SMART TOURISM

Managing Technological Revolution in the Tourism Industry

Perspectives and Anthropology in Tourism and Hospitality (PATH)

SMART TOURISM

Managing Technological Revolution in the Tourism Industry

Santus Kumar Deb, PhD

Editor

AAP APPLE ACADEMIC PRESS

First edition published 2027

Apple Academic Press Inc.
1265 Goldenrod Circle, NE,
Palm Bay, FL 32905 USA

760 Laurentian Drive, Unit 19,
Burlington, ON L7N 0A4, Canada

CRC Press
2385 NW Executive Center Drive,
Suite 320, Boca Raton FL 33431

4 Park Square, Milton Park,
Abingdon, Oxon, OX14 4RN UK

Apple Academic Press exclusively co-publishes with CRC Press, an imprint of Taylor & Francis Group, LLC

Library and Archives Canada Cataloguing in Publication

Title: Smart tourism : managing technological revolution in the tourism industry / Santus Kumar Deb, PhD, editor.
Other titles: Smart tourism (Palm Bay, Fla.)
Names: Deb, Santus Kumar, editor.
Series: Perspectives and anthropology in tourism and hospitality.
Description: First edition. | Series statement: Perspectives and anthropology in tourism and hospitality | Includes bibliographical references and index.
Identifiers: Canadiana (print) 20260130443 | Canadiana (ebook) 20260130451 | ISBN 9781779641083 (hardcover) | ISBN 9781779641090 (ebook)
Subjects: LCSH: Tourism—Information technology—Management. | LCSH: Tourism—Data processing—Management. | LCSH: Tourism—Technological innovations—Management.
Classification: LCC G156.5.I5 S63 2026 | DDC 338.4/7910285—dc23

Library of Congress Cataloging-in-Publication Data

Names: Deb, Santus Kumar editor

Title: Smart tourism : managing technological revolution in the tourism industry / Santus Kumar Deb, editor.

Description: First edition. | Palm Bay, FL : Apple Academic Press, 2026. | Series: Perspectives and anthropology in tourism and hospitality (PATH) | Includes bibliographical references and index. | Summary: "An emerging concept in tourism, smart tourism is the integration of smart technology and tourism for seamless, personalized, and sustainable travel experience. This new volume, Smart Tourism: Managing Technological Revolution in the Tourism Industry, delves into the realm of smart tourism and explores how advances in technology is shaping the landscape of travel and hospitality. This book unravels the complexities of smart tourism, examining the integration of artificial intelligence, big data analytics, augmented reality, and other cutting-edge technologies into the tourism ecosystem. The book presents case studies and highlights emerging trends that illustrate how smart tourism is not just a buzzword, but a transformative force with the power to redefine the way travelers experience the world. Focusing on technological advancement in tourism, the chapters explore myriad applications and aspects of digital technologies in tourism services, such as the integration of artificial intelligence, IoT (Internet of Things), big data, chatbots, QR codes, etc., for sustainable destination development, for improving hospitality customer service, for effective digital marketing, for value creation, for empowering citizen engagement in tourism, for gauging hospitality employees mental health, and more. The volume also addresses the use smart tools for crisis recovery in the tourism industry. Drawing on insights from industry experts, academics, and innovators, this book provides a comprehensive overview of the opportunities and challenges presented by employing smart tools in tourism management and promotion. Whether you are a seasoned professional in the tourism sector, a technology enthusiast, or simply a curious traveler, this book offers valuable perspectives on the intersection of technology and tourism in the 21st century"-- Provided by publisher.

Identifiers: LCCN 2026003145 (print) | LCCN 2026003146 (ebook) | ISBN 9781779641083 hardcover | ISBN 9781779641090 ebook

Subjects: LCSH: Tourism--Information technology | Tourism--Data processing | Tourism--Technological innovations

Classification: LCC G155.A1 S5637 2026 (print) | LCC G155.A1 (ebook) | DDC 338.4/791--dc23/eng/20260323

LC record available at https://lccn.loc.gov/2026003145

LC ebook record available at https://lccn.loc.gov/2026003146

ISBN: 978-1-77964-108-3 (hbk)
ISBN: 978-1-77964-109-0 (ebk)

ABOUT THE BOOK SERIES: PERSPECTIVES AND ANTHROPOLOGY IN TOURISM AND HOSPITALITY (PATH)

SERIES EDITOR:

Dr. Anukrati Sharma
Head & Associate Professor, Department of Commerce and Management
University of Kota, Rajasthan, India

In spite of the adverse effects of the pandemic, the tourism and hospitality sectors have grown more rapidly than any other. The purpose of this series is to highlight the innovations in tourism and hospitality in recent years. Further, the book series will provide an overview of future opportunities, challenges, and perspectives in the tourism and hospitality sector worldwide. It is open to not only academics and researchers but also to policymakers, bloggers, YouTubers, and industry people to submit proposals. Involving them is intended to bridge the gap between industry and academia and to provide practical learning opportunities. The series encourages bloggers and YouTubers to submit book proposals since they are better at telling stories about destinations, cuisine, events, festivals, etc.

This series aims to highlight tourism and hospitality themes that are yet to be explored and untapped. A key component of the series is the anthropology element in tourism since this topic appears to be one that will remain in demand for research indefinitely. Anthropology encompasses a wide spectrum of subjects, from human relationships to social, cultural, environmental, and community engagement and ethics to changing tourist behavior.

This series provides a critical analysis of the below-mentioned topics but is not limited to them:

- Tourism and the Anthropological Perspectives
- Growth and Sustainability of Livelihoods in the BRICS Countries
- Antecedents, Consequences and Future of Tourism Potential
- Tourism and Entrepreneurship
- People, Psychology and Peace
- Exchange, Emotions and Experiences
- Socio-Cultural Dimensions of Tourism
- The Historical and Cultural Phenomenon of Tourism

- Tourism in Smart Societies
- Hospitality and Anthropology
- Back to Basics in Hospitality
- Tourist Insights and Behavior
- Anthropologists and Skills
- Strategic Repositioning in Tourism and Hospitality
- Cross-Cultural Exchange and Communication
- Ethics and Responsible Behavior
- Environment and Climate Change

BOOKS IN THE SERIES

- **Anthropology of Tourism: Exploring the Social and Cultural Intersection**
  Editors: Maximiliano E. Korstanje, PhD, and Vanessa GB Gowreesunkar, PhD
- **Smart Tourism: Managing Technological Revolution in the Tourism Industry**
  Editor: Santus Kumar Deb, PhD
- **Artificial Intelligence in Hospitality and Tourism: Transforming the Future**
  Editors: Divya J. Thakur, PhD, Amna Mirza, PhD, Arsheed Ganie, PhD, and Ghousia Khatoon, PhD
- **Intercultural Communication in Tourism**
  Editors: Sangeeta Tripathi, PhD, Islam Elbayoumi Salem, PhD, and Andreia de bem Machado, PhD

ABOUT THE EDITOR

Santus Kumar Deb, PhD
Professor, Department of Tourism and Hospitality Management, Faculty of Business Studies, University of Dhaka, Bangladesh

Santus Kumar Deb, PhD, is a Professor in the Department of Tourism and Hospitality Management at the Faculty of Business Studies, University of Dhaka, Bangladesh, and a visiting Research Fellow at INTI International University, Malaysia. As a Visiting Professor, he has conducted classes at Vincent Pol University in Lublin, Poland (Europe). He serves as the head of research at the Centre for Tourism Studies (CTS), a renowned tourism research firm in Bangladesh. He supervises PhD and MPhil research students and actively contributes to academic journals. He serves as a moderator of the FBS Debating Club at the University of Dhaka, Bangladesh.

Dr. Deb has served as the Chairman of the Department of Tourism and Hospitality Management at the University of Dhaka (Bangladesh) and as the Director of the MBA (Evening) Program at the Department of Tourism and Hospitality Management, University of Dhaka (Bangladesh). Dr. Deb has served as a member of the Bangladesh Tourism Board and worked as a consultant for the National Tourism Master Plan, a national initiative designed by IPE Global, India.

In 2025, Dr. Deb was awarded the "DU Research Excellence Recognition-2025" (in recognition of exemplary contribution to scholarly excellence through high-quality research publication) from the University of Dhaka, Bangladesh. He has also received the Best Research Poster Award (Second Prize) from the University of Dhaka at the Centennial Research and Publication Fair in 2022. He has been awarded the Erasmus Mundus scholarship for the PhD researcher mobility program. Dr. Deb is serving as an editorial board member in a recognized SCOPUS indexed Q2 rank journal, 'Sport and Tourism Central European Journal,' and is also serving as one of the editors of the Journal of Digital Marketing and Communication (JDMC) published by Techno Scientifica (Indonesia). He is a regular reviewer of outcome-based education curricula on tourism for the University Grants Commission (UGC) of Bangladesh. He is an ad hoc reviewer for several publishers, including Emerald, Springer, Inderscience, Routledge, and Taylor & Francis Group.

He served as an expert member of the Federation of Bangladesh Chambers of Commerce and Industry (FBCCI) Standing Committees named 'Resorts and Tourism Development,' 'Civil Aviation and Tourism,' 'Youth Entrepreneurship Development,' and 'Sport Tourism' for promoting and formulating policy in Bangladesh.

He has completed his BBA and MBA with a major in Marketing, a Postgraduate Diploma (PgD) in Supply Chain Management, an Executive Program in Strategic Digital Marketing, and a PhD in e-Tourism Service Management. He has written two books titled *Tourism in Bangladesh* and *Fundamentals of Tourism and Hospitality Management* for undergraduate students in Bangladesh. His Edited (Scopus Index) book, Tourism and Hospitality Management: A Mordern Fusion of Sectors has been published by Bentham Science (Singapore). His areas of research interest include sustainable tourism, smart tourism, digital marketing, community-based tourism, crisis management, and emotional intelligence.

He has delivered keynote speeches and spoken at numerous international conferences held in Germany, Italy, Poland, Latvia, Thailand, Brunei, India, Indonesia, Sri Lanka, and Bangladesh. He has published more than 35 peer-reviewed articles in renowned journals related to tourism and marketing, most of which are indexed in SCOPUS (Q1/Q2, ABS, and ABDC-ranked journals) and Web of Science.

CONTENTS

CONTRIBUTORS

Sanjoy Kumar Acharjee
Department of Tourism and Hospitality Management, Noakhali Science and Technology University, Noakhali, Bangladesh

Tanvir Ahmed
Department of Tourism and Hospitality Management, Noakhali Science and Technology University, Noakhali, Bangladesh

Sara Alaviyeh
International School of Hospitality, Sports, and Tourism Management, Silberman College of Business, Fairleigh Dickinson University, Vancouver, Canada

Sadia Afrin Ananya
Department of Tourism and Hospitality Management, Gopalganj Science and Technology University, Gopalganj, Bangladesh; and University of South Australia, Adelaide, South Australia, Australia

Shahrin Ashraf
Department of Marketing, University of Dhaka, Dhaka, Bangladesh

Arup Barman
Department of Business Administration, Assam University, Silchar, Assam, India

Chhanda Biswas
Department of Tourism and Hospitality Management, Gopalganj Science & Technology University, Gopalganj, Bangladesh

Susanta Ranjan Chaini
Faculty of Hospitality and Tourism Management, Siksha 'O' Anusandhan (Deemed to be University), Bhubaneswar, Odisha, India

Iva Rani Das
Daffodil Institute of IT, Dhaka, Bangladesh

Santus Kumar Deb
Department of Tourism and Hospitality Management, University of Dhaka, Dhaka, Bangladesh

Md. Abdul Halim
Department of Tourism and Hospitality Management, Rangamati Science and Technology University, Rangamati, Bangladesh

Md. Tanvir Alam Himel
Department of Marketing, University of Dhaka, Dhaka, Bangladesh

Md. Manower Hossain
Department of Tourism and Hospitality Management, Gopalganj Science & Technology University, Gopalganj, Bangladesh

Pritom Kundu Joy
School of Graduate Studies, Trent University, Peterborough, Ontario, Canada

Tohomina Khatun
Department of Anthropology, Comilla University, Cumilla, Bangladesh

Sanjeev Kumar
School of Hotel Management and Tourism, Lovely Professional University, Phagwara, Punjab, India

Bapon Chandra Kuri
Department of Tourism and Hospitality Management, Gopalganj Science & Technology University, Gopalganj, Bangladesh

Patita Paban Mohanty
Faculty of Hospitality and Tourism Management, Siksha 'O' Anusandhan (Deemed to be University), Bhubaneswar, Odisha, India

Shohel Md. Nafi
Department of Tourism and Hospitality Management, Noakhali Science and Technology University, Noakhali, Bangladesh

Hiran Roy
Associate Professor, School of Management, Canadian University Dubai, Dubai, UAE

Sanjoy Kumar Roy
Department of Tourism and Hospitality Management, Noakhali Science and Technology University, Noakhali, Bangladesh

Shadia Sharmin
Department of Tourism and Hospitality Management, Gopalganj Science & Technology University, Gopalganj, Bangladesh

Abubakar Siddik
Department of Tourism and Hospitality Management, Gopalganj Science & Technology University, Gopalganj, Bangladesh

Mohammad Badruddoza Talukder
College of Tourism and Hospitality Management, International University of Business Agriculture and Technology, Dhaka, Bangladesh

Manika Devi Tanchangya
Department of Tourism and Hospitality Management, Gopalganj Science & Technology University, Gopalganj, Bangladesh

Marco Valeri
Faculty of Economics, Niccolò Cusano University, Rome, Italy

Liton Chandra Voumik
Department of Economics, Noakhali Science and Technology University, Noakhali, Bangladesh; and Central Queensland University, Melbourne, Victoria, Australia

A. S. Sochipem Zimik
Assistsnt Professor, Department of Management, SRM University, Sikkim, India

ABBREVIATIONS

4IR	fourth industrial revolution
AI	artificial intelligence
AR	augmented reality
AWS	Amazon Web Services
B2C	business-to-consumer
BDAI	big data and artificial intelligence
CAGR	compound annual growth rate
CASA	computers as social actors
CCPA	California Consumer Privacy Act
CE	constructed environment
CMC	computer-mediated communication
CRM	customer relationship management
CRS	computerized reservation systems
CSR	corporate social responsibility
DMO	destination management organization
DMS	destination management systems
DS	digital simulation
EAC	emotionally aware chatbot
ECM	expectation-confirmation model
ECST	European Capital of Smart Tourism
EDEN	European Destination of Excellence
ERP	enterprise resource planning
FGD	focus group discussion
FoG	Fog of Things
GAI	generative artificial intelligence
GDPR	General Data Protection Regulation
GDS	global distribution systems
H&T	hospitality and tourism
HDFS	Hadoop distributed file system
HDM	human-driven method
HMD	head-mounted display
ICT	information and communication technology
IoT	Internet of Things

ISS	information system success
IT	information technology
ITC	information technology communication
IV-TRS	virtual travel advisory method
KPIs	key performance indicators
LBSM	location-based social media
MGC	marketer-generated content
ML	machine learning
NFC	near-field communication
NLP	natural language processing
NLU	natural language understanding
OSF	Open Semantic Framework
OTA	online travel agency
OTA	online travel aggregators
P2P	peer-to-peer
PCAs	pedagogical conversational agents
PCI	Payment Card Industry
DSS	Data Industry Standard
PIR	post-incident review
PPC	pay-per-click
QR	quick response code
RFID	radio-frequency identification
RM	revenue management
ROI	return on investment
RQ	research question
SCO	shared content object
SDG	Sustainable Development Goal
SDL	service-dominant logic
SEM	search engine marketing
SEO	search engine optimization
SMM	social media marketing
SPG	Starwood Preferred Guest
SST	self-service technology
STD	smart tourism destination
STT	smart tourism technology
SWOT	Strengths, Weaknesses, Opportunities, and Threats
TAM	Technology Acceptance Model
TAS	tourism as a service

TDM	technology-driven method
TIES	The International Ecotourism Society
TQM	Total Quality Management
TSA	Tourism Satellite Account
UAM	Urban Air Mobility
UGC	user-generated content
UNWTO	United Nations World Tourism Organization
UTAUT	Unified Theory of Acceptance and Use of Technology
VPN	virtual private network
VR	virtual reality
WHO	World Health Organization
WSN	wireless sensor networks
WTO	World Tourism Organization

PREFACE

In an era where technology permeates every aspect of our lives, the tourism industry stands at the forefront of innovation. The dawn of the digital age has revolutionized the way we travel, plan vacations, and experience destinations. From booking flights and accommodations to discovering local attractions and sharing experiences on social media, technology has transformed every step of the journey. At its core, smart tourism aims to optimize efficiency, enhance the traveler's experience, foster sustainability, and empower local communities—ultimately shaping a more connected, responsible, and enjoyable future of travel.

The essence of smart tourism lies in harnessing the power of technology to create seamless, personalized, and sustainable travel experiences. It encompasses leveraging advancements such as artificial intelligence (AI), big data analytics, augmented reality (AR), and the Internet of Things (IoT) to enhance every aspect of the tourism journey—from trip planning and destination exploration to accommodation and local engagement. As we traverse an increasingly interconnected world, smart tourism technologies continue to transform the way we discover new places. It's more than just convenience and personalization; it's about establishing a sustainable and secure path to unforgettable travel experiences. The future of tourism is clearly smarter and more promising than ever before, offering a glimpse into a world where travel seamlessly blends the best of technology with the spirit of adventure. So, whether you're a seasoned explorer or a curious traveler, smart tourism is your ticket to a future full of great trips and lasting memories.

This book delves into the realm of smart tourism and explores how advancements in technology are shaping the landscape of travel and hospitality. As we embark on this exploration, we will unravel the complexities of smart tourism and examine the integration of artificial intelligence (AI), big data analytics, augmented reality (AR), and other cutting-edge technologies into the tourism ecosystem. But beyond the gadgets and algorithms, smart tourism is fundamentally about enhancing the traveler's experience, fostering sustainable practices, and empowering local communities.

The book focuses on case studies, practical applications of the smart tourism concept, and emerging trends that illustrate how smart tourism is not just a buzzword, but a transformative force with the power to redefine the way travelers experience the world. Drawing on insights from industry experts, academics, and innovators, this book aims to provide a comprehensive overview of the opportunities and challenges presented by smart tourism.

Whether you are a seasoned professional in the tourism sector, a technology enthusiast, or simply a curious traveler, this book offers valuable perspectives on the intersection of technology and tourism in the 21st century. Join us on a journey into the future of travel, where innovation and imagination converge to create unforgettable experiences. That's why we say—welcome to the world of smart tourism.

This edited book encompasses a wide range of technological advancements in tourism and hospitality and focuses on various aspects of smart tourism strategies rather than a single focal point. It exclusively highlights innovations such as technological advancement in tourism, Tourism 4.0 for the integration of digital technologies, applications of the Internet of Things (IoT) in travel and tourism, big data and smart tourism, the impact of generative AI on hospitality employees' mental health, the emergence of AI chatbots for 24/7 travel assistance, and the transformative impact of artificial intelligence in hospitality and tourism.

It also explores smart tourism for sustainable destination development, creating environments for smarter tourism, eco-smart destinations, smart tourism for empowering citizen engagement, e-safety technologies driving innovation in the tourism industry, value creation through smart tourism services, smart hotels leading with service excellence, digital marketing as a crucial tool in smart tourism development, and smart tourism for crisis recovery through strategic responses to challenges.

The practical implications of smart tourism are far-reaching and impact various stakeholders within the tourism ecosystem, including tourists, businesses, destinations, and local communities. In addition, smart tourism plays a key role in enhancing travelers' experiences through smart tourism technologies, enabling tourists to enjoy more seamless and personalized services. AI-powered chatbots provide real-time assistance, while mobile apps offer personalized recommendations and itineraries. Travelers benefit from improved convenience, efficiency, and satisfaction throughout their journey. Smart tourism also improves operational efficiency in tourism businesses, as smart technologies streamline operations and optimize resource allocation. This includes automated booking and reservation systems, data-driven marketing strategies, and predictive analytics for demand forecasting. Furthermore, by leveraging technology, tourism businesses can reduce costs, increase productivity, and enhance their comparative advantages.

To ensure sustainable practices in tourism, one of the key strategies is smart tourism, which promotes sustainability by encouraging responsible environmental and cultural stewardship. Through smart initiatives such as smart destination management systems, eco-friendly infrastructure, and community-based tourism projects, destinations can minimize their ecological footprint and preserve natural and cultural heritage for future generations. Smart tourism empowers local communities to actively participate in and benefit from the tourism industry. Digital platforms and online marketplaces enable local businesses, artisans, and service providers to showcase their offerings and reach a global audience. This promotes economic development, cultural preservation, and social inclusion within destination communities. By harnessing the power of big data analytics, smart tourism enables stakeholders to make informed decisions based on real-time insights and trends.

Destination marketers can better understand visitors' behavior, preferences, and satisfaction levels, allowing them to tailor their offerings and digital marketing strategies accordingly. Tourism businesses can also optimize their operations and improve customer service by leveraging data-driven insights, along with smart tourism technologies, which play a crucial role in ensuring the safety and security of travelers.

Moreover, from biometric authentication and contactless payments to real-time monitoring and emergency alert systems, smart tourism initiatives promote inclusivity and accessibility for travelers of all backgrounds and abilities. By implementing accessible infrastructure, digital platforms, and inclusive marketing campaigns, destination marketers can cater to diverse audiences and ensure that everyone can enjoy meaningful travel experiences.

Overall, smart technology applications in the tourism and hospitality industry are driving innovation, efficiency, and customer satisfaction. While also promoting sustainability and leveraging technology responsibly and collaboratively, stakeholders can create a more sustainable, inclusive, and enjoyable tourism experience for all.

This book is a meticulous effort of minds from different countries, including Bangladesh, India, Australia, Italy, and Canada. I express special thanks to each author. I feel genuinely grateful for each author's professionalism, enthusiasm, and willingness to share insights and valuable contributions. Your contributions have undoubtedly made a significant impact in enriching the quality of this book, and I am confident that your chapters will be well received by readers in the tourism and hospitality sector, especially in the area of technological applications in tourism.

Thank you once again for your outstanding work and for being a valuable contributor to this book. I look forward to the opportunity to collaborate with you again in the future.

CHAPTER 1

TOURISM 4.0: THE INTEGRATION OF DIGITAL TECHNOLOGIES AND THE TRANSFORMATIVE NATURE OF THE CONTINUOUS CHANGES IN THE TOURISM INDUSTRY

CHHANDA BISWAS*, MANIKA DEVI TANCHANGYA, and ABUBAKAR SIDDIK

Department of Tourism and Hospitality Management, Gopalganj Science & Technology University, Gopalganj, Bangladesh

*Corresponding author

ABSTRACT

In the dynamic era of modern technology, the need to integrate digital technologies, which led to "Tourism 4.0," refers to a paradigm shift characterized by the proper integration of digital technologies and software developments. The prime goal of this study is to figure out how aspects of Tourism 4.0 affect the changing nature due to the digital integration of the traveling industry. The framework of Tourism 4.0, the integration of digital technologies, was comprehensively investigated through qualitative research, using the snowball sampling method from tourism industries to collect samples from 10 groups. The potential outcomes provide valuable information to explore the challenges of Tourism 4.0 and harness the continued growth and innovation of tourism. In addition, these results help us understand the complex relationship between digital technologies and constant changes in travel patterns. This book chapter gives knowledge about the contribution of ICT and digital technologies transformation in

Smart Tourism: Managing Technological Revolution in the Tourism Industry. Santus Kumar Deb (Ed.)

DOI: 10.1201/9781779641090-1

experiencing tourism facilities, including the understanding of Tourism 4.0 technologies on the travel experience. Therefore, this chapter contains both theoretical and practical contributions to tourism literature.

1.1 INTRODUCTION

The tourism sector is at the forefront of major changes driven by the introduction of digital technologies in the changing era of modern technology. This paradigm shift, called Tourism 4.0, will allow the convergence of travel demand with digital innovation (Pencarelli, 2020). The needs of travelers around the world and the continuous advancement of technology have changed, which has created a period of constant change in the travel industry's facilities. Tourism 4.0 refers to the terms that bring the significant effects of innovative digitization that will bring revolutionary changes in tourism activities (Gretzel et al., 2015). This integrative pathway will ensure significant transformation in tourism management and signature service delivery to relevant sectors. Destination managers, industrial planners, and management companies are embracing technological changes and the implementation of Tourism 4.0 in this new context (Neuhofer et al., 2015). This integration goes beyond adopting innovative technological development and has a dynamic and constantly growing workforce in the tourism sector. Traveling participants are interested in the digital technology opportunities needed to improve operational efficiency while developing new, efficient, and sustainable methods of managing and providing memorable tourism experiences (Stankov and Gretzel, 2020).

In addition, smartphones are becoming an integral part of every aspect of our lives, including travel. Tourists' experiences are changing by using smartphones, and they feel secure in those ways (Kim et al., 2018; Kramer et al., 2022; Paris, 2012). Today, approximately 5.3 billion people use the Internet, and the number is increasing every day (International Telecommunication Union, 2022). People use these digital technologies in all sectors for social media, travel planning, vacations, and work. There will be a major change in the tourism sector that encourages visitors to search for their travel information needs through various mobile guide applications, technologies, and virtual websites, online ticket booking, and so forth, and that number is increasing daily (Natalia et al., 2020). Internet of Things (IoT) is an application that is recognized for being used to refer to everything that users look for when searching for information and data.

The tourism sector has great opportunities due to digital technologies and innovative concepts brought by Tourism 4.0 (Kramer et al., 2022). This chapter presents an example of a virtual travel advisory method (IV-TRS). This is a form that presents a virtual tour of a place using visual information such as photos, videos, and audio, which is known as the IV-TRS model. The model has three layers: website development, application, and calendar. Travel Advice is managed by IV-TRS, which uses Image Based Mapping and the SCORM concept (TRIPS) to integrate relevant tourism data stored in Shared Content Object (SCO) format. The stored data was saved as SCO that enhanced the web layer, confirmed by CORDRA, and created according to the SCORM. SCORM refers to the Sharable Content Object Reference Model, which is a model that metadata focuses on to describe the process of storing, referencing, and using information. CORDA searches on the web that is an emphasis on finding and referencing SCORM repositories. The second layer is cognitive, which informs the system's decision engine, expert system technologies, or artificial neural networks, which can be developed on semantic web tools. This research creates a calendar database via an emphasis on Semantic Web technology. The application layer consists of two parts: Search and display. The CORDRA feature module is to crawl, which is used to search and index SCOs on the web. The presentations section allows users who propose to customize presentations according to their needs during user interactions. Virtual travel is experienced by software applications like the IV-TRS system, which provides users with better travel selection and planning (Kramer et al., 2022). Elsewhere, the reforming of various digital tools like artificial intelligence (AI), virtual reality (VR), and augmented reality (AR) transform the overall traveling experience. These digital technologies provide interactive and convenient experiences, chatbots provide personalized experiences to assist with information search and booking services, and AI provides travel recommendations based on analysis of historical tourist data and dynamic pricing. Virtual reality can help with pre-trip decisions, enable virtual exploration of desired locations, and serve as a training tool. AR creates opportunities to explore real-world experiences, offer cultural tours, and promote better experiences. According to Gretzel et al. (2015); Neuhofer et al. (2015), these new technologies represent the essence of more personalized, incremental, and progressive tourism experiences for both tourists and tourism agents.

Smart tourism is described as a new phase to narrate the growing involvement in revolutionary ICT merge with tourism sectors, industries, and the hospitality market that can turn into the gather information indicators of value. However, the definition as a concept remains poorly defined, which makes theory development difficult. This chapter introduces smart tourism, introduces the current trends of smart tourism, and then presents its technical and business fundamentals. The prospects and weaknesses of smart tourism are briefly discussed below. This chapter also addresses the critical need for research to inform the development and management of smart tourism. Applications provide customer reviews, recommendations, detailed information, and destination information with just one click, and you will board the plane because the trip is canceled through this app. This determining factor is forced to the range of memorable experiences through technological advancement (Stankov and Gretzel, 2020). The app should provide feedback, customer recommendations, and detailed location-related information in one click, and since roaming is excluded in this app, it is important to determine the extent of the technology-enhanced experience (Gretzel et al., 2015; Neuhofer et al., 2015). The app should provide feedback, customer recommendations, and detailed location-related information in one click, and since roaming is excluded in this app, it is important to determine the extent of the technology-enhanced experience (Neuhofer et al., 2015).

The tourism sector faces challenges, opportunities, and problems due to the impact of the latest industrial revolutions. Disabled people can get assistance from advanced technological facilities in tourism-related events, thus raising important concerns. People benefit from tourism in many ways. Getting around can be difficult for people with disabilities and the elderly. People with motor and cognitive disabilities often face difficulties when they cannot get where they need to go for many reasons, including sightseeing, walking, sitting, and carrying. Because of this situation, they cannot go to explore the beauty of nature (Teixeira et al., 2020). However, security concerns are another psychological aspect of digital technology that should be given more consideration when planning the introduction of these technological advances.

However, the integration of digital technologies in Tourism 4.0 represents more than just a mere evolution; it signifies a revolutionary redefinition of the industry's fundamental paradigms. This transformation is not confined to the adoption of innovative tools and platforms; rather, it encompasses a holistic and interconnected approach that permeates every facet of the tourism experience (Neuhofer et al., 2015).

From personalized customer interactions to the optimization of operational efficiency, the impact of digitalization is shaping a tourism landscape that is both dynamic and adaptive. This exploration delves into the intricacies of how digital technologies are not only reshaping the industries to reform Tourism 4.0 in the operational landscape but also catalyzing a transformative nature characterized by continuous change (Kononova et al., 2020; Kumar et al., 2023). Few relevant studies are found regarding Tourism 4.0 along with this present chapter, where there are plenty of studies that consider the knowledge of long-term viability, which increases the quality of tourism practices (Pencarelli, 2020). Tourism 4.0 forces the adoption of technologies on generating potential benefit, which occurs as a positive tourist experience (Stankov and Gretzel, 2020), a challenging adaptation of digital tools in developing tourism market growth (Peceny et al., 2019a), providing a new shape to the future tourism industry by digital technologies (Steyn, 2010). For this reason, this chapter fills the literature gap in Tourism 4.0 and involves key stakeholders, including tourism management organizations, local communities, tourists, and government organizations, to realize responsible tourism and ensure safety issues, including experts and affiliates and related fields.

1.2 LITERATURE REVIEW

1.2.1 CONCEPT OF TOURISM 4.0

The widespread use of cutting-edge digital technology has caused a paradigm change in the tourist sector, which is embodied in the idea of Tourism 4.0 (Abdurakhmanova et al., 2022). This revolutionary method aims to improve and change every facet of the travel experience, from booking and preparation to on-location events and post-trip interaction. The phrase "Tourism 4.0" highlights how digitization, connectivity, and data-driven insights are altering old methods, and it bears similarities to the more general notion of Industry 4.0 (Kantamalajaroen and Summaniti, 2021).

1.2.2 COMPONENTS OF THE TOURISM 4.0

Tourism 4.0 reflects a vision for a technologically empowered and adaptive tourism industry, where digital innovations are harnessed to create more personalized, efficient, and sustainable travel experiences (Peceny et al., 2020). This evolution holds the potential to reshape not only the way people travel but also how destinations are managed and promoted in an increasingly interconnected and data-driven global landscape. However, the components of Tourism 4.0 are explained below subsections (Figure 1.1).

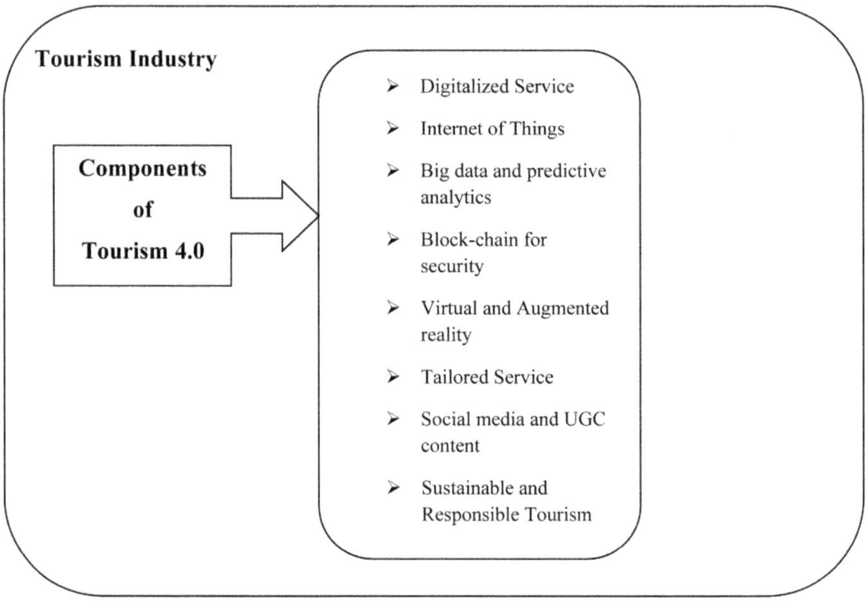

FIGURE 1.1 Components of tourism 4.0.

Source: Developed by authors.

1.2.2.1 DIGITALIZATION OF TOURISM SERVICES

Transformation of tourist services by adopting digitalized facilities is one of the most important aspects of the industry, which represents a shift from old techniques to a more tech-savvy and connected approach. This idea entails utilizing cutting-edge digital technology to optimize and improve several travel-related factors. Here are key components of the digitalization of tourism services in the context of Tourism 4.0.

1.2.2.2 ONLINE BOOKING AND RESERVATIONS

Online platforms are adopted by travel companies, hotels, airlines, and tour operators, introducing travelers to plan itineraries, make bookings, and book lodging through websites and mobile applications with ease (Amir et al., 2020). Travelers can make knowledgeable decisions quickly thanks to digital systems' real-time information on accessibility, pricing, and promotions (Stankov and Gretzel, 2020). The service of mobile applications is generating facilities that provide access to the users, including guided tours, flight reservations, and e-booking while they're on the go, increasing flexibility and convenience. The traveler's present position informs contextualized information, tailored suggestions, and navigation support using GPS and location-based technology (Kramer et al., 2022). User profiles and preferences using user data analyzed by sophisticated analytics and machine learning algorithms, comprehensive profiles are enabled, enabling highly customized travel suggestions based on individual tastes and behavior (Stankov and Gretzel, 2020). With the help of technological tools, travel agencies can now provide customized itineraries, advice, and add-on activities to suit the particular requirements of each visitor. Virtual Previews: Travelers may explore and experience locations before making reservations thanks to AR/VR technology, which provides virtual previews of locations, lodging options, and activities (Pencarelli, 2020). Through impressive storytelling, historical information overlays, and interactive guides, AR improves on-site experiences, making visitors' visits richer and more interesting.

1.2.2.3 E-TICKETING AND DIGITAL CHECK-INS

Paperless Transactions are used for digital check-ins and electronic tickets instead of paper tickets, the check-in procedure is streamlined, and the environmental effect is reduced. On the other side, contactless Solutions are used for electronic tickets and digital check-ins to make operations more efficient and contactless, particularly when it comes to health and safety issues (Kramer et al., 2022).

1.2.2.4 INTERNET OF THINGS (IOT) INTEGRATION

1. **Smart Hospitality:** adding IoT gadgets in cabins on cruise liners, hotels, and other lodgings adds to a networked hospitality experience. Smart entertainment, lighting, and temperature controls improve visitor comfort. Additionally, connected attractions which include Internet of Things-enabled attractions offer individualized and interactive experiences while gathering information on visitor preferences and behavior to enhance offerings (Pencarelli, 2020).

1.2.2.5 CUSTOMER RELATIONSHIP MANAGEMENT (CRM)

CRM platforms let travel agencies stay in touch with clients on a personal level by providing them with customized offers, loyalty plans, and ways to express feedback. Therefore, CRM system analytics offer insightful information about customer behavior and preferences, which helps guide strategic choices for improving marketing and service offerings (Neuhofer et al., 2015).

1.2.2.6 BLOCKCHAIN FOR TRUST AND SECURITY

Blockchain technology guarantees transparent and safe transactions, promoting confidence between travelers and service providers. By using blockchain technology to securely verify identities, fraud in reservations and transactions may be decreased, and security can be improved (Otebolaku and Lee, 2018). Tourism 4.0's digitalization of tourism services is an all-encompassing strategy for utilizing technology to the advantage of industry stakeholders and tourists alike. The tourism industry wants to be responsive to the changing demands and expectations of contemporary tourists while delivering a more effective, customized, and immersive travel experience by embracing these digital innovations (Otebolaku and Lee, 2018). Thus, digital technological facilities can affect tourists' overall experiences, positive destination image, and tourists' preferences. Travel service has been identified below based on the impact of digital technology and travel experience via a conceptual model (Figure 1.2).

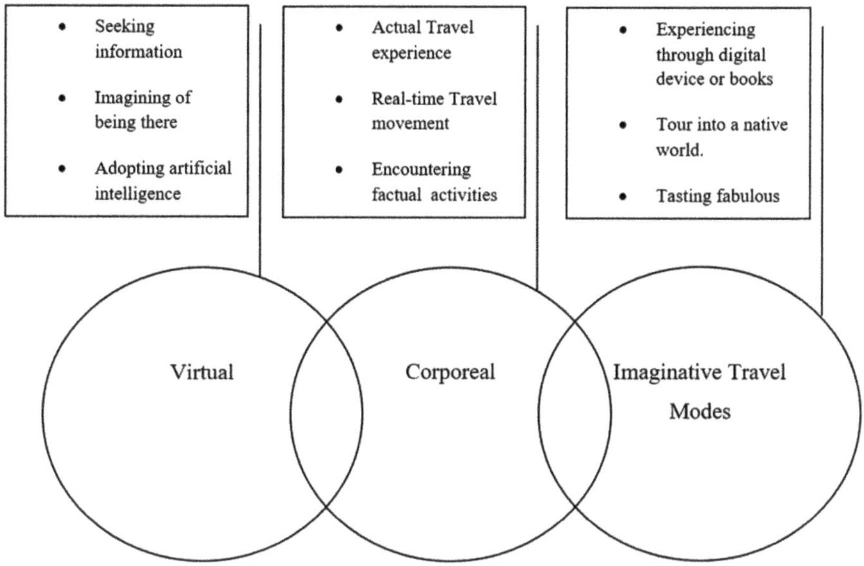

FIGURE 1.2 An interconnection between corporeal, virtual, and imaginative travel modes.
Source: Developed by authors.

1.2.3 PERSONALIZATION AND CUSTOMIZATION

Within the framework of Tourism 4.0, personalization and customization in the travel industry play a critical role. Cutting-edge digital technologies are utilized to customize travel experiences and cater to each visitor's unique preferences and requirements. This idea goes beyond conventional one-size-fits-all strategies in an attempt to provide highly customized and interesting trips. Here's a deeper look at how Tourism 4.0 uses customization and personalization.

1.2.3.1 USER PROFILING AND DATA ANALYTICS

1. **Data-driven Understanding:** To comprehend passenger behavior, preferences, and demographics, Tourism 4.0 uses advanced data analytics. In-depth user profiles are produced by analyzing past data

and current interactions. Predictive modeling: By using this data to forecast future preferences, algorithms enable travel agencies to anticipate each visitor's wants and preferences (Steyn, 2010).

1.2.3.2 TAILORED RECOMMENDATIONS AND ITINERARIES

Digital platforms employ user profiles to provide tailored suggestions for places to stay, things to do, and places to visit. These recommendations take into account the interests of the visitor, making for a more pleasurable trip. Travel 4.0 allows for the development of personalized itineraries that take into account individual preferences, preferred activities, and travel speed. Travelers may create custom itineraries that fit their preferences such as (Otebolaku and Lee, 2018):

1. **Personalized Pricing:** State-of-the-art pricing algorithms modify prices according to booking history, unique user activity, and preferences. Travelers will receive competitive offers that are tailored to them thanks to our dynamic pricing method (Abdurakhmanova et al., 2022). Individual tastes are taken into account when creating unique promotions, discounts, and loyalty awards, which boost consumer loyalty and encourage repeat business.
2. **On-the-Go Personalization:** Based on a traveler's location and preferences, mobile applications serve as personal travel assistants, providing real-time recommendations, navigation support, and customized notifications. By enabling users to assess experiences, offer comments, and modify preferences while on the road, apps make sure that the travel experience is flexible and sensitive to shifting tastes (Otebolaku and Lee, 2018).
3. **Smart Accommodations:** IoT gadgets in lodging facilities enable guests to tailor their stay by establishing their own lighting and temperature preferences. Tourists gather information about tourism sites that focus on visitor behavior to allow for individualized experiences with immersive storytelling, guided tours, and interactive exhibitions via the IoT (Pencarelli, 2020). AR technology enables tourists to virtually explore and personalize their travel experiences before arrival by offering previews of popular sites and activities (Otebolaku and Lee, 2018).

1.2.3.3 IOT (INTERNET OF THINGS) INTEGRATION

Within the scope of Tourism 4.0, the idea of IoT integration in the travel sector centers on the thoughtful integration of intelligent networked gadgets and technologies to optimize and transform a variety of travel-related experiences (Abdurakhmanova et al., 2022). IoT integration in Tourism 4.0 goes beyond individual apps and gadgets; it entails building an ecosystem in which many components of the infrastructure for tourism cooperatively and smoothly exchange information to improve productivity, security, and the overall experience for tourists. Here are the key elements of the concept discussed in the subsections.

1.2.3.4 SMART DESTINATIONS

A smart tourism destination (STD) is a concept that figures out a way to develop areas and transform them into smart cities. Since technology permeates all businesses and entities, destinations will leverage the synergies between their social components and ubiquitous sensing technologies to allow the development of tourist experiences such as:

1. **Interconnected Infrastructure:** IoT devices are integrated into the infrastructure of tourist attractions, such as waste management, smart transportation systems, and environmental monitoring. The entire efficiency of operations is improved by this link.
2. **Internet of Things (IoT)-Enabled Facilities:** IoT devices are used in hotels and other lodging establishments for lighting, temperature control, and room amenities. In addition to improving the energy economy, these clever innovations also accommodate individual visitor preferences. According to Kansakar et al. (2019), the hospitality service platform can be redeveloped by embracing IoT and advanced smart devices. The hotel industry offers to combine the same network to activate the services that guests want—integrated services like set-top boxes, thermostats, light switches, automated door locks, electric bills, and other devices—by applying the IoT. The hospitality personnel may get to bring back both the efficiency and provide the visitor experience by making faster and more tailored services (Sharma et al., 2021).

1.2.3.5 ENHANCED VISITOR SAFETY

1. **Real-time Monitoring:** IoT devices aid in the real-time monitoring of congested locations by revealing information about the density of visitors and any security risks. This helps to manage and ensure visitor safety in a proactive manner (Agarwal et al., 2021). By identifying and reacting to unique preferences and behaviors, IoT makes it possible to provide context-aware services (Otebolaku and Lee, 2018). This can include tailored suggestions for places to see, eat, or do activities. The Internet of Things is helping tourists to provide some suggestions based on locations and their culture (Dorcic et al., 2019).

1.2.3.6 INTERACTIVE ATTRACTIONS

1. **Immersive Experiences:** Interactive exhibitions and immersive experiences are made possible by IoT in tourist sites. By giving visitors relevant information tailored to their interests, smart sensors and gadgets increase visitor engagement. When smart cities improve traveler experiences, offer sophisticated platforms for information collection and dissemination, enable the effective use of tourism resources, and combine macro and micro-level tourism suppliers, they become smart tourist destinations (Cepeda-Pacheco and Domingo, 2022).

1.2.3.7 LOCATION-BASED SERVICES

1. **Direction and Guidance:** Software application services such as the IoT provide users with real-time guidance, details about neighboring sites of interest, and context-specific information in specific locations about their surroundings. Some applications such as sensor devices, IoT, and Smartphones provide a new location-based service (Almobaideen et al., 2017). These technologies are relatively new. These services might provide a cutting-edge suite of services that make traveling safer and simpler for mobile tourists or tourists who are currently visiting many locations (Jeong and Shin, 2020).

1.2.3.8 EFFICIENT TRANSPORTATION

1. **Smart Transportation Systems:** Internet of Things integration enhances transportation networks with dynamic route planning, real-time tracking, and predictive vehicle maintenance, improving the effectiveness of transit both inside and between locations. The energy used in digitalized areas is attributed to digitalized mobility and transportation by greater than 25% in general (Nguyen et al., 2020). These are important components of smart cities. The seven main categories of smart transportation are special transport, commercial, private, public, freight, and product distribution.
2. **Environmental Monitoring:** IoT devices help with responsible resource management by monitoring environmental aspects including waste management, energy usage, and the quality of the air and water. This promotes sustainable tourism. It helps the environment and the natural resource use in a sustainable way (Abdurakhmanova et al., 2022).
3. **Analytics and Insights:** Data-driven decision-making important insights into visitor behavior, preferences, and operational efficiency may be gained from the data produced by IoT devices. This data helps to force the industry stakeholders in the tourism sector to make potential decisions. AI can be the future of big data-driven technologies in the tourism sector as well as the technology of AR which makes virtual tours, machine learning (ML), and VR open doors for accessible tourist facilities as mix up with technology, where traveler can use the IoT that provide location-based services. These technologies set up high-valued management techniques at the tourism businesses that increase standardized service for recreational activities (Deb et al., 2023; Nguyen et al., 2020).

1.2.3.9 HEALTH AND WELLNESS MONITORING

Wellness-oriented Features: IoT devices, which offer services like contactless check-ins, health monitoring, and access to wellness amenities, may track metrics connected to health within tourist establishments. Establishing a networked, intelligent, and adaptable tourist ecosystem is the main objective of IoT integration in Tourism 4.0 (Almobaideen et al., 2017).

The tourist sector wants to improve guest experiences, and long-term growth, raise the standard of services, and participate in technologically sophisticated recreational activities by utilizing IoT devices and digital technology (Pai et al., 2020).

1.2.4 BIG DATA AND PREDICTIVE ANALYTICS

Big Data and Predictive Analytics are dynamic and crucial in transforming the tourist sector within the framework of Tourist 4.0 and the use of digital technology. With this strategy, large volumes of data produced by several digital touchpoints are gathered, and powerful analytics are applied to extract insightful information (Stankov and Gretzel, 2020). This is an examination of the idea within the context of the tourist industry's ongoing revolutionary change.

1.2.4.1 DATA AGGREGATION AND VARIETY

1. **Big Data Sources:** Tourism 4.0 is welcoming digital tools including social media interactions, mobile app usage, IoT devices, and online reservations to get a large scale of information. The foundation is built with big data due to the sheer amount and diversity of this data (Li et al., 2018). Massive amounts of organized and unstructured data are being created, recorded, saved, and gathered due to the quick advancement of computer science and Internet technology. This creates big data, ushering in a new era (Almobaideen et al., 2017).
2. **Dynamic Insights:** Real-time data analysis via big data analytics gives tourism stakeholders dynamic insights into consumer behavior, industry trends, and operational performance. This gives companies the ability to quickly and decisively adjust to the constantly shifting business environment (Cepeda-Pacheco and Domingo, 2022).

1.2.4.2 PERSONALIZATION AND CUSTOMER EXPERIENCE

1. **Tailored Services:** Personalized services are made possible by the building of comprehensive client profiles made possible

by Big Data and Predictive Analytics (Nguyen et al., 2020). Through awareness of personal preferences and behaviors, tourist operators may customize experiences and offerings to match the specific requirements of passengers in the context of tourism 4.0. Intelligence expansion can minimize the difficulties through the adoption of big data in vacation spots and traveling businesses (Huang et al., 2017). Hence, they are planning based on their data and environment.

2. **Predictive Modelling:** Predictive analytics forecasts future trends in tourism demand by analyzing patterns and data from the past. This skill contributes to improved management of visitor flows by assisting with efficient capacity planning, resource allocation, and peak travel period prediction (Cepeda-Pacheco and Domingo, 2022).

1.2.4.3 OPERATIONAL EFFICIENCY

1. **Resource Optimization:** By offering insights into service demand, traffic patterns, and resource use, big data analytics maximizes operational efficiency. This makes it possible to respond to ongoing developments in the tourist sector with proactive modifications that improve operational performance (Nguyen et al., 2020).
2. **Identifying Trends and Risks:** Potential hazards and new trends in the travel industry can be found with the use of predictive analytics. This proactive strategy assists in risk avoidance and strategic planning, whether it is by forecasting changes in consumer preferences or anticipating external variables that affect the tourism market (Otebolaku and Lee, 2018).

1.2.4.4 MARKETING AND DESTINATION MANAGEMENT

1. **Targeted Campaigns:** Agarwal et al. (2021) stated that marketing initiatives may be more focused and successful because of big data. Tourism 4.0 makes use of predictive analytics to pinpoint niche markets, which makes it possible to develop customized marketing plans that appeal to a range of tourist demographics.

2. **Continuous Learning:** Proper analysis of big data helps tourist organizations learn continuously by enabling them to modify and improve their strategy in response to real-time feedback and shifting market conditions. This flexibility is essential in a field where tastes and trends change quickly (Deb and Nafi, 2021; Otebolaku and Lee, 2018).
3. **Strategic Advantage:** In the cutthroat world of Tourism 4.0, companies that leverage Big Data and Predictive Analytics will get a competitive edge. Organizations are positioned at the forefront of industry change when they possess the capacity to anticipate trends and respond proactively (Sustacha et al., 2023).

To summarize, tourist 4.0 is a data-driven strategy to tackle the transformational character analysis of the big data of ongoing changes and predictive analytics in the tourist sector (Otebolaku and Lee, 2018). Businesses may improve client experiences, acquire a deeper insight into market dynamics, streamline processes, and position themselves to prosper in the dynamic and ever-changing world of contemporary tourism by utilizing sophisticated analytics.

1.2.5 AUGMENTED REALITY (AR) AND VIRTUAL REALITY (VR)

The ability to interact with visitors has significantly advanced in the introduction of tourism 4.0 by utilizing AR immersive technologies along with VR. The business can improve experiences and adjust to the transformational nature of constant change (Deb et al., 2022; Stankov and Gretzel, 2020). The following is an examination of how AR and VR relate to digital technology and the changing tourist industry landscape.

1.2.5.1 IMMERSIVE DESTINATION EXPLORATION

1. **Virtual Previews:** Travelers may virtually experience places before they arrive thanks to virtual reality (Pencarelli, 2020). This immersive experience helps with travel planning and decision-making by giving a feel of the environment, attractions, and cultural features of the place.

2. **Augmented Information:** It helps to provide memorable experiences by accepting digital technologies and gives real-time excellence to the tourist. This might provide tourists with a better and more educational experience by including historical information, specifics about sites, or interactive guides (Almobaideen et al., 2017).

1.2.5.2 VIRTUAL TOURS AND ATTRACTIONS

The development of digital technology has been essential in recent years to draw the attention of modern tourists and maintain a competitive edge. Digital technology utilized in public areas introduces augmented and virtual reality to the tourist such as:

1. **Immersive Attractions:** VR makes it easier to create virtual tours and attractions, which lets visitors explore locations, museums, and landmarks from the comfort of their own homes. This gives attractions additional sources of income in addition to being a marketing tool (Almobaideen et al., 2017).
2. **Augmented Reality (AR)-enhanced Exhibits:** AR creates dynamic, interactive experiences out of static exhibitions at museums and other cultural venues. Utilizing AR applications, visitors may access multimedia content, interactive components, and more, encouraging learning and engagement (Agarwal et al., 2021).

1.2.5.3 PERSONALIZED NARRATIVES

1. **Tailored Storytelling:** AR and VR make customized narratives possible. Narratives that are customized to the interests, tastes, and historical background of travelers can enhance the significance and engagement of their travel experience.
2. **Augmented Reality (AR) Wayfinding:** By giving real-time, context-aware information about sites of interest, routes, and other facts, AR helps travelers navigate. The efficiency with which visitors may tour a location is improved by this technology (Brunyé et al., 2018).

3. **Virtual Meetings and Events:** VR enables online conferences, meetings, and events, which eliminates the need for in-person participation. This is especially important in an environment where travel limitations and unpredictability might affect face-to-face meetings (Romão, 2021).

1.2.5.4 CRISIS PREPAREDNESS AND TRAINING

Catastrophes, whether man-made, natural, or hybrid, are mostly to blame for the immense loss of life and property. A disaster's appropriate preparation may have a profound impact on society. Hence, VR helps the staff to train and prepare for the disaster such as (Baggio et al., 2020):

1. **Simulated Training:** VR may be used for staff training and crisis preparedness. Tourism workers may better adjust to unanticipated obstacles by participating in scenarios that are simulated, such as emergency response or customer service simulations (Pencarelli, 2020).
2. **Interactive Advertising:** Users can engage with virtual items relevant to vacation locations or services by integrating AR into advertising campaigns. This participatory method increases interest and might affect travel choices (Brunyé et al., 2018).

1.2.5.5 ACCESSIBILITY AND INCLUSIVITY

1. **Virtual Accessibility:** According to Agarwal et al. virtual reality opens up travel experiences to those with physical disabilities. An inclusive approach is made possible by virtual travel, which makes it possible for a larger audience to discover and experience locations. AR and VR approach tourism facilities that redefine passenger engagement with destinations and attractions, adding to the industry's revolutionary character in a reformation of tourism 4.0. In an era marked by constant innovation, these immersive technologies provide new dimensions to the tourism experience, adjust to ongoing changes, and offer creative solutions for destination marketing and visitor engagement (Almobaideen et al., 2017).

1.2.6 BLOCKCHAIN FOR SECURITY AND TRANSPARENCY

Reforming security in several tourism-related areas and building transparency is known as "Blockchain for Security and Transparency in Tourism 4.0." Blockchain is a technological tool of fundamental component that provides several important advantages and a transformative nature of driven development in the tourist industry.

1.2.6.1 SECURE TRANSACTIONS

1. **Decentralized Ledger:** A system that identifies the nodes of transactions that focus on the spreadable and impenetrable ledger. This network guarantees the tourism stakeholders safe, open, and fraud-proof financial transactions (Baralla et al., 2021).
2. **Traceability:** End-to-end service supply chain visibility in the tourist industry is made possible by blockchain technology (Deb et al., 2024). Tracking the provenance of goods and services, guaranteeing authenticity, and encouraging sustainable and ethical tourist activities all depend on this openness (Agarwal et al., 2021).
3. **Self-executing Contracts:** Blockchain makes it possible to see the whole service supply chain to the stakeholders. To monitor the origin of goods and services, guarantee authenticity, and encourage sustainable and responsible tourist behaviors, this openness is essential (Demirel, 2023).

1.2.6.2 IDENTITY VERIFICATION

1. **Secure Identity Management:** Blockchain technology may be applied to decentralized and safe identity verification. In the tourist sector, this is especially important for confirming visitors' identities, guaranteeing safe service access, and lowering the possibility of identity theft (Banerji et al., 2021).
2. **Immutable Records:** Because of the immutability of blockchains, data cannot be changed or removed once it is recorded. By safeguarding critical data including financial transactions, travel schedules, and personal information, this function improves data security (Agarwal et al., 2021).

3. **Reduced Friction:** Blockchain opens up quicker and more affordable cross-border payments by interconnecting and lowering transaction activity. This may help the international travel and tourism sector's financial ecology become smoother and more effective (Demirel, 2023).

1.2.6.3 CUSTOMER LOYALTY PROGRAMS

1. **Transparent Reward Systems:** Blockchain improves the security and transparency of tourist loyalty schemes. Travelers may earn and redeem loyalty points in a decentralized and verifiable manner using tokenized incentives on a blockchain (Banerji et al., 2021).
2. **Distributed Architecture:** Blockchain is decentralized; it is naturally resistant to cyber-attacks. In the context of Tourism 4.0, where data breaches and cyber-attacks are major concerns, blockchains can provide a strong layer of protection (Otebolaku and Lee, 2018).
3. **Trustworthy Feedback:** Blockchain technology may be utilized to guarantee the legitimacy of reviews and ratings on travel-related websites. Travelers may make more trustworthy decisions by relying on user-generated material that has been verified by an unchangeable ledger (Gaki et al., 2016).

1.2.6.4 SUSTAINABLE AND RESPONSIBLE TOURISM

1. **Tracking Provenance:** In the tourism sector, blockchain technology makes it possible to track the origin of environmentally beneficial and sustainable activities. This openness promotes responsible tourism by empowering tourists to make well-informed decisions that are consistent with their beliefs (Banerji et al., 2021).

In conclusion, security and transparency in Tourism 4.0 signifies a revolutionary change that will make the sector safer, more open, and more productive (Otebolaku and Lee, 2018). Through the utilization of blockchain technology, the tourist industry can effectively tackle security issues, establish credibility, optimize workflows, and adjust to the ever-evolving digital terrain, ultimately forming a more robust and reliable tourism network.

1.2.7 SUSTAINABILITY AND RESPONSIBLE TOURISM

In the framework of Tourism 4.0, the ideas of sustainability and responsible tourism represent a dedication to reducing environmental effects, supporting ethical behavior in the face of ongoing change, and building community well-being. Transformative ideas are used in this digital age to build a more responsible and sustainable tourist sector. The following are the main ideas of this concept.

1.2.7.1 DIGITAL SUSTAINABLE PRACTICES

1. **Smart Resource Management:** IoT and data analytics are two examples of digital technologies that Tourism 4.0 uses to manage resources intelligently. This entails maximizing resource use as well as minimizing waste disposal in lodging and attractions (Pencarelli, 2020).
2. **Digital Mobility Solutions:** Eco-friendly mobility choices are emphasized in sustainable and responsible tourism. To lessen the carbon impact of travel, digital platforms encourage the use of ridesharing, electric or hybrid cars, and other environmentally friendly transportation options (Otebolaku and Lee, 2018).

1.2.7.2 BLOCKCHAIN FOR SUSTAINING SUPPLY CHAIN

1. **Certification and Provenance:** Blockchain is used to track and confirm the sustainability credentials of goods and services in the tourist supply chain system. This openness guarantees the maintenance of ethical and environmentally friendly behavior (Banerji et al., 2021).
2. **Digital Channels for Regional Companies:** Travel 4.0 platforms offer a digital marketplace to assist small companies. In addition to ensuring equitable economic distribution and encouraging

responsible use of regional goods and services, this also empowers communities (Prasetyo et al., 2020).

1.2.7.3 CROWD-SOURCED SUSTAINABILITY INITIATIVES

1. **Engagement through Apps:** Travelers are involved in crowd-sourcing sustainability projects using mobile applications and platforms. Instilling a sense of responsibility in visitors involves encouraging them to report environmental concerns, take part in clean-up activities, and contribute to nearby conservation programs (Banerji et al., 2021).
2. **Augmented Reality (AR) for Environmental Education:** AR is used to deliver engaging and instructive experiences on regional ecosystems, protecting species, and environmentally friendly behaviors. This improves travelers' comprehension of the significance of the site for the environment.
3. **Blockchain for Carbon Credits:** Transparent carbon offsetting methods are made easier by blockchain technology. By using digital payments, travelers may make contributions to carbon offset programs, guaranteeing that their travel plans are in line with sustainability objectives (Demirel, 2023).

1.2.7.4 DATA-DRIVEN CONSERVATION EFFORTS

1. **Internet of Things (IoT) for Wildlife Monitoring:** IoT devices gather information about animal behavior, habitat quality, and any dangers, which helps with wildlife conservation. Making educated decisions to safeguard biodiversity is made easier with the use of this data (Pencarelli, 2020).
2. **Digital Collaboration Tools:** With the help of digital cooperation and remote work, tourism 4.0 lessens the need for regular business travel. As a result, the carbon emissions from travel-related activities are reduced.
3. **Blockchain-backed Certifications:** The integrity of certificates for sustainable tourism is guaranteed by blockchain. Verifiable evidence of compliance is provided via certificates, such

as eco-labels and responsible tourism badges, which are safely kept on the blockchain (Dutta et al., 2020). In Tourism 4.0, the notion of sustainability and responsible tourism acknowledges the relationship between digital technology and the need to develop an ecologically, socially, and culturally conscious tourist sector (Pencarelli, 2020). Through the integration of digital advances, the sector can effectively adjust to ongoing developments while bringing beneficiaries through simultaneously the preservation of natural resources and the welfare of nearby populations.

1.2.8 USER-GENERATED CONTENT AND SOCIAL MEDIA

Stankov and Gretzel (2020) stated that User-Generated Content (UGC) in conjunction with Social Media in Tourism 4.0 highlights the noteworthy influence of internet platforms and user-generated material. Modern-day social media is now recognized as a digital tool and the ever-changing nature of the tourism industry's communication, marketing, and community development. The following are the main ideas of this concept.

1.2.8.1 DESTINATION MARKETING AND PROMOTION

1. **Real-time Engagement:** Social media platforms enable places to rapidly highlight attractions, events, and unique experiences to passengers in real time.
2. **Influencer Partnerships:** Tourism 4.0 embraces influencer marketing, which uses influencers to produce and disseminate interesting content that attracts a wide audience and highlights experiences and places (Deb and Mallik, 2023).
3. **Digital Travel Communities:** Social media helps to create online communities where travelers may exchange advice, suggestions, and experiences. These groups offer insightful information as well as a feeling of camaraderie among like-minded people.
4. **Interactive Content:** By utilizing tools like surveys, tests, and interactive maps, users may become more engaged, fostering a vibrant and active online travel community (Pencarelli, 2020).

1.2.8.2 REAL-TIME CUSTOMER FEEDBACK

1. **Instant Reviews and Ratings:** Travelers may instantly submit feedback via social media channels by leaving reviews, ratings, and comments. Businesses in the tourism industry may utilize this input to enhance offerings and quickly resolve client complaints (Kitsios et al., 2022).
2. **Data Analytics for Insights:** To help businesses adjust to the shifting expectations of their customers, Tourism 4.0 uses data analytics on social media platforms to gather insights into customer attitudes, preferences, and trends.
3. **Live Streaming:** Locations and businesses may provide instantaneous experiences, virtual tours, and updates by using social media capabilities such as live streaming. Travelers are informed and kept interested in this real-time material (Stankov and Gretzel, 2020).
4. **User-Generated Live Content:** By sharing their experiences in real-time, travelers may add to the live content, helping to create a lively and genuine representation of the places they visit (Pencarelli, 2020).

1.2.8.3 INTERACTIVE MARKETING CAMPAIGNS

1. **User-Generated Campaigns:** User-generated material is welcome in Tourism 4.0 marketing initiatives. Travelers may become brand ambassadors by sharing their experiences through contests, challenges, and hashtag campaigns (Teixeira et al., 2020).
2. **Augmented reality (AR) Effects and Filters:** AR capabilities are integrated into social media platforms, enabling users to add location-specific effects and filters to their material, therefore promoting destination branding.
3. **Personalized Contact:** Social media provides a means of facilitating individualized contact between travelers and travel companies. Businesses may customize offers, promotions, and messaging for users depending on their preferences by using CRM tools on social media sites.
4. **Chatbots and Direct Messaging:** Chatbots and direct messaging tools on social media offer real-time, tailored customer assistance, improving the entire customer experience (Teixeira et al., 2020).

5. **Trend Analysis:** Tourism firms may examine current trends in social media and customer behavior. Companies can remain relevant and flexible due to the use of the strategy in the face of constant change (Peceny et al., 2019b).

1.2.8.4 CRISIS COMMUNICATION AND MANAGEMENT

In times of crisis, social media platforms play a crucial role as communication channels. Businesses in the tourism industry can exhibit accountability and transparency by addressing issues, managing the narrative, and giving timely updates (Baralla et al., 2021). User-generated content and Social Media are essential elements of Tourism 4.0, impacting tourist decisions and forming the industry narrative. Teixeira et al. (2020) stated that by utilizing social media and user-generated content, the tourist industry can remain flexible, establish connections with a worldwide audience, and adjust to the always-changing digital landscape.

1.3 METHODOLOGY

A qualitative method is followed to gather important information to integrate digital technologies and the changing nature of this transformative environment. In research, strategies are needed to identify the experiences, perspectives, and challenges faced by key actors. This research was enhanced with a snowball sampling technique to gather information and feedback from participants. However, to ensure a large number of observations and to collect data for this new work, we recruited 10 participants to conduct interviews. Before starting the interview session, the researcher will make sure that the participant can easily answer the questions related to the research. Six questions were asked of the participants to help researchers understand the implementation of modern technology in the travel business, to collect in-depth data, and enrich the results.

This study is conducted by collecting data from key tourism industry stakeholders, including tourism industry leaders, technology innovators, experts, and researchers. Selected from the leaders of the tourism industry, the three interviewees are managers and decision-makers of major organizations and tourism companies who can explain the strategic direction of the industry and have a great opportunity to create and influence digital change. Two interview questions were asked to understand management approaches, challenges, and opportunities for integrating digital tools. These two questions: (1) "How does your organization approach the integration of digital technologies in the context of Tourism 4.0, and what overarching vision guides these initiatives?" and (2) "What are the primary challenges and opportunities your organization faces in adapting to the continuous changes brought about by Tourism 4.0? (Figure 1.3)."

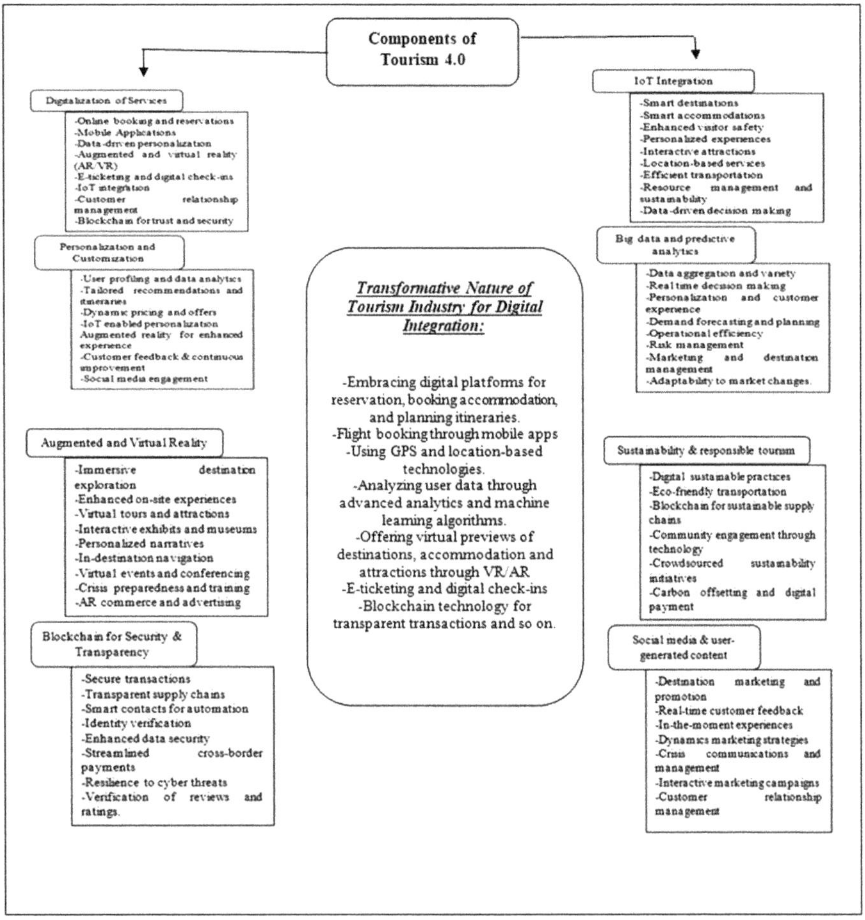

FIGURE 1.3 Model of tourism 4.0 for integrating digital technologies to create a transformative environment through its continuous change.

Source: Developed by authors.

On the other hand, two respondents were selected from the technology innovators who were from technology companies specializing in solutions and play pivotal roles, including making reservations, ticketing, IoT providers, and AI-driven services in providing tourism facilities. Similarly, they were asked two interview questions as well: (1) "What innovative technologies or solutions does your company provide to address the specific needs and challenges of the tourism industry in the era of Tourism 4.0?" and (2) "What emerging trends in technology do you foresee having a significant impact on the tourism industry in the coming years?" Lastly, five respondents were selected from academics and researchers who were experts in the concept of tourism management, technology adoption in tourism, or digital transformation, and they had an understanding of the dynamics in the hospitality sector. However, two questions were asked, and the questions were: (1) "What specific aspects of Tourism 4.0 and digital transformation in the tourism industry have been the focuses of your research?" and (2) "Based on your insights, what areas within the intersection of tourism and digital technologies do you believe warrant further research attention in the future?" By asking these six questions to a total of 10 participants, the researchers tried to understand the activities of tourism organizers' search through in-depth information and continuous changes in the travel business. However, chapter findings are found by the researchers in which they describe what respondents are asked interview questions.

1.4 FINDINGS OF THE STUDY

Six questions in all were posed to the respondents, who provided their opinions, convictions, and experiences on the incorporation of digital technologies and their disruptive character in the tourist sector, which reshaped Tourist 4.0. The first respondents were leaders in the tourist sector; there were a total of three, and they were asked two questions in all.

The first question was asked, *"How does your organization approach the integration of digital technologies in the context of Tourism 4.0, and what overarching vision guides these initiatives?"*

Sustainability in the tourist business is a priority, according to the leaders of the industry who responded to the survey. To increase the sustainability and value of the tourist sector, they also seek to draw in and satisfy customers. They reduce detrimental effects to promote the proper use of digital technology in the tourism business. While another respondent wants to place more emphasis on the strategic use of resources with the aid of digital technologies and embracing the ongoing changes in the hospitality industry, one of them also feels that collaboration and adaptability with the stakeholder parties are necessary when using digital and innovative technologies from government bodies, technology providers, industry parties, and so forth. Additionally, this formulation starts with the proper use of Digital Simulation (DS) and AI for accessible tourists.

The second question was asked, *"What are the primary challenges and opportunities your organization faces in adapting to the continuous changes brought about by Tourism 4.0?"*

Leaders in the tourist sector confront the difficulty of maintaining cultural integrity, which is a challenging endeavor that requires striking a balance between contemporary technology and cultural or historic authenticity. Respondents, on the other hand, are addressing the lack of expertise in industry strategic initiatives and development parties, emphasizing the necessity of group training sessions, data privacy implementation, and security considerations while using digital technologies in the travel and tourist sectors. The options include hints about collaborating with other groups and taking concrete steps to implement sustainable practices in the tourist sector, where stakeholders with relevant training may be attracted. Additionally, the respondents say that strategic decision-making and the appropriate adoption of changes related to digital technology may help the sector realize its intended vision and purpose.

The next respondents were technology innovators; there were a total of two respondents. They were asked a total of two questions, and they answered the questions very strategically. The first question was asked, *"What innovative technologies or solutions does your company provide to address the specific needs and challenges of the tourism industry in the era of Tourism 4.0?"*

In Tourism 4.0, the respondent discusses the imposition of cutting-edge solutions and the implementation of novel technology. The subject matter covered included destination management systems (DMS), the need for AR, smart technologies, VR, AI platforms, as well as IoT, mobile concierge apps, and other related topics. The authors highlight the significant threat associated with the industry stakeholders to gather information through digital devices.

The next question was asked, *"What emerging trends in technology do you foresee having a significant impact on the tourism industry in the coming years?"*

The improvement of digital pioneers, which are regarded as Tourism 4.0, hopes to foresee several new developments in the travel and tourism sector. Positive aspects of AI, robotics implications, drone technology, autonomous vehicles, neuro-tourism, and so forth, are covered. Technology innovators who responded to the survey anticipate significant changes as a whole and the integration of tourism's fourth revolution in the upcoming years. A significant impact arises on technological advancements by the kind of modification that could be both beneficial and harmful depending on human ethics.

The final respondents were from academia and research. There were a total of five participants who were asked a total of two questions. The first question asked was, *"What specific aspects of Tourism 4.0 and digital transformation in the tourism industry have been the focuses of your research?"*

Our first respondent is tourism executives, where researchers have primarily concentrated on having prioritized sustainable methods with them. The integration of digital technologies with ethics, crisis management, and its applications, the implications of digital travel, DMS, and so forth, is this chapter's prime objective. One of the researchers brought attention to the urgent need for digital technology practices in the hospitality sector, emphasizing resource sustainability through the use of transformational technology. The second answer concentrated on the drawbacks of integrating digital technologies, stating that our dependence on technology has made our survival as a species untenable and that we risk losing our culture, authenticity, and history. A further field of tourism executives stated that embracing positive technological changes that will elevate the passion of all key stakeholders and practices culture, history, values, and so forth, in the transformative effects of sustainability depends not only on modern technology but also on undergoing authentic culture, picturesque surroundings, and historic sites.

The final question asked to the respondent was, *"Based on your insights, what areas within the intersection of tourism and digital technologies do you believe warrant further research attention in the future?"*

The respondents brought up several topics they felt needed more investigation in the future. First, they adopted tourist 4.0 and thought about the moral implications of technology usage in the travel and tourism sector. Additionally, VR maximizes accessibility; executives also intend to incorporate AR, AI, and other technologies that can help travelers practice using intangible resources by using digital technology during real trips to historical sites, cultural centers, and other locations. In their future study, a few respondents highlighted the cross-cultural section and acknowledged the significance of connectedness among all parties involved in the tourist industry. A portion of the participants expressed faith in the capacity of digital technologies to foster global connections, foster mutual understanding, advance culture and history, close gaps, and improve responsible interactions among all parties involved.

To adopt Tourist 4.0 in the tourist sector, the majority of respondents sought to employ sustainable practices in the transformative digital technology where cooperation, adaptability, strategic implementation, avoiding bad behaviors, and intact norms, values, and culture are vital.

1.5 CONCLUSION

"Tourism 4.0" demonstrated an evolution of travel that is made simple for the travel industry by using digital technology. Better travel experiences for all travelers—including the elderly and the disabled—will also be gathered thanks to access. In 2020, Teixeira et al., stated that foreign travel is evolving in the new landscape of digital technologies and seems to provide popular venues and experiences that will revitalize (Peceny et al., 2019b). The tourist business may take advantage of the digital revolution by embracing various chances and adaptable offerings, training visitor workers to perform exceptionally well, and promptly managing significant customer conversions.

However, it underscores the value of innovation in providing a thrilling and genuine travel experience by the beneficiary use of digital technology in the travel industries. Key stakeholders may meet a range of technology issues and grow tourism in the global tourist business by utilizing the unique features of Tourism 4.0. The survey also emphasizes stakeholders' privacy and security issues, which technology management teams should take seriously. People with physical limitations who are unable to travel but still like to visit their favorite places may now use all tourism amenities thanks to Tourism 4.0. In response to shifts in technological advancement, sustainable activities and the preservation of social culture, prestige, and standing are highly regarded.

Travel management companies may quickly interact with consumers and provide search information for preferred places. These prospects boost the tourist industry's sustainability and encourage long-term travel. As part of the revitalization of Tourism 4.0, one of the features of Tourism 4.0 is the opportunity to enhance agents' environmental and tourism resource knowledge and competencies by adjusting to the ongoing changes in digital technologies. As per the World Tourist Organization (WTO) of the United Nations, there is an increasing apprehension over Tourism 4.0. This initiative seeks to create links between individuals globally, maintain customs and cultural heritage for their benefit, and employ digital technology in creative ways to minimize tourist resources. It assures that global travel will be detailed and constrain long-viability (Ramos and Brito, 2020). The primary stakeholders should not be the only focus of innovation and digital technology development but also sustainability; the tourist sector also has to embrace strategic and lasting adjustments. In light of the tourism business's need to embrace transformative technology and revolutionary character in continuing changes of the tourist business, we also value the ethical innovation and continual development of Tourism 4.0.

1.6 IMPLICATION OF THE STUDY

1.6.1 THEORETICAL IMPLICATION

This research covers the potential for digital integration in the tourist sector, which promotes revolutionary developments in the sector. Tourism 4.0, sustainability, competitiveness, and comprehension of the requirements of the implication in academic research that carries the re-evaluation of the dynamic Tourism 4.0 frameworks in the contemporary tourism industry are prioritized by transformative digital technologies (Gretzel et al., 2015). The current chapter offers a model or conceptual structure that shows how elements of Tourism 4.0 might use and integrate digital technology to bring about revolutionary changes in the tourist sector. In contrast, this research suggests working with DMS and combining with ICT as ways to evolve the new era, which is not only timely but also a vital component of the constantly changing environment (Peceny et al., 2020). This model will contribute to the literature on tourism. Future research studies will benefit greatly from it as it will make it easier to verify the gap in the dynamics of the current period and comprehend the role that integration plays in the tourist sector of emerging areas (Pahrudin et al., 2023). Additionally, Tourism 4.0 is reforming tourism by enabling the application of sustainable heritage preservation practices and beneficial growth for local communities. This promotes the essentiality of theoretical frameworks that influence changes in social influences and socioeconomic well-being (Neuhofer et al., 2015).

Additionally, the theoretical framework of the current study will make it easier to examine destination management organizations (DMOs) and identify potential information technology-based activities under Tourism 4.0, where offering a wide range of amenities promotes and facilitates accessible travel. It highlights potential effects due to TDPS and Enterprise Resource Planning (ERP) and focuses on ICT sophistication and offers a suitable remedy for newly arriving operational workforces, regarding security concerns and access to a variety of digital facilities (Abdurakhmanova et al., 2022). Finally, this study connects the many sectors and highlights the good developments and theoretical understanding in the growing sector in the form of academics. Future academic endeavors, the incorporation of sustainable practices, and the improvement of operational efficiency in associated industries will all be impacted by the model. Understanding digital transition is emphasized to prepare researchers to embrace the theoretical difficulties associated with evolutionary processes. These factors may be extremely important for future studies and procedures using the dynamic technological advances used in the travel and tourism sector.

1.6.2 PRACTICAL IMPLICATION

This study shows the needs of tourism industries and adopting new technology as well as how tourism facilities can develop their competency and industries can meet them strategically to transform into the continuing tourism industry. Effective outcomes include developing personalized experiences, verifying data, recording tourist preferences, and providing services and recommendations (Xiang et al., 2017). This research can explore the needs of key stakeholder collaboration, adapting digital technologies, understanding dynamic change, managing resources, and creating broad approaches for key stakeholders.

The potentiality of the sector is not yet fully recognized and requires greater collaboration, participation, and implementation, which is a challenge for companies. Increasingly, researchers can build knowledge, and understand key stakeholders, and the potentiality of the research. Aiming to provide accessible tourism facilities by using different kinds of technological applications example AI, AR, VR the IoT, and DMS to tourists (Aburakhmanova et al., 2022). Technological advancement involves the era of tourism 4.0 which starts with a strategic understanding of these digital tools to improve the overall tourist perception, expectations, and optimal use of resources, sustainable management practices, and long-term sustainability. Investment in artificial intelligence, AR, and VR will bring about profound changes in tourism companies that drive positive developments and ensure the lengthy success of business organizations (Neuhofer et al., 2015).

Therefore, this chapter provides an opportunity to collaborate with trained key industry stakeholders, including local communities, tourists, and local government, on the implementation of sustainability and addressing safety in tourism issues. Gretzel Sigala et al. (2015) prefer to blend with IoT will increase smart tourism spot richness and optimize the overall quality, infrastructure, and resource use according to tourists' expectations. The practical application which concerned with enriching the traveler experience associated with AR (Ganguly, 2018). The future is to focus on satisfying experiences adapted to the needs of tourists. This personalized approach will enable the most relaxed and safest tour which makes memorable and enjoyable travel highlighting customized recommendations, signature services, and prominent itineraries to the tourist (Xiang et al., 2017). These digital transformations align new opportunities to the dynamic tourism world that focuses on and encourages sustainable tourism practices, minimizing waste, and smart hotels with automated services. Embracing these digital technologies will increase efficiency and effective tourism, personalized service, and quality tourism which support accessible and reliable tourism and create long-lasting memories for tourists (Neuhofer et al., 2015; Stankov and Gretzel, 2020). In conclusion, this study about the reform tourism industry and building accessible tourism by adapting digital tools with a transformative environment for easing future visits, encompassing personalized services, Smart Destinations Management Systems, augmented reality, block-chain security, data-driven, decision-making, and the implications of emerging digital technologies. These developments highlight industry insights and point towards a future implementation where technology enhances efficiency and security for travelers but also interconnects stakeholders with evolving societal values and expectations where need to stay updated on the latest technology and reports for continued insights into this dynamic landscape.

KEYWORDS

- **augmented reality**
- **digital technologies**
- **Internet of Things (IoT)**
- **modern technology**
- **qualitative method**
- **social media**
- **stakeholders**
- **Tourism 4.0**
- **transformative nature**
- **virtual reality**

REFERENCES

Abdurakhmanova, G. K., Astanakulov, O. T., Goyipnazarov, S. B., & Irmatova, A. B. (2022). Tourism 4.0: Opportunities for applying Industry 4.0 technologies in tourism. *ACM International Conference Proceeding Series*, 33–38. https://doi.org/10.1145/3584202.3584208 (accessed on 8th July 2025).

Agarwal, S., Page, S. J., & Mawby, R. (2021). Tourist security, terrorism risk management, and tourist safety. *Annals of Tourism Research*, *89*, 103207. https://doi.org/10.1016/j.annals.2021.103207 (accessed on 8th July 2025).

Almobaideen, W., Krayshan, R., Allan, M., & Saadeh, M. (2017). Internet of Things: Geographical routing based on healthcare centers vicinity for mobile smart tourism destination. *Technological Forecasting and Social Change*, *123*(March 2016), 342–350. https://doi.org/10.1016/j.techfore.2017.04.016 (accessed on 8th July 2025).

Amir, S., Dura, N., Yusof, M. A., Nakamura, H., & Nong, R. A. (2020). Challenges of smart tourism in Malaysia eco. *Journal of the Malaysian Institute of Planner*, *18*(4), 442–451.

Baggio, R., Micera, R., & Del Chiappa, G. (2020). Smart tourism destinations: A critical reflection. *Journal of Hospitality and Tourism Technology*, *11*(3), 407–423. https://doi.org/10.1108/JHTT-01-2019-0011 (accessed on 8th July 2025).

Banerji, D., Rashideh, W., Arora, B., & Pratihari, A. R. (2021). Application potential of blockchain technologies in the travel and tourism industry. In *Blockchain Applications in IoT Ecosystem* (pp. 289–299). Cham: Springer International Publishing.

Baralla, G., Pinna, A., Tonelli, R., Marchesi, M., & Ibba, S. (2021). Ensuring transparency and traceability of food local products: A blockchain application to a smart tourism region. *Concurrency and Computation: Practice and Experience*, *33*(1), 1–18. https://doi.org/10.1002/cpe.5857 (accessed on 8th July 2025).

Brunyé, T. T., Gardony, A. L., Holmes, A., & Taylor, H. A. (2018). Spatial decision dynamics during wayfinding: Intersections prompt the decision-making process. *Cognitive Research: Principles and Implications*, *3*(1), 1–19. https://doi.org/10.1186/s41235-018-0098-3 (accessed on 8th July 2025).

Buhalis, D., & Amaranggana, A. (2013). Smart tourism destinations. In *Information and Communication Technologies in Tourism 2014* (pp. 553–564). Springer International Publishing Switzerland 2013. https://doi.org/10.1007/978-3-319-03973-2 (accessed on 8th July 2025).

Cepeda-Pacheco, J. C., & Domingo, M. C. (2022). Deep learning and Internet of Things for tourist attraction recommendations in smart cities. *Neural Computing and Applications*, *34*(10), 7691–7709. https://doi.org/10.1007/s00521-021-06872-0 (accessed on 8th July 2025).

Choi, Y., Hickerson, B., Lee, J., Lee, H., & Choe, Y. (2022). Digital tourism and wellbeing: Conceptual framework to examine technology effects of online travel media. *International Journal of Environmental Research and Public Health*, *19*(9), 1–13. https://doi.org/10.3390/ijerph19095639 (accessed on 8th July 2025).

Deb, S. K., & Mallik, N. (2023). Effects of social media in tourism marketing: Outlook on user generated content. *Journal of Digital Marketing and Communication*, *3*(2), 49–65. https://doi.org/10.53623/jdmc.v3i2.316 (accessed on 8th July 2025).

Deb, S. K., & Nafi, S. M. (2021). Assessing critical success factors of ICT-based learning: An analysis. *The Cost and Management*, *49*(04).

Deb, S. K., Kuri, B. C., & Nafi, S. M. (2024). Application of knowledge management in tourism and hospitality industry: A sustainable approach. In *Knowledge Management and Knowledge Sharing: Business Strategies and an Emerging Theoretical Field* (pp. 99–116). Cham: Springer Nature Switzerland.

Deb, S. K., Nafi, S. M., & Valeri, M. (2022). Promoting tourism business through digital marketing in the new normal era: A sustainable approach. *European Journal of Innovation Management*.

Deb, S. K., Nafi, S. M., Mallik, N., & Valeri, M. (2023). Mediating effect of emotional intelligence on the relationship between employee job satisfaction and firm performance of small business. *European Business Review*, *35*(5), 624–651.

Demirel, E. (2023). Application of blockchain-based smart contract in sustainable tourism finance. In *Blockchain for Tourism and Hospitality Industries* (pp. 122–138). Routledge.

Dorcic, J., Komsic, J., & Markovic, S. (2019). Mobile technologies and applications towards smart tourism–State of the art. *Tourism Review*, *74*(1), 82–103. https://doi.org/10.1108/TR-07-2017-0121 (accessed on 8th July 2025).

Dutta, P., Choi, T. M., Somani, S., & Butala, R. (2020). Blockchain technology in supply chain operations: Applications, challenges and research opportunities. *Transportation Research Part E: Logistics and Transportation Review*, *142*(2), 102067. https://doi.org/10.1016/j.tre.2020.102067 (accessed on 8th July 2025).

Gaki, E., Kostopoulou, S., Parisi, E. D., & Lagos, D. G. (2016). The evaluation of tourism satisfaction in island destinations: The case of the Ionian Islands of Greece. *56th Congress of the European Regional Science Association: "Cities & Regions: Smart, Sustainable, Inclusive.* https://www.econstor.eu/bitstream/10419/174645/1/Paper0295_EleniGaki.pdf (accessed on 8th July 2025).

Ganguly, P. (2018). Role of information and communication technologies (ICTs) to sustainable tourism development in Santiniketan, West Bengal–India. *International Journal of Management Studies, V* (Special Issue 2), 54. https://doi.org/10.18843/ijms/v5is2/07 (accessed on 8th July 2025).

Gretzel, U., Sigala, M., Xiang, Z., & Koo, C. (2015). Smart tourism: Foundations and developments. *Electronic Markets, 25*(3), 179–188. https://doi.org/10.1007/s12525-015-0196-8 (accessed on 8th July 2025).

Huang, C. D., Goo, J., Nam, K., & Yoo, C. W. (2017). Smart tourism technologies in travel planning: The role of exploration and exploitation. *Information and Management, 54*(6), 757–770. https://doi.org/10.1016/j.im.2016.11.010 (accessed on 8th July 2025).

International Telecommunication Union. (2019). Global and regional ICT data. https://www.itu.int/en/ITU-D/Statistics/Pages/stat/default.aspx (accessed on 8th July 2025).

Jeong, M., & Shin, H. H. (2020). Tourists' experiences with smart tourism technology at smart destinations and their behavior intentions. *Journal of Travel Research, 59*(8), 1464–1477. https://doi.org/10.1177/0047287519883034 (accessed on 8th July 2025).

Kansakar, P., Munir, A., & Shabani, N. (2019). Technology in the hospitality industry: Prospects and challenges. *IEEE Consumer Electronics Magazine, 8*(3), 60–65. https://doi.org/10.1109/MCE.2019.2892245 (accessed on 8th July 2025).

Kantamalajaroen, A., & Summaniti, L. (2021). Development of tourism potential for Tourism 4.0 case study of Kuedchang Sub District, Mae Taeng, Chiang Mai. http://202.28.38.45/dspace/handle/123456789/452 (accessed on 8th July 2025).

Kim, D. Y., Park, J., & Morrison, A. M. (2018). A model of traveler acceptance of mobile technology. *International Journal of Tourism Research, 10*(5), 393–407. https://doi.org/10.1002/jtr.669 (accessed on 8th July 2025).

Kitsios, F., Mitsopoulou, E., Moustaka, E., & Kamariotou, M. (2022). User-generated content in social media: A twenty-year bibliometric analysis in hospitality. *Information & Computer Security, 13*(12), 558–574.

Kononova, O., Prokudin, D., & Tupikina, E. (2020). From e-tourism to digital tourism: Terminologically review. *CEUR Workshop Proceedings, 34*(4), 164–177.

Kramer, R., Modsching, M., & De, R. H. (2022). Information and communication technologies in tourism. *Information and Communication Technologies in Tourism 2007, 32*(3), 1–13. https://doi.org/10.1007/978-3-211-69566-1 (accessed on 8th July 2025).

Kumar, S., Kumar, V., Kumari Bhatt, I., Kumar, S., & Attri, K. (2023). Digital transformation in tourism sector: Trends and future perspectives from a bibliometric-content analysis. *Journal of Hospitality and Tourism Insights, 12*(3), 1–14. https://doi.org/10.1108/JHTI-10-2022-0472 (accessed on 8th July 2025).

Li, J., Xu, L., Tang, L., Wang, S., & Li, L. (2018). Big data in tourism research: A literature review. *Tourism Management, 68*, 301–323. https://doi.org/10.1016/j.tourman.2018.03.009 (accessed on 8th July 2025).

Natalia, N., Liliya, S., Aigul, B., Nazgul, K., & Dolores, P. (2020). Digital technologies as a driver for the development of the tourism industry. *E3S Web of Conferences, 04002*(4), 1–12. https://doi.org/10.1051/e3sconf/20201590400 (accessed on 8th July 2025).

Neuhofer, B., Buhalis, D., & Ladkin, A. (2015). Technology as a catalyst of change: Enablers and barriers of the tourist experience and their consequences. *Information and Communication Technologies in Tourism 2015, 14*(1), 789–802. https://doi.org/10.1007/978-3-319-14343-9_57 (accessed on 8th July 2025).

Nguyen, D. D., Rohács, J., Rohács, D., & Boros, A. (2020). Intelligent total transportation management system for future smart cities. *Applied Sciences (Switzerland)*, *10*(24), 1–31. https://doi.org/10.3390/app10248933 (accessed on 8th July 2025).

Osondu, J. (2018). Tourism digitization using the main components of Industry 4.0. *Journal of International Economy and Business*, *6*(63), 12–19.

Otebolaku, A., & Lee, G. M. (2018). A framework for exploiting the Internet of Things for context-aware trust-based personalized services. *Mobile Information Systems*, *23*(3), 1–25. https://doi.org/10.1155/2018/6138418 (accessed on 8th July 2025).

Pahrudin, P., Hsieh, T. H., & Liu, L. W. (2023). Information, communication, and technology in the field of tourism and hospitality: A bibliometric approach. *Engineering Proceedings*, *38*(1), 1–15. https://doi.org/10.3390/engproc2023038004 (accessed on 8th July 2025).

Pai, C. K., Liu, Y., Kang, S., & Dai, A. (2020). The role of perceived smart tourism technology experience for tourist satisfaction, happiness and revisit intention. *Sustainability (Switzerland)*, *12*(16). https://doi.org/10.3390/su12166592 (accessed on 8th July 2025).

Paris, C. M. (2012). Flashpackers: An emerging sub-culture? *Annals of Tourism Research*, *39*(2), 1094–1115. https://doi.org/10.1016/j.annals.2011.12.001 (accessed on 8th July 2025).

Peceny, U. S., Urbančič, J., Mokorel, S., Kuralt, V., & Ilijaš, T. (2019a). Tourism 4.0: Challenges in marketing a paradigm shift. In *Consumer Behavior and Marketing*. IntechOpen.

Peceny, U. S., Urbančič, J., Mokorel, S., Kuralt, V., & Ilijaš, T. (2019b). Tourism 4.0: Challenges in marketing a paradigm shift. In *Consumer Behavior and Marketing*. IntechOpen.

Peceny, U. S., Urbančič, J., Mokorel, S., Kuralt, V., & Ilijaš, T. (2020). Tourism 4.0: Challenges in marketing a paradigm shift. *Consumer Behavior and Marketing*, *22*(1), 1–19. https://doi.org/10.5772/intechopen.84762 (accessed on 8th July 2025).

Pencarelli, T. (2020). The digital revolution in the travel and tourism industry. *Information Technology & Tourism*, *22*(3), 455–476.

Prasetyo, B. D., Kurniadi, R., & Setyowati, R. (2020). Industry 4.0 and the future challenges for the local community to manage eco-tourism (A case study of the participatory eco-tourism management in Menipo Island, West Timor, Indonesia). *IOP Conference Series: Earth and Environmental Science*, *533*(1), 0–13. https://doi.org/10.1088/1755-1315/533/1/012002 (accessed on 8th July 2025).

Ramos, C. M., & Brito, I. S. (2020). The effects of Industry 4.0 in tourism and hospitality and future trends in Portugal. In *The Emerald Handbook of ICT in Tourism and Hospitality* (pp. 367–378). Emerald Publishing Limited.

Romão, J. (2021). Tourism, smart specialization, and sustainable development. In *Tourism, Smart Specialization and Sustainable Development*. https://doi.org/10.3390/books978-3-03943-672-9 (accessed on 8th July 2025).

Sharma, S., Rishi, O. P., & Sharma, A. (2021). IoTeST: IoT-enabled smart tourism—Shaping the future of tourism. In *Rising Threats in Expert Applications and Solutions: Proceedings of FICR-TEAS 2020* (Vol. 1187, pp. 175–186). Springer Singapore. https://doi.org/10.1007/978-981-15-6014-9_20 (accessed on 8th July 2025).

Stankov, U., & Gretzel, U. (2020). Tourism 4.0 technologies and tourist experiences: A human-centered design perspective. *Information Technology & Tourism*, *22*(3), 477–488.

Steyn, M. (2010). *Innovative Issues and Approaches in Social Sciences (IIASS). Innovative Issues and Approaches in Social Sciences*, *3*(3), 50–82.

Sustacha, I., Baños-Pino, J. F., & Del Valle, E. (2023). The role of technology in enhancing the tourism experience in smart destinations: A meta-analysis. *Journal of*

Destination Marketing and Management, *30*(August), 1–12. https://doi.org/10.1016/j.jdmm.2023.100817 (accessed on 8th July 2025).

Teixeira, P., Teixeira, L., & Eusébio, C. (2020). Accessible@Tourism 4.0: An exploratory approach to the role of Industry 4.0 in accessible tourism. In *Handbook of Research on Social Media Applications for the Tourism and Hospitality Sector* (pp. 192–211). IGI Global.

Xiang, Z., Du, Q., Ma, Y., & Fan, W. (2017). A comparative analysis of major online review platforms: Implications for social media analytics in hospitality and tourism. *Tourism Management*, *58*(3), 51–65. https://doi.org/10.1016/j.tourman.2016.10.001 (accessed on 8th July 2025).

CHAPTER 2

APPLICATIONS OF IOT (INTERNET OF THINGS) IN THE TRAVEL AND TOURISM INDUSTRY

MD. ABDUL HALIM[1*] and PRITOM KUNDU JOY[2]

[1]*Department of Tourism and Hospitality Management, Rangamati Science and Technology University, Rangamati, Bangladesh*

[2]*School of Graduate Studies, Trent University, Peterborough, Ontario, Canada*

*Corresponding author

ABSTRACT

The tourism industry has been undergoing substantial expansion for the last few decades, which makes it one of the most rapidly developing sectors. The pandemic has significantly impacted the industry, leading to new business practices and opportunities, mainly through the implementation of the Internet of Things (IoT). The IoT facilitates convenient accessibility and engagement with diverse information. The tourism industry is continuously evolving with the advancement of IoT technologies, which have unprecedentedly revolutionized operations and customer experiences. The chapter introduces, first, the fundamental IoT ideas and their existing application to the travel and tourism industry. Then, the real-world usages of IoT have been examined in the components and distribution channels of the travel and tourism industry, ranging from attraction and lodging to transportation, activities, and destination management. Furthermore, the chapter demonstrates how IoT contributes to developing Smart Cities and Smart Tourism, and the advantages and challenges of adopting IoT

Smart Tourism: Managing Technological Revolution in the Tourism Industry. Santus Kumar Deb (Ed.)

DOI: 10.1201/9781779641090-2

have also been addressed. Secondary sources have been used to compile this chapter's information. The findings reveal that the tourism industry is the pioneer consumer of adopting IoT applications for their customers to improve tourist feedback and provide customized, smooth, and engaging travel experiences, boosting operational excellence. Data privacy and security concerns are the significant challenges in adopting these IoT technologies.

2.1 INTRODUCTION

Humans are naturally curious about technology, which makes them want to make their lives easier and dependent on technology. Technological advancements have transformed humanity in the last few decades, with recent breakthroughs in Industry 4.0 and the innovation of IoT, Big Data, and AI. The Internet of Things (IoT) is a significant development in the ever-evolving realm of technology, representing a paradigm shift reshaping several businesses globally. IoT is recognized as a vital area of cutting-edge technology and is generating considerable attention from several industries (Lee and Lee, 2015). IoT enables the interconnection of devices, systems, and services beyond machine-to-machine (M2M) communication, which serves a wide range of applications in several sectors (Sharma et al., 2019). Integrating the IoT in the travel and tourism industry has sparked a significant shift, revolutionizing our encounters and reshaping how we engage with the world. IoT is an extensive and linked network that encompasses many publicly and privately owned businesses, educational institutions, and federal networks. These networks, which may span from micro to macro scales, are interconnected by utilizing a diverse array of connectivity methods such as wired, wireless, and optical networks (Nunberg, 2012). IoT is an interconnected system of material or physical items, including different gadgets, devices, automobiles, households, and other objects, where they collect and exchange data among themselves to enhance users' experiences.

The notion of a network including intelligent devices was first deliberated in 1982, whereby a Coca-Cola vending machine at Carnegie Mellon University was retrofitted to establish the first instance of an appliance linked to the internet. However, the idea behind the IoT was developed over time due to the growth of several technologies and concepts rather than having a single creator. The first proposal for the IoT was put out by Kevin Ashton in 1999. Ashton defined the IoT as a network of traceable interconnected things via radio-frequency identification (RFID) technology.

The IoT is a distinctive worldwide network architecture that can be constructed autonomously based on interoperable communication protocols. IoT system design must link physical and virtual worlds via networking, communication, and procedures where adaptable, scalable, and device-operable should enable dynamic interaction and communication, and IoT should be diverse and decentralized for real-time engagement (Gokhale et al., 2018). The components of the IoT may be classified into three tiers in the following manner: Level 0 encompasses sensors, controllers, objects, and gateways; Level 1 encompasses data storage and data analytics; and Level 2 encompasses data visualization and interpretation (Sharma and Gupta, 2021).

The concept of the IoT enables the distant detection and management of devices via the utilization of current network infrastructure. This technology has the potential to provide a smoother incorporation of real-life situations with systems powered by computers, resulting in improved productivity and accuracy. The tourism business has consistently observed and adapted to developments, continuously improving its practices and appearance to meet changing market needs (Deb et al., 2022). According to the International Data Corporation, it is projected that there will be 41.6 billion interconnected IoT devices, often referred to as "things," by 2025, and this substantial increase in IoT devices will result in the generation of around 79.4 zettabytes of data (Balandina et al., 2015). According to projections (Statista, 2022), the number of IoT-linked devices in use is expected to exceed 75 billion by 2025. This would represent an almost threefold surge compared to the number of IoT devices deployed in 2019. IoT applications are revolutionizing travel by providing visitors with precise directions and quiet times, and these services enhance decision-making and local economic balance, which also enable local enterprises to compete for market share and improve service quality based on implicit feedback, ensuring a pleasant journey for all visitors (Wise and Heidari, 2019). Major tourist and hospitality corporations have recognized the significance of technology and the extensive prospects it presents for contemporary firms (Car et al., 2019).

Due to the widespread use of mobile devices globally, social mobile crowd sensing via the IoT enables tourists to access information about famous events in a specific location, which also facilitates interactive communication with other tourists in various places, offers insights into the optimal locations to visit at specific times, provides climate predictions, and estimates journey durations all day long (Wise and Heidari, 2019). Simultaneously, tourism suppliers are adopting IoT to enhance their customer experiences through product development and personalization. The concept of smart home, a prominent application of IoT, is adopted by hotels in their different department like security, housekeeping for waste management and so forth.

This chapter explores the use of IoT in the travel and tourism sector, focusing on its practical applications in attractions, accommodations, transportation, activities, and destination management. The chapter also illustrates the evolution of IoT and its different significant components and challenges. The entire chapter is arranged into three major segments. The first segment discusses the ideas of IoT and its evolution; in the second segment, applications of IoT have been discussed, dividing into the different components of tourism, and in the last segment, opportunities and challenges have been demonstrated with its other implications in the field of smart cities, smart tourism, and destination management.

2.2 LITERATURE REVIEW

2.2.1 UNDERSTANDING INTERNET OF THINGS (IOT): EVOLUTION OF THE FOUNDATION AND CONCEPTS

The internet facilitates the connection of many devices, such as computers, printers, and fax machines. However, technological progress has expanded its capabilities, including recognizing and controlling ordinary things via RFID tags, sensors, and smartphones. The IoT denotes a network system of tangible entities which are able to detect, interact, and connect to the Internet, becoming an essential component of this interconnected system (Sharma et al., 2019). The IoT, sometimes acknowledged as the Internet of Everything or the Industrial Internet, is a novel tech framework conceived as a global network of interconnected devices and gadgets with the ability to communicate and engage with one another (Lee and Lee, 2015). When commonplace tangible things, for example, detectors, actuators, label badges, smartphones, and other similar items, are linked together so they can communicate with each other, either directly or indirectly, via cable networks or the web, this is what we refer to as IoT (Munir et al., 2017). Five main groups make up the business IoT ecosystem: developers of software and hardware platforms, network technologies, applications or solutions, and lastly, users and customers (Lee, 2019). A broad spectrum of IoT applications has surfaced, and the eagerness of organizations to adopt them is increasing quickly. Sundmaeker et al. (2010) have discussed the prime evolution of IoT.

Table 2.1 summarizes the evolution of diverse fundamental technologies of IoT development in the domain of fundamental IoT devices, including architecture, system, applications and procedures, physical parts, and data analysis. Multiple industries, including retailing, production, medical and policies, household goods, machinery for industrial use, travel companies, and administration of logistics, are being penetrated by the IoT (Lee and Lee, 2015).

2.2.2 ESSENTIAL INTERNET OF THINGS (IOT) TECHNOLOGIES

Lee and Lee (2015) have discussed five frequently used IoT technologies for the implementation of effective IoT-based goods and services: RFID, wireless sensor networks (WSN), middleware, cloud computing, and IoT application software. RFID technology enables the automatic detection and data extraction of objects through employing radio signals, a label, and an interpreter; WSN are self-regulating devices spread in space to monitor physical or environmental variables; Middleware is an application tier that connects applications programs, facilitating interaction and input/output operations for developers; cloud computing allows users to access a repository comprising various customizable resources shared among users, including hardware devices, applications, networks, servers, and storages, and the IoT applications enable the creation of several industry-focused and user-tailored IoT applications.

According to Madakam et al. (2015), the prerequisites for successful execution of the IoT include (i) changing needs for resources; (ii) instantaneous requirements; (iii) accelerated increase of demand; (iv) apps' accessibility; (v) confidentiality of the data and user; (vi) energy usage efficiency of the software; (vii) running apps close to final users; and (viii) entree to a cloud platform that is open and can be used by anyone.

TABLE 2.1 Key Internet of Things (IoT) Technology Evolution

	Before 2010	2010–2015	2015–2020	Beyond 2020
Innovation in the field of IoT	• IoT infrastructure Blueprint; • Context-aware middleware; • Platforms for smart thinking.	• IoT infrastructure advancement; • IoT infrastructure in the Finance sector; • Network of networks design; • Interoperability across Fog of Things (FoG) platforms.	• Dynamic, situation-based design; • Self-properties.	• Intellective framework; • Experimental architectures.
Technology for Network	• Sensor networks.	• Self-aware and self-organizing networks; • Sensor network Location transparency; • Delay-tolerant networks; • Storage and power networks; • Assorted networking technologies.	• Network context awareness.	• Network cognition; • Self-learning, self-repairing networks.
Software and Algorithms	• Relational database incorporation; • IoT-oriented RDBMS; • Event-based system; • Sensor and Sensor networks middleware; • Proximity/Localization algorithms.	• Massive, Open Semantic Framework (OSF); • Breakable and combinable algorithms; • Next generation IoT: social and business software.	• Outcome-based application; • Distributed intelligence, problem solving; • Things-to-Things association environments.	• Consumer-based application; • The invisible IoT; • Easy-to-deploy IoT software; • Things-to-Humans collaboration; • IoT 4 All.

TABLE 2.1 *(Continued)*

	Before 2010	2010–2015	2015–2020	Beyond 2020
Hardware	• RFID tags and select sensors; • Mobile device-integrated sensors; • NFC in mobile phones; • More compact and cost-effective; • MEMs technology.	• Readers that support multiple protocols and standards; • Additional sensors and actuators; • Low-cost, secure identifiers (such as silent tags).	• Intelligent biochemical sensors; • Additional actuators and sensors (tiny sensors).	• Novel materials and nanotechnology.
Data and Signal Processing Technologies	• Processing of serial data; • Processing of parallel data; • Service quality.	• Energy and frequency spectrum-aware data processing; • Context-adaptive data processing.	• Context-aware data processing and responses to data.	• Processing of cognitive information and optimization.

Source: Sundmaeker et al., 2010, P. 74.

According to Miorandi et al. (2012), IoT has three primary system-level attributes: (i) all objects can communicate: smart entities have the capability to establish wireless communication with one other, so creating impromptu networks of linked items; (ii) everything can be identified: intelligent objects are assigned a digital identifier: connections between things may be defined in the digital realm when physical linkage is not possible; and (iii) all objects have the ability to interact: intelligent objects may engage with the immediate surroundings by using their sensing and actuation capabilities anytime they are present. Girau et al. (2016) categorized Cloud IoT architecture into four levels: Application, Service, Virtualization, and Real World. The Application level deploys user-oriented applications, while the Service layer handles application requests and connects them to atomic services. The Virtualization layer enhances objects' functionality, and the Real-World layer contains physical Smart City devices. Two additional cross-layers manage application quality and secure communication.

In their study, Nitti et al. (2017) examined the essential criteria for an IoT platform in a smart city setting, focusing on security, adaptability, and data needs as distinct areas of consideration. They suggested four levels of IoT platforms consisting of a hierarchical design: Application Layer, Service Layer, Virtualization Layer, and Real-World Layer. In order to meet the quality criteria and ensure secure communication in real-world devices of the Smart City, it is necessary to add two more cross layers while still retaining trustworthiness and security.

2.2.3 EVOLUTION OF THE INTERNET OF THINGS (IOT) IN THE TRAVEL INDUSTRY

The IoT is broadly acknowledged as a crucial emerging technology domain and generates substantial attention from several enterprises. Technological developments such as the IoT have gained prominence within the tourist sector, presenting a burgeoning trend that has the potential to impact and reshape the whole business significantly by emphasizing customization, securitization, and optimization within the travel business for airlines, hotels, and cruises (Koteshov, 2023). Different sectors, like tourism, have adopted IoT. Smart home is one of the most emerging sectors which has applied IoT massively. The global number of smart homes is projected to be over 300 million and 450 million by 2025, and in 2023, smart home devices will be actively used in around 60.4 million households in the

United States (Curls, 2023). Smart home devices are also directly connected with the tourism accommodation industry, where hotels and resorts use devices to customize, improve, and collaboratively develop tourist services during their stay (Rahman et al., 2023; Sharma and Gupta, 2021).

The hotel industry has seen a significant transformation due to the IoT. When visitors leave a room, electronic doorbells use infrared sensors to detect body heat and discreetly scan the space, allowing the staff to clean the room without disturbing the hotel guests (Bauer et al., 2014). Utilizing technology intelligently will enhance current services by providing real-time updates, such as localized weather forecasts via smartphone applications. Additionally, it will provide personalized services to enhance the entire travel experience, and tourism involves the consumption of information in a timely manner, by offering suitable services to the relevant individuals (Wise and Heidari, 2019). The information will be supplied depending on users' geographical placement, eliminating the need for users to search for suitable websites. This enables immediate and real-time delivery and authentication of data (Wise and Farzin, 2018). Smart tourism is an emerging kind of tourism that utilizes advanced engineering such as IoT, cloud computing, Big Data, Artificial Intelligence, and others to enhance the visitor experience, support industrial growth, and improve administrative oversight (Wu, 2017). The latest applications of IoT in the tourist industry include implementing computerized parking systems, intelligent swimming pools, and managing hotel temperatures (Kansakar et al., 2019). An important implementation in hotels is the digital room key, which is provided to the guest's phone to access their room (Sharma and Gupta, 2021) (Figure 2.1).

Zheng et al. (2021) addressed the application of IoT in recreational travel for sports, where they discussed its limitations: existing operational inefficiency and extensive resource requirements for implementing IoT in sport tourism for its conventional operational method. They contend that it is essential to focus on combining resources and improving the entire procedure of managing the sports tourist service business, which will achieve benefits in search time, fusion malfunction, and power usages. The use of an IoT service model was discussed for creating smart hotel rooms by Lee (2019). Lee explored the potential for developing IoT services to bring innovation to hotel rooms by discussing an IoT service model with four crucial factors: value proposition, networked activities, resources, and sustainability. A smart city IoT framework called iTour was designed to facilitate the autonomous movement of visitors by Tripathy et al. (2018) by addressing the necessity and importance of all stakeholders. iTour offers a range of services via smart devices, such as hotel booking, taxi reservation, and searching for key services like banking, guides, and restaurants. In addition, intelligent gadgets provide user authentication, enhance security, and assist in urgent scenarios such as theft, road rage, and medical crises. Crowdsourcing may provide individuals with information like the optimal times to visit sights or restaurants, alternate routes to consider, strategies to prevent unexpected inconveniences while traveling, suitable parking options, and the most efficient public transportation alternatives (Wise and Heidari, 2019).

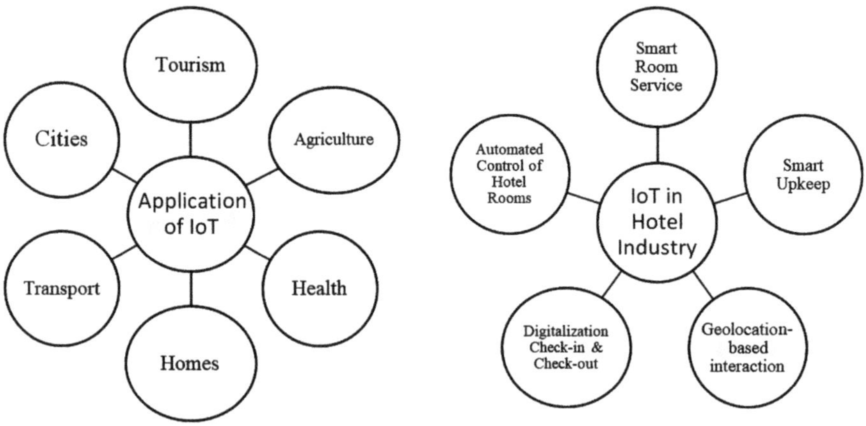

FIGURE 2.1 Application of Internet of Things (IoT) and IoT in hotel industry.
Source: Sharma and Gupta, 2021.

Kansakar et al. (2019) described and identified many crucial improvements of IoT thsat must be included in the hotel business to adapt its service platform to the contemporary technology environment. They proposed that the entire hospitality service operation benefit both hotel guests and hotel owners, and in-room IoT sensors will work to obtain, store, and analyze data for them. There is a solid link between smart cities and smart tourism. The technologies developed for smart cities facilitate tourist services, profoundly enhancing their travel experiences. The study conducted by Nitti et al. (2018) examines the convergence of Smart Cities and Accessible Tourism, emphasizing the significance of integrating the needs and specifications of individuals with impairments in the establishment of intelligent urban settings. The key requirements for developing IoT applications in a smart city context were addressed as security, flexibility, and data.

The research conducted by Bauer et al. (2014) highlights the top 10 sectors investing in sensors, where Asia is at the forefront. The sectors most impacted by the rapid proliferation of IoT technology are those where many sensors need to be produced. These industries include energy and excavation, power and utilities, automobiles, manufacturing, tourism, medical care, retailing, leisure, technology, and finance.

Table 2.2 Sharma and Gupta (2021) shows the information on sensors that may be included to improve hotel operations, which was adopted (Nizetic et al., 2020; Verma and Shukla, 2019). During the coronavirus pandemic in 2020, Walt Disney World adopted precautionary steps to safeguard the health and well-being of its visitors (Deb and Nafi, 2023). The implementation of the MagicBand RFID technology facilitated park reservation, capacity control, and contact tracking. This infrastructure enables Disney to optimize return on investment by enhancing visitor service by implementing additional services, such as FastPass+ (Podeschi, 2023). Bob Iger, the head of The Walt Disney Company, used a cutting-edge approach to include RFID technology in resort and theme park environments. Organizations are increasingly using IoT technology in non-traditional applications to enhance their consumers' experience in the tourism industry. Airlines have used RFID baggage monitoring technology to provide passengers with enhanced real-time tracking of their bags throughout the whole flight process (Kang, 2016). Customers may track the whereabouts of their baggage on a map using a smartphone application and get real-time push alerts from important checkpoints along their bags' trip (Podeschi, 2023). IoT in the tourism industry will shift from scheduled service to real-time service, possibly improving the alignment between tourists' wishes, requests, and consumption requirements (Wise and Heidari, 2019).

2.3 METHODOLOGY

The prime intention of this book chapter is to study IoT's application and implications in the field of travel and tourism. There is a significant lack of research articles on IoT in travel and tourism. However, the application of IoT in other businesses has impacted the tourism industry directly and indirectly. To this end, an exploratory research methodology was used

TABLE 2.2 Sensor and it's Operation in Hotel Industry

Position of Sensor	Type of Sensor	Operation
Inside the hotel premises	Location Sensor	Offer meals and other amenities to guests across the whole hotel premises.
Storeroom	Inventory tag	Check the expiration date and the par-stock level. Identify the characteristics and whereabouts of the object.
Hotel exterior	Thermal sensor	Monitor the ambient temperature and control energy use
Hotel room for guest	Temperature Sensor, Light sensor, voice sensor, door sensor	Offers a preferable experience for guests inside the confines of the hotel.
Garden area of Hotel	Moisture Sensor	Measure the soil's moisture level and implement a computerised method for irrigating plants.

Source: Sharma and Gupta (2021).

to carry out this study. The data were obtained from secondary sources such as journal papers, pertinent literature, website data, and published materials. Secondary data analysis involves analyzing data from published publications, text, tables, graphs, and appendices or using the original data without initial planning or gathering, and the responsible party is not involved in data collection (Church, 2002). In the study, the authors tried to demonstrate the IoT evolution practices in travel and tourism, dividing them into different components. The search initially focused on the keyword "IoT" and its relevance to tourism, specifically "smart tourism." This search was conducted within the title, abstract, and keywords of chapters in the Scopus and Google Scholar databases. Subsequently, the keywords "hospitality," "hotel," and "tourist" were employed to refine the search and identify the most pertinent papers for the study. The chapter thoroughly examines how the IoT has significantly changed the travel and tourist business. It recognizes the challenges of IoT and discusses the effects in developing smart cities to smart tourism. Additionally, it establishes a foundation for future study and progress in this field.

2.4 INTERNET OF THINGS (IOT) IN TRAVEL AND TOURISM

According to Car et al. (2019), the IoT has immense potential, particularly as a technology that may provide novel and enhance current solutions in the development of tourist goods and services. The tourism sector can enhance operational efficiency and provide more tailored visitor experiences by using IoT technologies. The tourism industry is anticipated to see significant expansion due to the increasing use of IoT technologies. Mimos (2014) presented many instances of IoT applications in the field of (smart) tourism (Table 2.3).

IoT can help the hotel business offer bundled services with devices that are controlled by apps and automatic mechanisms, like door locks, TVs, thermostats, phones, light switches, voice-activated controls, electric blinds, and other items linked to a central network to provide the services that guests desire (Car et al., 2019).

2.5 INTERNET OF THINGS (IOT) IN ACCOMMODATION

The pervasive influence of IoT has orchestrated a profound transformation within the accommodation sector, redefining the standards of guest experience. Anchored in the connectivity of devices and sensors to the Internet, IoT facilitates a seamless amalgamation of data collection, analysis, and automation, offering unprecedented opportunities for innovation and optimization (Pappas et al., 2021; Vermanen and Harkke, 2019). Accommodations stand as a focal point for IoT applications within the dynamic landscape of the travel and tourism industry. IoT emerges as a linchpin technology in elevating the customer experience, streamlining operational processes, and fortifying safety and security (Deb et al., 2024; Shah et al., 2024). This discourse delves into the nuanced facets of IoT's impact on accommodations, scrutinizing its role in the evolution of smart rooms, operational efficiency optimization, and implementing cutting-edge safety measures.

The concept of smart rooms is at the core of IoT's influence in the accommodation sector. Here, IoT extends a personalized canvas for guests, empowering them to exert control over their environment through the conduits of mobile devices or voice assistants. The control spectrum encompasses nuanced adjustments – from room temperature and lighting to entertainment systems and curtains, all meticulously tailored to individual preferences (Arapou and Kapiki, 2023; Sharma, 2016; Voumik, et al., 2023). The conventional realm of physical key cards dissipates with the advent of smartphones as digital keys, ushering in a frictionless check-in process that amplifies overall convenience (Bharwani and Mathews, 2021). IoT is

TABLE 2.3 Internet of Things (IoT) Implications in the Field of Tourism

Field	Internet of Things (IoT) Implications	Advantages/ Examples
Geo-location technologies	• Monitor the location and gather data that enhance travel experiences. • Route mapping, accommodation coordination, and itinerary development.	• Widespread devices provide passengers convenience and protection by capturing data on time, velocity, distance, position, and altitude, enabling immediate contact with other travelers on the way or in the same place.
Medical and Wellness tourism	• Manage waiting times and the supply of highly skilled medical staff. • Remote monitoring and continuity of care. • Creating customized travel insurance.	• Tourists prioritize Healthcare facilities, wait times, and medical staff. To minimise difficulties, they seek cheap treatments, spas, and post medical vacations. Flights, transportation, accommodations, and language guides have risen to meet this demand.
Ecotourism, Green and Heritage Tourism	• Green tourism practices through recycling or renewable tech which generates opportunities for other business. • IoT quantification to make ecotourism, green, and heritage tourism more appealing and meaningful.	• -Through direct measurement and social media engagement, IoT devices may boost tourist locations' appeal. Communicators and environmental designers may supply e-demonstrators and e-rangers/tour guides to enhance visitor experiences and ensure safety.
Customs and Immigration Verification	• Simplified verification process. • Mitigate human error in unlawful entry and exit.	• IoT-linked airport communication terminals can concurrently handle several entry/exit requests, automating customs and immigration decisions and reducing human error. Verification is quicker, planes are missed less, and airport transit times are shorter.
Retail Tourism Services	• To create distinctive shopping experiences. • Geolocation services like virtual tour to the destination. • Providing interactive travel experiences through gamification. • A smooth shopping experience.	• The amalgamation of IoT technology into mobile devices allows tourist service providers to effectively advertise, market, and cultivate client loyalty by comprehending consumer preferences and providing location-specific services.

TABLE 2.3 *(Continued)*

Field	Internet of Things (IoT) Implications	Advantages/ Examples
Digital Valets and Virtual Concierges	• Personalised services by using digital valets from check-in to check-out. • Concierge services provided virtually such as track baggage.	• IoT-enabled digital valets offer personalized amenities like room configuration, climate control, illumination, music selection, beverages, and heated tubs, and can schedule morning calls for efficient guest wake-up.
Travel Disease Surveillance	• Educating and preventing of diseases infections. • Secure and healthy travel by monitoring visitors.	• Wearable devices can help travelers monitor time, distance, and location data, enabling track-and-trace capabilities. This helps identify illnesses, notify healthcare authorities, and aid medical professionals in diagnosing and treating diseases.

Source: Mimos, 2014.

not merely a conduit for guest-centric enhancements; it metamorphoses the operational bedrock of accommodations. Embedded sensors within rooms function as vigilant sentinels, monitoring occupancy, energy consumption, and maintenance requisites in real time. This aligns with the imperatives of energy efficiency and raises the bar on service quality (Minoli et al., 2017). Inventory management, a herculean task in its traditional guise, metamorphoses into an automated symphony as IoT tracks and replenishes items, ranging from towels and toiletries to minibar supplies, with unparalleled precision (Minoli et al., 2017).

IoT assumes a pivotal role in safety and security, leveraging its capabilities to detect potential hazards – smoke, fire, gas leaks, or unwarranted intrusions. The swift communication of such events in real-time to authorities or guests underscores the potency of IoT in fostering a secure environment. Contactless check-in and check-out processes, coupled with secure digital payment systems, reduce operational friction and serve as formidable bulwarks against infection or fraud (Mercan, 2021; Yang et al., 2021). Strategic alliances with tech behemoth Google signify a marriage of innovation and convenience. Voice-controlled smart rooms, orchestrated through Google Assistant, empower guests with seamless command over room settings, access to hotel services, and immersive entertainment experiences (Lee et al., 2013). The digital keys on smartphones not only obviate the need for physical artifacts but also exemplify the convergence of security and convenience. Concurrently, the proliferation of smart thermostats, attuned to guests' preferences, crystallizes the industry's unwavering commitment to cutting-edge technology (Lee et al., 2013).

An exemplary initiative, the Backyard program stands at the avant-garde of innovation. By conceptualizing and constructing smart homes for rental or sale, Backyard seamlessly integrates IoT devices and sensors attuned to the nuanced preferences of both guests and hosts. This initiative places a premium on environmental friendliness, modular adaptability, and synthesis of customization and flexibility, thereby encapsulating the zeitgeist of modern technological trends (Alam et al., 2012; Dingli and Seychell, 2015). As the accommodation industry navigates the transformative landscape sculpted by IoT, it becomes clear that this technological vanguard is reshaping paradigms (Alam et al., 2012). The confluence of personalization, efficiency, and safety unraveled by IoT, coupled with strategic collaborations with tech giants and avant-garde initiatives like

Backyard's smart homes, portends a future where accommodation spaces cease to be mere venues but metamorphose into experiential ecosystems meticulously curated by the prowess of IoT.

2.6 INTERNET OF THINGS (IOT) IN TRANSPORTATION

The integration of IoT within transportation has emerged as a pioneering revolution in mobility by using transformative force, orchestrating heightened automation, robust data collection, and intricate analysis (Bojan et al., 2014; Rodney, 2020; Yang et al., 2021). This technological paradigm shift permeates various transportation modes, spanning automobiles, public transit, and logistics, with the promise of substantial advancements in efficiency, safety, and the overall travel experience (Rodney, 2020; Vermanen and Harkke, 2019). At the core of IoT's impact on transportation lies the genesis of intelligent, interconnected systems. In the automotive sector, IoT catalyzes the evolution of smart vehicles endowed with advanced sensors and connectivity features (Chand and Karthikeyan, 2018). These interconnected vehicles establish seamless communication amongst themselves and with smart infrastructure, culminating in improved traffic management, reduced congestion, and elevated road safety (Jadaan et al., 2017; Xu, et al., 2017). Public transit systems harness IoT's potential to optimise operations and enrich the passenger journey. The deployment of smart transportation systems, fueled by real-time data from sensors, facilitates the meticulous monitoring of vehicle locations, route optimization, and the provision of precise arrival times. Passengers, in turn, gain access to this dynamic information through user-friendly mobile applications, fostering an informed and convenient travel experience (Vemuri et al., 2024; Xu, et al., 2017). In logistics and supply chain management, IoT plays a pivotal role in revolutionizing the tracking and monitoring of goods. Connected sensors furnish instantaneous insights into shipments' location, condition, and status (Bojan et al., 2014; Deb et al., 2023a). This elevates the precision of inventory management and empowers proactive decision-making to mitigate potential challenges within the supply chain (Vermanen and Harkke, 2019; Witkowski, 2017).

Safety, a paramount concern in the transportation domain, is being robustly addressed through the substantial contributions of IoT. Connected vehicles, equipped to detect and communicate potential hazards such as accidents or road obstacles in real-time, exemplify the transformative power of IoT in ensuring road safety. Furthermore, IoT-enabled infrastructure, particularly adaptive traffic management systems, dynamically responds to traffic conditions, fostering safer roads (Deb et al., 2023b; Witkowski, 2017). Strategic collaborations between transportation entities and technology innovators have borne witness to the genesis of cutting-edge solutions. Noteworthy partnerships with industry leaders like Microsoft have resulted in the seamless integration of IoT technologies into navigation systems and the conception of intelligent transportation applications, epitomizing the convergence of innovation and excellence (Shah et al., 2024). Looking ahead, the trajectory of IoT in transportation continues to be transformative. From the evolution of smart vehicles and the implementation of adaptive traffic control to the infusion of data-driven logistics, IoT technologies are reshaping the transportation landscape, promising efficiency gains, safety enhancements, and an overall elevated travel experience (Yang et al., 2021). The road ahead is marked by a paradigm shift where mobility becomes not just a journey but an orchestrated symphony of technology-driven efficiency and safety.

2.7 INTERNET OF THINGS (IOT) IN FOOD AND BEVERAGE

The global market share of the Food and Beverage (F&B) industry is $281.14 billion in 2020, which makes the industry a prominent participant in the global economy (Culot et al., 2020; Statista, 2021). The employees in this sector should enhance their efficiency, making use of advanced technologies like IoT, due to the growing pressure from customers for a steady supply of F&B, efficient transportation of fragile food items, and prioritization of general protection (Ali and Aboelmaged, 2022; Mano et al., 2020; Parsad et al., 2020). Extensive research has been carried out on the widespread adoption of technology in the F&B industry by using the technology adoption model like the UTAUT model. Ali et al. (2023) have discovered that the behavioral intention to embrace IoT for the digital shifting of the F&B is strongly influenced by performance expectation, anticipated cooperative benefit, expected effort, interpersonal impact, and favorable circumstances. Moreover, organizational indolence and elevated estimated expenses are frequently cited as barriers to acquiring cutting-edge technology. The results also indicate that, whereas gender is an insignificant moderator, seniority and willingness significantly moderate the majority of the relationships (Ali et al., 2023; Ramadas and Satish, 2018). Demir and Dincre (2020) mentioned the importance of cloud computing, visual technologies, robotic technology, cyber-physical systems, 3D printers, big data in the F&B industry, and smart factory applications. However, proper investment is needed to adopt these new technologies and make innovations in the F&B sector.

The use of IoT in F&B helps to increase productivity, performance, task achievement, mutual problem solving and support, and positivity about work that triggers managers to adopt IoT. In order to ensure a seamless transition from conventional to cutting-edge technologies, managers should take into consideration the training requirements, skill development, and capacity building of senior staff when implementing the IoT adoption initiative. Apart from the F&B production and delivery process, the supply chain plays a crucial role in the overall success of this sector. Ali and Aboelmaged (2022) and Culot et al. (2020) studied the use of IoT on supply chain management in the F&B industry. These studies examine four key factors—reduction of supply-demand misalignment, rapidly evolving customer needs, avoidance of legal penalties, and cost optimization—that may encourage businesses to adopt supply chain 4.0. Furthermore, findings revealed three significant obstacles that firms are currently facing when implementing supply chain 4.0: a lack of supply chain collaboration, organizational inertia, and a lack of awareness.

2.8 OPPORTUNITIES AND CHALLENGES OF INTERNET OF THINGS (IOT) IN TRAVEL AND TOURISM

Iansiti and Lakhani (2014) say that companies will be able to gain a competitive edge by using the IoT in today's market. Crucially, establishing an internet connection for a gadget does not automatically result in the widespread availability and sharing of its information. The owner retains a certain level of control over determining who may access the information (Wise and Heidari, 2019). According to Sharma et al. (2019), the IoT continues to extend its presence in residential, commercial, and industrial settings. An increasing number of devices will be interconnected to facilitate data sharing, real-time communication, and data analytics. From the perspective of the user, here are some advantages of using smart products: (i) enhanced energy efficiency; (ii) enhanced safety; (iii) enhanced security; and (iv) enhanced product quality.

The proliferation of IoT devices with constrained functionality has expanded due to their connectivity, use of cloud services, and analytics integration. Obstacles continue to hinder the adoption of IoT, requiring a step-by-step approach to eventually achieving distant and ambiguous goals. Car et al. (2019) has recognized many difficulties in the adoption of IoT, including expensive initial investment, scalability, self-organisation, security and privacy, and data quantities. They also suggested that hotel professionals develop proficiency and effectiveness in adopting emerging IoT applications. The authors Angelova et al. (2017) have examined the beneficial effects of the IoT on the travel and tourism industry. These effects include automation, improved communication control, enhanced customer satisfaction, increased revenue, and cost savings in various areas such as time, energy, resources, money, transportation, security, asset utilization, and quality.

Table 2.4 demonstrates the summary of the challenges of IoT in adopting the tourism industry. Moreover, Zeinab and Elmustafa (2017)

TABLE 2.4 Challenges of Internet of Things (IoT)

Challenges	Authors
Security of data, protection of data, quality of data, management of data, accurate analysis of data, privacy in the absence of standards, privacy protocols, privacy connectivity, and legal problems.	Angelova et al. (2017)
Security of data, Security breaches	Pate and Adegbija (2018)
Obstacles related to technology and regulations in regards to exchanging data	Kaur and Kaur (2016)
Installation expenses, compatibility of the devices, the potential need for device replacement, the challenging and unregulated environment, the capacity of the sensors to operate independently, and the risk of excessive reliance on the sensors	Evans (2011)
Data security, volume issues related to data management, data validity, and the absence of necessary expertise before and after training.	Brou and Janssen (2015)
Expandability, arrive and operate, compatibility, investigation, algorithmic complexity, amount of data, data explanation, regarding protection and privacy, tolerance for faults, power provision	Mattern and Floerkemeier (2010); Zeinab and Elmustafa (2017)
Privacy and data protection, security, radio spectrum, standards and interoperability	Davies (2015)

have further identified that the IoT encounters two significant obstacles in ensuring uninterrupted network access. The first difficulty pertains to the coexistence of many networks, while the second challenge is associated with the large scale of IoT's big data.

2.9 INTERNET OF THINGS (IOT) IN DEVELOPING SMART CITIES, SMART HOMES, SMART TOURISM, AND DESTINATION MANAGEMENT

Novera et al. (2022) discussed that implementing IoT in tourism will enhance its appeal to a larger audience, including a technologically advanced hotel equipped with a smart communication system and a comprehensive management system that covers all essential aspects. Mastering this skill has already become effortless for industrialized nations in the current era, while emerging countries are striving to enhance their proficiency in this area. According to Kaur and Kaur (2016), the concept of smart tourism encompasses several technologies that collaborate to provide intelligent tourist services, including online and cell phone services, cloud computing, the IoT, and Big Data analysis. Developing smart cities helps tourists to get advantages from the presence of digital infrastructures, consolidated urban data, platform services, regulations that facilitate operations, and city residents who provide information on many aspects such as the economy, community, culture and entertainment, mobility and transportation, as well as urban locations and spaces (Koo et al., 2017). Concepts of smart cities may be expanded upon by smart tourism, which aims to improve destination quality and experience based on value generated, traded, and consumed in addition to improving urban quality of life (Gretzel, 2011). The rise of smart homes is now extending to smart hotels, presenting hoteliers with the chance to enhance guest experiences, provide additional services, and optimize operations with unprecedented efficiency (Car et al., 2019). The advancements in smart gadgets and IoT are propelling the transformation of technology used in the hospitality service platform (Kansakar et al., 2019). Verma and Shukla (2019) addressed that IoT-enabled smart cities in Singapore provide autonomous taxi services, reducing costs and enhancing travel efficiency. Intelligent parking systems and traffic lights decrease travel and waiting durations, while live navigation notifications decrease fuel use and financial gains. As a result, this leads to improved traffic flow, reduced ambiguity, and enhanced selection of destinations.

This technological innovation in developing smart cities, directly and indirectly, improves tourist experiences. The Internet has significantly augmented the capacity to promote locations and stimulate visitors (Soteriades, 2012). Today, all destinations have an online presence to ensure the information is available for potential tourists. IoT has stimulated this presence by connecting devices with the internet. Utilizing digital connection enables the identification and positioning of visitors within their unfamiliar environment, facilitating the targeted dissemination of information by a destination (Borseková et al., 2017). The standards for being competitive are constantly shifting; therefore, travel destinations must keep up with technological advancements (Wise and Heidari, 2019). Expanding upon the 6As of prosperous destinations (Attraction, Accessibility, Amenities, Availability, Activities, and Ancillaries) as outlined by Buhalis and Amaranggana in 2014, it is suggested that the 6As should be formulated taking into account interaction and up-to-date data, as proposed by Brandt et al. (2017). To construct compelling smart tourist destinations, it is necessary to establish a conceptual framework for components. Essential elements of smart data include the standard of facts, the source's legitimacy, the level of interaction, and the ease of accessibility (Yoo et al., 2017). Advanced IoT sensors integrated into physical objects in the surroundings provide valuable data and insights into several intricate sensory systems. This information may pertain to offering additional transit options or giving guidance to the closest hotel, dining establishment, or tourist site, for instance (Wise and Heidari, 2019).

2.10 CONCLUSION

IoT has changed the entire business shape with drastic transformations in business and human life simultaneously. The development of IoT has been adopted in every business, including tourism. Hotels, resorts, restaurants, amusement parks, and so on have adopted IoT services to enhance tourist services and experiences. Consumers are also interested and excited to experiment with emerging IoT services. With the development of IoT innovations, the challenges are rising as data security and data volume are prime concerns for tourists and enterprises. For that reason, businesses are investing a lot in data privacy and storing large data volumes. With the innovation of IoT, homes have become smart homes, cities have become smart cities, and tourism is becoming smart tourism. Big data plays a crucial role in IoT technologies, where enormous amounts of data are analyzed to better function these devices. The Fourth Industrial Revolution (4IR) contributed to every business, including tourism, and will continue to do so.

The chapter thoroughly demonstrates and enlightens the application of IoT, especially in the tourism sector, categorizing the main supply components of the tourism system, which will be beneficial for policymakers, government, and other stakeholders. The tourism industry is adopting IoT devices to meet the demand and attract tourists by offering them smooth service experiences. The findings of this chapter will guide tourism enterprises, especially from low-income countries where tourism organizations are less interested in investing in new IoT infrastructure, to adopt existing IoT applications from the success tale of the industry. In the future, blockchain technology may take over the financial transaction method as a data security mode, and web.03 will give consumers opportunities to protect their data, which may impact the IoT industry as web.03 talks about data protection and gives preference to which data to share and which not. As the tourism sector adapts to the digital era, stakeholders are significantly impacted by this innovation and must adapt to cutting-edge technologies.

KEYWORDS

- **accommodation**
- **data privacy**
- **food and beverage**
- **Internet of Things (IoT)**
- **smart cities**
- **smart tourism**
- **tourism industry**
- **transportation**

REFERENCES

Alam, M. R., Reaz, M. B. I., & Ali, M. A. M. (2012). A review of smart homes—Past, present, and future. *IEEE Transactions on Systems, Man, and Cybernetics, Part C (Applications and Reviews)*, *42*(6), 1190–1203.

Ali, I., & Aboelmaged, M. G. S. (2022). Implementation of Supply Chain 4.0 in the food and beverage industry: Perceived drivers and barriers. *International Journal of Productivity and Performance Management*, *71*(4), 1426–1443.

Ali, I., Aboelmaged, M., Govindan, K., & Malik, M. (2023). Understanding the key determinants of IoT adoption for the digital transformation of the food and beverage industry. *Industrial Management & Data Systems*, *12*(7), 1887–1910. https://doi.org/10.1108/IMDS-02-2022-0082 (accessed on 8th July 2025).

Angelova, N., Kiryakova, G., & Yordanova, L. (2017). The great impact of Internet of Things on business. *Trakia Journal of Science*, *15*, 406–412. https://doi.org/10.15547/tjs.2017.s.01.068 (accessed on 8th July 2025).

Arapou, D., & Kapiki, S. (2023). Enhancing the guest experience in hospitality through technology: An outlook of the future. *ITHGC 2023: IV International Tourism, Hospitality & Gastronomy Congress*.

Balandina, E., Balandin, S., Koucheryavy, Y., & Mouromtsev, D. (2015). IoT use cases in healthcare and tourism. In *2015 IEEE 17th Conference on Business Informatics* (Vol. 2, pp. 37–44). IEEE.

Bauer, S., Burkitt, F., Curran, C., & Gioia, L. (2014). Sensing the future of the Internet of Things. *PwC 6th Annual Digital IQ*. https://www.pwc.com/gx/en/services/advisory/assets/digital-iq-sensors-iot-snapshot-final.pdf (accessed on 8th July 2025).

Bharwani, S., & Mathews, D. (2021). Techno-business strategies for enhancing guest experience in luxury hotels: A managerial perspective. *Worldwide Hospitality and Tourism Themes*, *13*(2), 168–185.

Bojan, T. M., Kumar, U. R., & Bojan, V. M. (2014). An Internet of Things based intelligent transportation system. In *2014 IEEE International Conference on Vehicular Electronics and Safety* (pp. 174–179). IEEE.

Borseková, K., Vaňová, A., & Vitálišová, K. (2017). Smart specialization for smart spatial development: Innovation strategies for building competitive advantages in tourism in Slovakia. *Socio-Economic Planning Sciences*, *58*(1), 39–50.

Brandt, T., Bendler, J., & Neumann, D. (2017). Social media analytics and value creation in urban smart tourism ecosystems. *Information & Management*, *54*(6), 703–713.

Brou, P., & Janssen, M. (2015). Effects of the Internet of Things (IoT): A systematic review of the benefits and risks. *ICEB 2015 Proceedings (Hong Kong, SAR China)*, *48*.

Buhalis, D., & Amaranggana, A. (2013). Smart tourism destinations. In *Information and Communication Technologies in Tourism 2014: Proceedings of the International Conference in Dublin, Ireland, January 21–24, 2014* (pp. 553–564). Springer International Publishing.

Car, T., Stifanich, L. P., & Šimunić, M. (2019). Internet of Things (IoT) in tourism and hospitality: Opportunities and challenges. *Tourism in South East Europe*, *5*, 163–175. https://doi.org/10.20867/tosee.05.42 (accessed on 8th July 2025).

Chand, H. V., & Karthikeyan, J. (2018). Survey on the role of IoT in intelligent transportation systems. *Indonesian Journal of Electrical Engineering and Computer Science*, *11*(3), 936–941.

Church, R. M. (2002). The effective use of secondary data. *Learning and Motivation, 33*(1), 32–45. https://doi.org/10.1006/lmot.2001.1098 (accessed on 8th July 2025).
Culot, G., Nassimbeni, G., Orzes, G., & Sartor, M. (2020). Behind the definition of Industry 4.0: Analysis and open questions. *International Journal of Production Economics, 226*, 1–15.
Curls, A. (2023). Top 35 smart home facts and statistics. *Today's Homeowner*. https://todayshomeowner.com/smart-home/guides/smart-home-facts-and-statistics (accessed on 8th July 2025).
Davies, R. (2015). The Internet of Things: Opportunities and challenges. *EPRS: European Parliamentary Research Service*.
Deb, S. K., & Nafi, S. (2023). Impact of pandemic, crises, and challenges on event tourism. In *Event Tourism and Sustainable Community Development* (pp. 231–245). Apple Academic Press.
Deb, S. K., Das, M. K., Voumik, L. C., Nafi, S. M., Rashid, M., & Esquivias, M. A. (2023b). The environmental effects of tourism: Analyzing the impact of tourism, global trade, consumption expenditure, electricity, and population on environment in leading global tourist destinations. *GeoJournal of Tourism & Geosites, 51*.
Deb, S. K., Kuri, B. C., & Nafi, S. M. (2024). Application of knowledge management in tourism and hospitality industry: A sustainable approach. In *Knowledge Management and Knowledge Sharing: Business Strategies and an Emerging Theoretical Field* (pp. 99–116). Cham: Springer Nature Switzerland.
Deb, S. K., Kuri, B. C., Roy, H., Roy, B., & Nafi, S. M. (2023a). The adaptability of tourism and hospitality supply chain management to manage COVID-19 crisis: Bangladesh perspective. *International Journal of Agile Systems and Management, 16*(4), 484–511.
Deb, S. K., Nafi, S. M., & Valeri, M. (2022). Promoting tourism business through digital marketing in the new normal era: A sustainable approach. *European Journal of Innovation Management, 27*(3), 775–799.
Demir, Y., & Dincer, F. I. (2020). The effects of Industry 4.0 on the food and beverage industry. *Journal of Tourismology, 6*(1), 133–145. https://doi.org/10.26650/jot.2020.6.1.0006 (accessed on 8th July 2025).
Dingli, A., & Seychell, D. (2015). Smart homes. In *The New Digital Natives: Cutting the Chord* (pp. 85–101). Springer, Berlin, Heidelberg.
Evans, D. (2011). The Internet of Things: How the next evolution of the Internet is changing everything. *Whitepaper, Cisco Internet Business Solutions Group (IBSG), 1*, 1–12.
Girau, R., Martis, S., & Atzori, L. (2016). A cloud-based platform of the social Internet of Things. In *Internet of Things. IoT Infrastructures: Second International Summit, IoT 360° 2015, Rome, Italy, October 27–29, 2015, Revised Selected Papers, Part I* (pp. 77–88). Springer International Publishing.
Gokhale, P., Bhat, O., & Bhat, S. (2018). Introduction to IoT. *International Advanced Research Journal in Science, Engineering and Technology, 5*(1), 41–44.
Gretzel, U. (2011). Intelligent systems in tourism: A social science perspective. *Annals of Tourism Research, 38*(3), 757–779.
Iansiti, M., & Lakhani, K. R. (2014). Digital ubiquity: How connections, sensors, and data are revolutionizing business. *Harvard Business Review, 92*(11), 19.
Jadaan, K., Zeater, S., & Abukhalil, Y. (2017). Connected vehicles: An innovative transport technology. *Procedia Engineering, 187*, 641–648.

Kang, A. (2016). Delta introduces innovative baggage tracking process. https://news.delta.com/delta-introduces-innovative-baggage-tracking-process (accessed on 8th July 2025).

Kansakar, P., Munir, A., & Shabani, N. (2019). Technology in the hospitality industry: Prospects and challenges. *IEEE Consumer Electronics Magazine*, *8*(3), 60–65. https://doi.org/10.1109/MCE.2019.2892245 (accessed on 8th July 2025).

Kaur, K., & Kaur, R. (2016). Internet of Things to promote tourism: An insight into smart tourism. *International Journal of Recent Trends in Engineering & Research*, *2*(4), 357–361.

Koo, C., Park, J., & Lee, J-N. (2017). Smart tourism: Traveler, business, and organizational perspectives. *Information & Management*, *54*(6), 683–686.

Koteshov, D. (2023). IoT in tourism and travel: Examples, trends and benefits for the travel industry. https://anywhere.epam.com/business/internet-of-things-in-travel-and-tourism (accessed on 8th July 2025).

Lee, H. A., Denizci Guillet, B., & Law, R. (2013). An examination of the relationship between online travel agents and hotels: A case study of Choice Hotels International and Expedia.com. *Cornell Hospitality Quarterly*, *54*(1), 95–107.

Lee, I. (2019). The Internet of Things for enterprises: An ecosystem, architecture, and IoT service business model. *Internet of Things*, *7*, 100078.

Lee, I., & Lee, K. (2015). The Internet of Things (IoT): Applications, investments, and challenges for enterprises. *Business Horizons*, *58*(4), 431–440. https://doi.org/10.1016/j.bushor.2015.03.008 (accessed on 8th July 2025).

Madakam, S., Ramaswamy, R., & Tripathi, S. (2015). Internet of Things (IoT): A literature review. *Journal of Computer and Communications*, *3*, 164–173. http://dx.doi.org/10.4236/jcc.2015.35021 (accessed on 8th July 2025).

Mano, A. P., Gouvea da Costa, S. E., & Pinheiro de Lima, E. (2020). Criticality assessment of the barriers to lean construction. *International Journal of Productivity and Performance Management*, *70*(1), 65–86. https://doi.org/10.1108/IJPPM-11-2018-0413 (accessed on 8th July 2025).

Mattern, F., & Floerkemeier, C. (2010). From the Internet of computers to the Internet of Things. In *Springer Berlin Heidelberg* (pp. 242–259).

Mercan, S., Cain, L., Akkaya, K., Cebe, M., Uluagac, S., Alonso, M., & Cobanoglu, C. (2021). Improving the service industry with hyper-connectivity: IoT in hospitality. *International Journal of Contemporary Hospitality Management*, *33*(1), 243–262.

Mimos, B. (2014). *IoT Idea Book*. Kuala Lumpur, Malaysia. https://www.mimos.my/wp-content/uploads/2016/01/IoT-Idea-Book-Experiential-Travel-and-Tourism.pdf (accessed on 8th July 2025).

Minoli, D., Sohraby, K., & Occhiogrosso, B. (2017). IoT considerations, requirements, and architectures for smart buildings—Energy optimization and next-generation building management systems. *IEEE Internet of Things Journal*, *4*(1), 269–283.

Miorandi, D., Sicari, S., De Pellegrini, F., & Chlamtac, I. (2012). Internet of Things: Vision, applications and research challenges. *Ad Hoc Networks*, *10*(7), 1497–1516. http://dx.doi.org/10.1016/j.adhoc.2012.02.016 (accessed on 8th July 2025).

Munir, A., Kansakar, P., & Khan, S. U. (2017). IFCIoT: Integrated Fog Cloud IoT—A novel architectural paradigm for the future Internet of Things. *IEEE Consumer Electronics Magazine*, *6*(3), 74–82.

Nitti, M., Giusto, D., Zanda, S., Di Francesco, M., Casari, C., Clemente, M. L., & Popescu, V. (2018). Using IoT for accessible tourism in smart cities. In *Assistive Technologies in Smart Cities*. IntechOpen.

Nitti, M., Pilloni, V., Giusto, D., & Popescu, V. (2017). IoT architecture for a sustainable tourism application in a smart city environment. *Mobile Information Systems, 2017*. https://doi.org/10.1155/2017/9201640 (accessed on 8th July 2025).

Nizetic, S., Solic, P., Gonzalez-de-Artaza, D. L. D., & Patrono, L. (2020). Internet of Things (IoT): Opportunities, issues and challenges towards a smart and sustainable future. *Journal of Cleaner Production, 274*.

Novera, C. N., Ahmed, Z., Kushol, R., Wanke, P., & Azad, M. A. K. (2022). Internet of Things (IoT) in smart tourism: A literature review. *Spanish Journal of Marketing-ESIC, 26*(3), 325–344.

Nunberg, G. (2012). The advent of the Internet: 12th April courses. *International Journal of Innovation and Research in Technology*.

Pappas, N., Caputo, A., Pellegrini, M. M., Marzi, G., & Michopoulou, E. (2021). The complexity of decision-making processes and IoT adoption in accommodation SMEs. *Journal of Business Research, 131*, 573–583.

Parsad, C., Mittal, S., & Krishnankutty, R. (2020). A study on the factors affecting household solar adoption in Kerala, India. *International Journal of Productivity and Performance Management, 69*(8), 1695–1720. https://doi.org/10.1108/IJPPM-11-2019-0544 (accessed on 8th July 2025).

Pate, J., & Adegbija, T. (2018). AMELIA: An application of the Internet of Things for aviation safety. In *2018 15th IEEE Annual Consumer Communications & Networking Conference (CCNC)* (pp. 1–6). IEEE. https://doi.org/10.1109/CCNC.2018.8319163 (accessed on 8th July 2025).

Podeschi, R. J. (2023). IoT in tourism. In *Disruptive Technologies in International Business: Challenges and Opportunities for Emerging Markets* (p. 29). Walter de Gruyter GmbH & Co KG.

Rahman, M. H., Voumik, L. C., Nafi, S. M., & Zimon, G. (2023). Effects of tourism and other macroeconomic variables on women's employment in agricultural, industry and service sectors: Evidence from African countries. *Current Issues in Tourism*, 1–21.

Ramadas, T., & Satish, K. P. (2018). Identification and modeling of employee barriers while implementing lean manufacturing in small- and medium-scale enterprises. *International Journal of Productivity and Performance Management, 67*(3), 467–486.

Rodney, B. D. (2020). Understanding the paradigm shift in education in the twenty-first century: The role of technology and the Internet of Things. *Worldwide Hospitality and Tourism Themes, 12*(1), 35–47.

Shah, S., Hussain Madni, S. H., Hashim, S. Z. B. M., Ali, J., & Faheem, M. (2024). Factors influencing the adoption of Industrial Internet of Things for the manufacturing and production small and medium enterprises in developing countries. *IET Collaborative Intelligent Manufacturing, 6*(1), e12093.

Sharma, D. (2016). Enhancing customer experience using technological innovations: A study of the Indian hotel industry. *Worldwide Hospitality and Tourism Themes, 8*(4), 469–480.

Sharma, N., Shamkuwar, M., & Singh, I. (2019). The history, present and future with IoT. In *Internet of Things and Big Data Analytics for Smart Generation* (pp. 27–51). https://doi.org/10.1007/978-3-030-04203-5_3 (accessed on 8th July 2025).

Sharma, U., & Gupta, D. (2021). Analysing the applications of Internet of Things in hotel industry. In *Journal of Physics: Conference Series* (Vol. 1969, Issue 1, pp. 012041). IOP Publishing.
Soteriades, M. (2012). Tourism destination marketing: Approaches improving effectiveness and efficiency. *Journal of Hospitality and Tourism Technology*, *3*(2), 107–120.
Statista. (2022). Internet of Things (IoT) connected devices installed base worldwide from 2015 to 2025. https://www.statista.com/statistics/471264/iot-number-of-connected-devices-worldwide/ (accessed on 8th July 2025).
Statista. (2021). Size of the worldwide functional food and beverage market from 2020 to 2021 with a forecast to 2028. https://www.statista.com/statistics/1264080/functionalfood-and-beverages-global-market-size/ (accessed on 8th July 2025).
Sundmaeker, H., Guillemin, P., Friess, P., & Woelfflé, S. (2010). Vision and challenges for realizing the Internet of Things. *Cluster of European Research Projects on the Internet of Things, European Commission*, *3*(3), 34–36.
Tripathy, A. K., Tripathy, P. K., Ray, N. K., & Mohanty, S. P. (2018). iTour: The future of smart tourism: An IoT framework for the independent mobility of tourists in smart cities. *IEEE Consumer Electronics Magazine*, *7*(3), 32–37.
Vemuri, N., Tatikonda, V. M., & Thaneeru, N. (2024). Enhancing public transit system through AI and IoT. *Valley International Journal Digital Library*, 1057–1071.
Verma, A., & Shukla, V. (2019). Analyzing the influence of IoT in tourism industry. In *Proceedings of International Conference on Sustainable Computing in Science, Technology and Management (SUSCOM)* (pp. 2083–2093). Amity University Rajasthan, Jaipur-India.
Vermanen, M., & Harkke, V. (2019). Findings from multipurpose IoT solution experimentations in Finnish SMEs: Common expectations and challenges. In *Proceedings of the 52nd Hawaii International Conference on System Sciences* (pp. 5246–5255).
Voumik, L. C., Rahman, M. H., Nafi, S. M., Hossain, M. A., Ridzuan, A. R., & Mohamed Yusoff, N. Y. (2023). Modelling sustainable non-renewable and renewable energy based on the EKC hypothesis for Africa's ten most popular tourist destinations. *Sustainability*, *15*(5), 4029.
Wise, N., & Farzin, F. (2018). See you in Iran on Facebook. In *Authenticity & Tourism* (Vol. 24, pp. 33–52). Emerald Publishing Limited.
Wise, N., & Heidari, H. (2019). Developing smart tourism destinations with the Internet of Things. In *Big Data and Innovation in Tourism, Travel, and Hospitality: Managerial Approaches, Techniques, and Applications* (pp. 21–29).
Witkowski, K. (2017). Internet of Things, big data, Industry 4.0–Innovative solutions in logistics and supply chains management. *Procedia Engineering*, *182*, 763–769.
Wu, X. (2017). Smart tourism based on Internet of Things. *Revita de La Faulted de Ingenieria*, *32*(10), 66–170.
Xu, H., Lin, J., & Yu, W. (2017). Smart transportation systems: Architecture, enabling technologies, and open issues. In *Secure and Trustworthy Transportation Cyber-Physical Systems* (pp. 23–49).
Yang, H., Song, H., Cheung, C., & Guan, J. (2021). How to enhance hotel guests' acceptance and experience of smart hotel technology: An examination of visiting intentions. *International Journal of Hospitality Management*, *97*, 103000.

Yoo, C. W., Goo, J., Huang, C. D., Nam, K., & Woo, M. (2017). Improving travel decision support satisfaction with smart tourism technologies: A framework of tourist elaboration likelihood and self-efficacy. *Technological Forecasting and Social Change*, *123*, 330–341.

Zeinab, K. A. M., & Elmustafa, S. A. A. (2017). Internet of Things applications, challenges and related future technologies. *World Scientific News*, *67*(2), 126–148.

Zheng, B., Mei, Z., Hou, L., & Qiu, S. (2021). Application of Internet of Things and edge computing technology in sports tourism services. *Security and Communication Networks*, 1–10.

CHAPTER 3

TECHNOLOGICAL INNOVATION FOR SMART TOURISM: FOCUS ON INTERNET OF THINGS (IOT), AUGMENTED REALITY (AR), AND VIRTUAL REALITY (VR)

SANJOY KUMAR ACHARJEE*

Department of Tourism and Hospitality Management, Noakhali Science and Technology University, Noakhali, Bangladesh

*Corresponding author

ABSTRACT

Driven by the integration of cutting-edge technologies like the Internet of Things, artificial intelligence, and augmented/virtual reality, the tourism industry is undergoing a profound transformation. This chapter provides a comprehensive analysis of these modern technologies and their profound impacts on the landscape of smart tourism. The integration of these disruptive technologies not only improves the overall tourism experience but also brings resource optimization, operational efficiency, and data-driven decisions. Data for this chapter is collected from secondary sources and in-depth interviews of smart tourism experts. However, issues like data privacy, cybersecurity risks, and resistance to the adoption of technology need to be addressed. The chapter also highlights strategies to balance innovation and the authentic experience of culture, mitigating environmental impacts, fostering sustainable tourism practices, and boosting local economies. In this chapter, insights from diverse case studies and academic research were synthesized to offer a holistic approach to harnessing smart

Smart Tourism: Managing Technological Revolution in the Tourism Industry. Santus Kumar Deb (Ed.)

DOI: 10.1201/9781779641090-3

technologies to shape the future of responsible and enriching travel experiences.

3.1 INTRODUCTION

The tourism industry is going through a profound transformation driven by the integration of modern technologies such as the Internet of Things (IoT), artificial intelligence (AI), and augmented/virtual reality (AR/VR) (Buhalis et al., 2019; Bulchand-Gidumal, 2022). These disruptive technologies not only enable operational efficiency, resource optimization, and data-driven decision-making but also improve the overall experience of tourism (Kontogianni and Alepis, 2020). The "smart tourism" concept has emerged, wherein cutting-edge technology is seamlessly woven into numerous aspects of travel, from personalized recommendations and virtual experiences to predictive maintenance, evaluation, and real-time monitoring (Gretzel et al., 2015).

A comprehensive analysis of modern technologies' impact on the landscape of smart tourism is provided in the chapter. It includes the functions of AI in personalization, recommendation systems, voice assistants, and forecasting (Claveria et al., 2015; Tussyadiah and Miller, 2019). The chapter also explores the role of the IoT in improving personalized hotel experiences, management of inventory, and real-time monitoring (Car et al., 2019; Elshaer and Marzouk, 2022). Moreover, it analyzes the applications and uses of AR and VR in virtual tours, destination management and marketing, and decision-making (Katkuri et al., 2019; Rainoldi et al., 2018). While these disruptive technologies provide various benefits, some issues also need to be addressed, such as data privacy, cybersecurity risks, and resistance to technology adoption (Ardito et al., 2019; Priyambodo and Artianingsih, 2022). The chapter sheds light on strategies for managing environmental impacts, generalizing sustainable tourism practices, and improving local economies (Buonincontri and Micera, 2016; McNaughton et al., 2020). Additionally, the chapter finds the balance between technology and authentic experiences, highlighting the preservation of local heritage and cultural integrity (Idris et al., 2016; Mowforth and Munt, 2015).

3.2 LITERATURE REVIEW

In tourism, AI refers to modern technologies that can mimic human intelligence for problem-solving (Lai and Hung, 2018). The functionality of

Artificial Intelligence in tourism includes personalization through recommendation systems, voice assistants, conversational chatbots, predicting demand of tourists, marketing strategies and trends, as well as applications for language translation (Tussyadiah and Miller, 2019). AI technologies have numerous functions in tourism. From the perspective of the consumer, it assists in getting relevant and reliable information while giving more mobility and makes the process of decision-making easier; ultimately, a great experience of tourism is provided (Tussyadiah and Miller, 2019).

From the service provider's perspective, they can use AI in almost every part of their management (Buhalis et al., 2019), particularly in the productivity and promotional areas (Tussyadiah and Miller, 2019). "Internet of Things" as a term was first used in 1999 by Kevin Ashton. IoT can be referred to as a network system that has the ability to connect with anything, anyplace, and anytime using technologies like REID (Radio Frequency Identification), WSN (Wireless Sensor Network), and 3g/4g/5g smartphone communication to locate, identify, monitor, track, and manage smart things or objects (Mingjun et al., 2012). While Virtual Reality creates environments that are simulated, Augmented Reality is used to overlay digital content onto the physical world (Samaddar and Mondal, 2023). These applications include virtual tours, education for professionals and tourists, preservation of heritage, in case of travel restrictions remote exploration, improving visitor experiences at sites of cultural importance, and enabling tailored planning of travel (Han et al., 2019). Effective mixes of marketing for travel destinations include programming, product, pricing, place, promotion, people, processes, sales support, and bundled packages (Mišúnová, 2015). Smart technologies enable itineraries tailored to personal interests, through behavioral data-targeted marketing, predictive workflow with seamless experiences, integrated mobile applications, and finally pre-trip exploration through immersive Artificial and Virtual Reality (Buhalis et al., 2015; Yavuz et al., 2018).

3.3 METHODOLOGY

The data were collected mainly from secondary sources. In the field of tourism and AI, a comprehensive review and search of secondary sources were conducted across academic databases, scholarly journals, conference papers, and reputable online repositories to find relevant literature. Keywords like "smart tourism," "tourism technology," "AI in tourism," "personalization in travel," and "IoT in hospitality" were used to filter search results. A few in-depth interviews with experts in tourism have been conducted for a better understanding of technological innovation in Tourism.

3.4 INTERNET OF THINGS (IOT) IN TOURISM

The IoT—a never-ending network of connected and electronic devices that not only communicate with a server but also each other—is turning into more than just a form of connectivity (Car et al., 2019). It represents a significant factor in the tourism and hospitality sector, opening unique ways of doing business.

"Internet of Things" as a term was first used in 1999 by Kevin Ashton. IoT can be referred to as a network system that has the ability to connect with anything, anyplace, and anytime using technologies like REID, WSN, and 3g/4g/5g smartphone communication to locate, identify, monitor, track, and manage smart things or objects (Mingjun et al., 2012).

Figure 3.1 demonstrates that with the IoT, any device can connect to the server at anytime from anywhere to provide service to anyone through the network. From this concept, new types of applications, including smart homes and smart vehicles, can be created to provide services such as notifications, security alerts, automation, and so forth. (Saranya and Nitha, 2015).

The IoT contains three main characteristics of system-level which includes (Miorandi et al., 2012):

1. **Anything Communicates:** Smart objects can wirelessly connect among themselves and via networks of interconnected things.
2. **Anything is Identified:** A digital name is the way to identify smart things: whenever a physical connection cannot be established, their relationships can be specified in the digital realm.
3. **Anything Interacts:** If actuation and sensing capabilities are present through them, smart objects can interact with the local environment.

3.4.1 FUNCTIONS OF INTERNET OF THINGS (IOT) IN TOURISM AND HOSPITALITY

1. **Personalized Hotel Rooms:** To experiment somewhat differently on the connected room concept, Hilton and Marriott enable the

user or guest to control the various features of the room from a provided tablet or their own smartphone. Moreover, these IoT technologies can even remember a guest's preferences using his previous history, which includes the temperature of the room, lights, shades, and TV channels, and set up the room automatically for the guest's next stay.

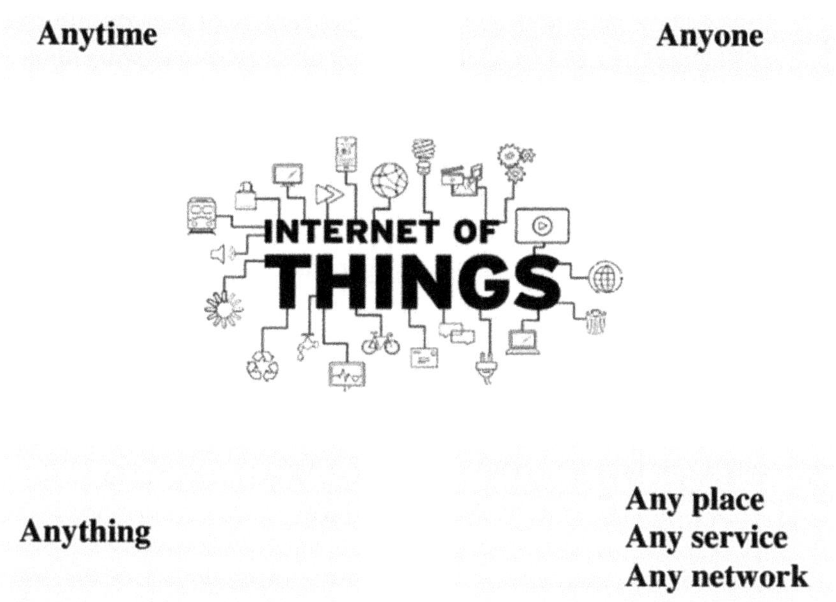

FIGURE 3.1 Concept of Internet of Things (IoT).

Source: Adapted from Car et al. (2019).

2. **Voice-Based Interaction:** Google Home or Amazon Alexa provides the user voice control on different smart and electronic home devices. It's like a guest's personal butler who always fulfills their needs without getting tired. These sorts of voice-control devices could be used by service providers to enable guests to adjust and control the room's smart electronics like thermostats, blinds, TVs, and lights.
3. **Inventory Management:** The way inventories are managed in hotels will be changed by IoT. IoT devices will keep track of the inventories and will manage the changes in the inventories automatically.
4. **Valpas – Autonomous Bed Bug Prevention:** An automated bed bug prevention system developed by Valpas allows guests to stay

relieved and carefree from bed bugs. Hotels can eliminate bed bugs caught by the system by exchanging the existing bed legs with Valpas' smart legs, which also enables the hotel to monitor and receive real-time notifications about bed bugs.

5. **Real-time Monitoring and Data Collection Applications:** IoT devices are not just about assistance and convenience, they are also collecting valuable information and data. By utilizing these data, sustainable and personalized tourism practices can be informed (Xiang et al., 2017). Optimized uses of resources, prevention of overcrowding, and data-driven decisions can be undertaken for sustainable tourism development by monitoring crowd levels at tourist destinations, tracking environmental parameters in ecosystems, and measuring traffic flow to a popular tourist destination (Elshaer et al., 2022). It is our responsibility to keep the natural wonder in existence for generations to come, and we can fulfill our duty by using monitoring and data collection applications like sensors to monitor wildlife populations and tracking systems that measure the impact of tourists on delicate ecosystems.

3.5 AUGMENTED REALITY (AR) AND VIRTUAL REALITY (VR) FUNCTION IN THE TOURISM INDUSTRY

AR and VR have emerged as transformative technologies in the travel and tourism industry, offering immersive and interactive experiences to passengers (Samaddar and Mondal, 2023). VR produces simulated environments, while AR applies digital content that is superimposed onto the physical environment. VR augments the perception of reality and provides distinctive avenues for user interaction (Mofokeng and Matima, 2018). Enterprises such as "Discovery VR" generate multimedia content that presents wildlife habitats and landscapes, thereby facilitating virtual travel experiences (Loureiro et al., 2020). To educate travelers about destinations and instruct tourism professionals, AR and VR are invaluable tools. Tour guides are educated via VR simulations, whereas VR field excursions supplement classroom instruction (Shen et al., 2022). In light of their ubiquitous ownership and accessibility, Jenny (2017) asserts that devices such as Samsung Gear VR, Google Cardboard, and smartphones augment the tourism industry with augmented reality and virtual reality.

The preservation of digital heritage is enhanced by virtual, mixed, and augmented realities, which provide immersive encounters with historical environments and interactive experiences. In times of travel restrictions, virtual tourism facilitated by VR/AR technologies offers an engaging and accessible means of investigating destinations, surpassing the constraints imposed by physical distance (Siddiqui et al., 2022). AR and VR are utilized by cultural tourism destinations to enhance the experiences of visitors through the provision of virtual enhancements, historical re-enactments, and remote destination exploration. Deliberations of visitor experience impacts, technology objectives, customer-centric design, and the transformational potential of AR in the context of cultural heritage engagement are noteworthy (Han et al., 2019). AR and VR are revolutionizing the tourism sector as a whole by providing novel approaches to engage visitors, promote destinations, and enhance their experiences. The ongoing development of these technologies is anticipated to have a greater influence on the tourism industry, thereby transforming the course of travel and hospitality. Using AR and VR, the travel experience can be customized. One could foresee the utilization of augmented reality to locate obscure attractions and indigenous encounters in the vicinity, or virtual reality to generate an individualized itinerary tailored to one's preferences such as (Kourouthanassis et al., 2015):

1. **QR Code:** While placed at various travel spots, QR codes offer instant access to historical facts, multimedia guides, and assistance for navigation via smartphones, enriching travelers' experience and understanding (Car et al., 2019). Additionally, QR codes can be served as digital tickets, which simplify entry processes and reduce wastage of paper.
2. **3D Video:** Integrating the technology of 3D video in tourism allows immersive previews of the destination, and also enables the potential visitor to virtually visit the tourist place or destination before their actual visit (Kapiki, 2021). These engaging videos offer more captivating experiences compared to the traditional media, assisting in making travel decisions.

3.6 DESTINATION MARKETING

To position and promote a destination as an attractive destination to potential visitors, marketing plays a significant role. Figure 3.2 illustrates the marketing mix suitable to promote a destination, encompassing various elements that are vital and need to be integrated and aligned carefully to make an appealing, comprehensive marketing strategy are:

FIGURE 3.2 Destination marketing mix.

Source: Adapted from Mišúnová (2015).

1. **Programming:** In shaping the attractiveness and appeal of a destination, activities, events, and experiences offered by the destination are instrumental. Programming that is well-designed can cater to various interests, from outdoor adventures to cultural performances and festivals, providing an engaging experience for the tourists while also creating a unique identity.
2. **Product:** The product is the core of destination marketing, which encompasses various tangible and intangible aspects. Natural attractions, accommodation, historical sites, culinary aspects, and the overall infrastructure of tourism are included in the tangible products. On the contrary, intangible products include ambiance, experiences, and emotional connections created by visitors with the destination.
3. **Price:** In determining the competitiveness and perceived value of a destination, strategies and policies of pricing play a significant role. Factors including accommodation costs, activity fees, transportation costs, and overall value for money must be crafted carefully.

Competitive pricing, along with special offers and attractive packages, can significantly enhance the destination's appeal and accessibility.

4. **Place:** Visitor decisions are critically influenced by the accessibility and location of a destination. This particular element covers several aspects, which include options for transportation like air, road, and rail, also travel routes, and channels of distribution through which tourists can purchase and book tourism services and products.
5. **Promotion:** To generate interest, create awareness, and finally attract tourists, it is vital to effectively promote a destination. This can be achieved by efforts of communication, such as social media campaigns, public relations initiatives, and participation in exhibitions and travel fairs. The strategy of promotion should be tailored to reach the target audience while highlighting the unique selling propositions.
6. **People:** To give the best experience of a destination to the audience, the human elements play a crucial role. Investment in training professionals and service providers in the tourism industry can enormously enhance the overall experience of the visitors.
7. **Processes:** To ensure a smooth and seamless experience for the visitor, efficient systems and processes are a must. Processes include various aspects like booking, information dissemination, and handling of complaints and inquiries. A properly managed process can improve convenience, reduce frustrations, and contribute to a better experience.
8. **Sales Support:** Collaboration with tour operators, travel agencies, and other channels of distribution is critical for selling and promoting a destination's products and services. These partners can have a better understanding and promote the destination to potential tourists if training, adequate sales support, and marketing materials are provided.
9. **Packages:** Bundling tourism offerings and services into packages can gain convenience and attract visitors by providing greater value. A package can be tailored to different budgets, interests, and travel styles, including the combination of activities, transportation, accommodation, and other offerings. A well-designed package can significantly simplify the planning for tourists and reduce time while increasing the appeal of the destination.

10. **Personalized Itinerary:** Customized itinerary is a vital aspect in tourism destination marketing, as it tailors the experiences to match the tourists' interest. By capitalizing on the feedback and data, marketers make personalized plans while showcasing the destination's unique aspect, attracting visitors, and driving engagement (Ozdemir, 2021). By digital means, these customized itineraries inspire tourists to explore and experience a destination fully.

3.7 IMPROVING CUSTOMER EXPERIENCE

Due to modern technological innovation, the tourism industry is transforming into smart tourism. Smart technology is changing the way the tourism industry interacts with its tourists and customers (Jasrotia et al., 2018). From personalized experiences and real-time interactions to automation and artificial intelligence, smart technology is playing a key role in changing customer demand and satisfaction. It explores the various ways in which smart technology constructs the tourism industry and creates customer experiences across industries, leading to enhanced efficiency, convenience, and proper interactions such as:

1. **Customized Travel Itineraries:** Choice exploration: tourist data, including departed blueprint visited, favored performance, tenement pick, and food preference, is resolved by intellectual processing using algorithms (Buhalis et al., 2015).
2. **Real-time Recommendations:** This process measures recommendations in real-time based on the impending situation of occurrence, such as regional things, airline attendance, and weather prophecy (Kapiki, 2021). Tourists can take in personalized admonishment via mobile apps or advertisements regarding airlines, deceleration itinerary interchange, or express close fascination risen their curiosity.
3. **General Loop:** By incessantly progressing and boosting afterlife excursion ideas or line drawings, machine science algorithms are acquainted from user comments to excellent expression.

3.8 SEAMLESS TRAVEL EXPERIENCES

1. **Integrating Artificial Intelligence (AI) and Internet of Things (IoT) to Ensure Smooth Operations:** Conjecturing upkeep:

simultaneous data on the agility of machinery is collected by an IoT detector, implanted in trains, airplanes, and other forms of transportation (Novera et al., 2022). By using AI to evaluate this data, conservation requisite can be predicted, minimizing unintentional delays. Refined responsibility: computational intelligence direction deconstruction of enormous volumes of input from universal deaminization tools, such as weather stations and traffic sensors, to create optimal routes that minimize travel postponement and increase spot-on-time accomplishment (Tsaih and Hsu., 2018). Magnified purchaser privilege: natural language processing-powered chatbots converse with travelers, providing personalized suggestions based on past data and preferences, real-time help, and updates on travel plans (Leung, 2022).

2. **Virtual Tour:** Virtual Reality Travel Experiences: VR provides 360° views of inns, visitor attractions, and even full towns, permitting consumers to virtually experience travel locations (Katkuri et al., 2019). This empowers visitors to experience a location practically before actually seeing it.
3. **Intensified Cultural Concentration:** As sightseers inspect, augmented reality (AR) applications cover real-time information about archaeological remains, landmarks, or areas of interest, furnishing a deeper knowledge of the area's ethnic significance (Rainoldi et al., 2018).
4. **Virtual Guides:** As visitors ramble through a region, AR-powered guides afford them up-to-date information on the local area, its historical background, or fascinating facts about nearby landmarks.
5. **Impact on Travel Decision-Making:** Emphasized pre-arranged simulated reality-based involvement assists passengers in making informed decisions by furnishing a more pragmatic foretaste of their potential resting place, influencing their choice of accommodations, exertion, and attractions (Katkuri et al., 2019).
6. **Commitment And Publication:** Artificial environment experiences in promotion movement implicit travelers, offering a more attractive way to showcase destinations, accommodations, and activities, ultimately influencing travel decisions (Kapiki, 2021).
7. **Approachable:** For individuals with motion challenges or limitations, brings the opportunity to exposure landing place that might be difficult to access physically, intensifying travel possibilities.

8. **Local Economy Boost Through Technology:** Technology increases the efficiency of systems, products, and services. It makes data flow, process tracking and optimization, and the maintenance of contacts and personnel records easier (Elshaer et al., 2022). In actuality, this increased operational efficiency reduces costs and enables rapid business expansion.

All industries have been profoundly affected by recent technological breakthroughs, either directly or indirectly (Rainoldi et al., 2018). Many sectors have had to change in order to stay competitive and prosper in the changing economy as a result of these technological breakthroughs. Automation and digitalization are two examples of technology-driven effects that have transformed business processes, upended established industries, and significantly reduced employment. However, the increasing significance of data and analytics has shown how important technology is to decision-making and how it is changing the face of business (Moiseeva et al., 2020).

Technological developments have not only changed industries but also had a significant impact on opening up new growth prospects for economic development. Technology-driven growth opportunities have evolved as businesses seek to stay competitive and prosper in the changing economy (Tsaih and Hsu., 2018). A new digital economy has emerged as a result of these options, which have also enhanced communication and distant work and opened up new markets and opportunities for collaboration.

3.9 AUGMENTED REALITY (AR) AND VIRTUAL REALITY (VR) IN TOURISM

Through offering immersive and interactive experiences that improve tour planning, site visits, and memory collection, the integration of VR and AR technologies has made a breakthrough in the industry of tourism. These disruptive modern techs have the potential to enhance the way travelers think and engage with an attraction, destination, and site of cultural heritage. (Guttentag, 2010).

3.9.1 AUGMENTED REALITY (AR) IN TOURISM

To create an interactive and enriched experience, AR overlays digital forms of information, like images, texts, or 3D models or objects, onto the real-world environment using a device with a camera like a smartphone, AR glasses, or tablet (Yovcheva et al., 2013). AR enhances and creates a new user perception of the real world by mixing virtual elements.

AR works by utilizing cameras and sensors to catch data about the nearby environment and then layer digital objects onto that view. This digital environment appears to be a part of the actual scene of the world, which enables users to interact with objects that are virtual just like they existed in the real space (Kourouthanassis et al., 2015). Technologies of augmented reality can identify surfaces, objects, and environments, allowing the digital content overlay to be accurate, adjusted, and placed according to the perception and movement of the users.

AR has been used in numerous industries, which includes education, gaming, retail, and last but not least, tourism, because it provides a unique way to improve the experiences of the users (Chung et al., 2018). In the industry of tourism, augmented reality proved its worth as a strong tool for improving travel and tour experiences, providing interactive services like guides and allowing immersive storytelling.

3.9.2 AUGMENTED REALITY (AR) APPLICATIONS IN TOURISM

1. **Augmented Reality (AR)-enhanced Travel Guides:** Displays mounted to the head and mobile applications have the capability to provide real-time info about the nearby destinations, attractions, local culture, and historical facts, which ultimately boosts tourist's appreciation and understanding of the destination (Yovcheva et al., 2013). For instance, the application TripAdvisorAR uses augmented reality to display reviews, ratings, and other necessary information about the closest attractions and restaurants (Tripadvisor, 2022).
2. **Augmented Reality (AR) Navigation:** It is capable of assisting tourists in navigating unknown environments through giving cues of direction, finding out points of interest, and providing real-time information on transportation and event updates (Kourouthanassis et al., 2015). One such application is the Google Maps AR navigation features, which use augmented reality to overlay directions and visual cues into the view of the real world, making it smoother for travelers to find their way (Google, 2019).
3. **Storytelling Through Augmented Reality (AR):** Cultural heritage destinations, especially historical sites, can use augmented

reality for recreating events of the past, characters, or architectural marvels, which allows a tourist to engage and visualize with the site's rich history (Chung et al., 2018). For example, the augmented reality app "Civilizations AR" by BBC gathers and brings age-old artifacts and figures of history to life through AR (BBC, 2021).

In Figure 3.3, the application of a smartphone overlays digital information, such as the names of restaurants, reviews, ratings, and directions, onto the real-world view of a town road taken through a phone's camera.

FIGURE 3.3 Example of an augmented reality (AR) application in a tourism setting.

Source: Author.

3.9.3 VIRTUAL REALITY (VR) IN TOURISM

VR technologies make a completely digital and immersive environment that users can explore and interact with just like they do in the real world. It is a computer-generated simulation that makes a completely interactive and immersive digital environment (Guttentag, 2010). The difference between AR and VR is that VR fully replaces the physical environment with a virtual one, cutting users off from the actual world.

VR includes wearing a headset or head-mounted display (HMD) that covers the eyes of the user and gives a view of the virtual environment in a 3D stereoscopic display (Huang et al., 2016). By tracking the movement of the user's head, the VR system adjusts the display accordingly, developing a sense of staying in the world created by the virtual reality. The experience of virtual reality can be made to simulate real-world surroundings, fictional worlds, or even scenarios that are supposed to be impossible (Huang et al., 2019). With the help of specialized controllers, gestures of the hand, and other input devices, a person can interact with objects in the virtual world, navigate the surroundings, and perform different actions.

VR has applications in numerous domains, which include entertainment, gaming, healthcare, education, and tourism (Bekele et al., 2018). In the tourism industry, virtual reality provides the capability to transport users to virtual attractions and destinations, enabling them to enjoy and explore those places in an interactive and immersive way before deciding to visit the destination physically.

3.9.4 VIRTUAL REALITY (VR) APPLICATIONS IN TOURISM

1. **Virtual Tours:** VR has the capability to allow potential tourists to experience an attraction or destination before they visit it, which provides them the opportunity to make informed decisions and ultimately increases their desire to travel (Huang et al., 2016). Many companies like YouVisit provide virtual tours of resorts, hotels, and attractions, enabling tourists to experience these places from the comfort of their houses (YouVisit, 2023).
2. **Heritage Experiences Virtually:** VR can recreate sites of historical importance, cultural heritage sites, and museums, which provides an engaging and immersive way to appreciate and learn

about cultural legacy (Bekele et al., 2018). For instance, the "Rome Reborn" project utilizes VR to recreate ancient Rome, enabling tourists to explore the city as it was during the time of the Roman Empire (Rome Reborn, 2020).

3. **Virtual Training:** VR can be utilized to train tourism professionals, like hotel staff or tour operators and guides, by creating a simulated environment, and enhancing their skills and readiness for real-world problems and situations (Huang et al., 2019). Several companies like Transfr offer VR training solutions for the hospitality and travel industry, by creating scenarios like the check-in process and interactions with customers or tourists (Transfr, 2023).

In Figure 3.4, it is shown that virtual reality can take users to a virtual world of attractions, enabling the users to experience and explore those places in an interactive and immersive way before making the decision to visit physically.

3.9.5 AUGMENTED REALITY (AR) AND VIRTUAL REALITY (VR) BENEFITS AND OPPORTUNITIES

The addition of technologies like AR and VR has transformed the tourism industry and also offers various opportunities and benefits:

1. **Enhanced Tourist Experience:** By increasing the satisfaction of tourists, retention of knowledge, and creating emotional connections with attractions and destinations, AR and VR can produce engaging and interactive experiences (Dieck and Jung, 2018). Research by Huang et al. (2020) showed that travelers and tourists who used VR and AR during their trip ended up with higher levels of satisfaction, learning, and enjoyment and also increased the intention to visit the attraction again.
2. **Marketing and Promotion:** AR experiences and virtual tours allow potential tourists to preview the attraction; in that way, AR and VR can be used as powerful marketing tools, increasing the number of tourists (Tussyadiah et al., 2018). It has been shown that VR marketing has been used by organizations of tourism to promote destination awareness and interest among all stakeholders, including travelers (Bogicevic et al., 2019).

FIGURE 3.4 Virtual reality (VR) application in a tourism setting.
Source: Author.

3. **Cultural Heritage Preservation:** VR and AR are capable of reconstructing and preserving sites of cultural heritages, artifacts, and traditions; these also ensure their longevity and accessibility for the generations to come (Bekele et al., 2018). For instance, the "Mosul Virtual Museum" project utilizes VR for preserving and reconstructing traditional heritage sites that have been destroyed and damaged (UNESCO, 2019).
4. **Accessibility and Inclusivity:** Augmented reality and virtual reality can transform tourism destinations, making them more accessible to those with mobility limitations or disabilities,

enabling them to experience attractions virtually or with improved features of accessibility (UNESCO, 2019). For instance, projects like "Access Mars" utilize VR to give an immersive experience of staying on the surface of Mars, making the exploration of space accessible and easy for persons with limitations or disabilities (NASA, 2018).

5. **Sustainable Tourism:** By reducing the negative impact of over-crowding and travel on the environment at popular destinations, AR and VR can contribute to the practice of sustainable tourism (Huang et al., 2020). AR experiences and virtual visits can provide alternative paths to enjoy an attraction without the necessity of physical travel, ultimately decreasing the emissions of carbon and mitigating issues of over-tourism.

As long as technologies of AR and VR continue to evolve, improve, and become more accessible and affordable, their uses in the industry of tourism will keep on expanding, providing new chances for engagement, innovation, and sustainable practices of tourism (Guttentag, 2010). Still, addressing potential challenges, like user experience problems, privacy concerns, and the need for regulation and standardization within the tourism industry, is essential (Marasco et al., 2018).

3.10 THE RAMIFICATIONS FOR ENVIRONMENTAL CONCERNS, LOCAL ECONOMY, AND VISITOR ENJOYMENT

3.10.1 POSITIVE IMPACT ON LOCAL BUSINESSES AND EMPLOYMENT

3.10.1.1 CASE STUDIES OF ECONOMIC UPLIFTMENT

The tourist business serves as a linchpin for global economic growth, playing a key role in national and regional development. The Maasai Mara Conservancies initiative illustrates a unique type of reciprocal interaction by cooperating with the indigenous Maasai people, a departure from usual tactics. This new way of doing things has not only boosted the community's income over five years, but it has also caused wildlife populations to rise again, showing that economic growth and conservation efforts can work together in harmony (Keane et al., 2020). Similarly, Penang, Malaysia, has successfully positioned itself as a "food paradise," leveraging local cooking customs to draw tourists. This focus on gastronomy has not only improved income for small eateries and street sellers but has also created work chances in related services, adding to both economic and cultural growth (Gani et al., 2017). In the face of difficulties brought by the 2015 earthquake in Nepal, the "Homestay Network Nepal" project emerged as a changing force. By teaching and helping rural communities in creating homestays, the program enabled income diversification, improved living standards, and worked as a driver for infrastructure development in remote areas. This project not only provided a stage for cultural exchange but also allowed tourists to experience genuine Nepali culture, promoting mutual understanding between guests and local communities (Daly et al., 2017).

3.10.2 SOCIAL AND CULTURAL BENEFITS

The diverse effects of tourist growth are obvious across economic, social, cultural, and natural aspects, as noted in various studies. Host communities' views toward tourism depend on the delicate balance between perceived benefits and costs, with support staying strong when benefits outweigh costs, as stressed by Dyer et al. (2007). Chiu and Yeh (2017) warn that despite the role of the tourist industry in boosting overall economic growth, a competitive loss within the industry may hinder such growth, especially when facing trade deficits. In their 2004 study, Gursoy and Rutherford observed a correlation between tourism development levels and the preferences of residents, revealing an inclination towards moderate to lower levels of development, suggesting that excessive development may lead to unhappiness among residents.

Capital investment, especially in human capital, land, and technology, appears as a major element affecting economic growth in the tourist industry, as stressed by Liu and Chen (2016). They argue for higher government budget allocation for fixed capital investment as a plan to boost the economy. Du et al. (2016) establish a positive link between tourism activities and GDP growth, urging partnerships between governments and NGOs to start diverse tourism goods and encourage increased spending on travel places. In Southeast Asia, the development of ecotourism is described as a shared skill and goal among partner countries (Ruekeith, 2014). However, the study of tourism's effect would be incomplete without considering its social and cultural aspects. Nyaupane et al. (2006) stress the contextual dependence of these effects on the specific social environment of the host area, emphasizing the need for a detailed understanding to grasp the full range of consequences coming from tourist growth (Figure 3.5).

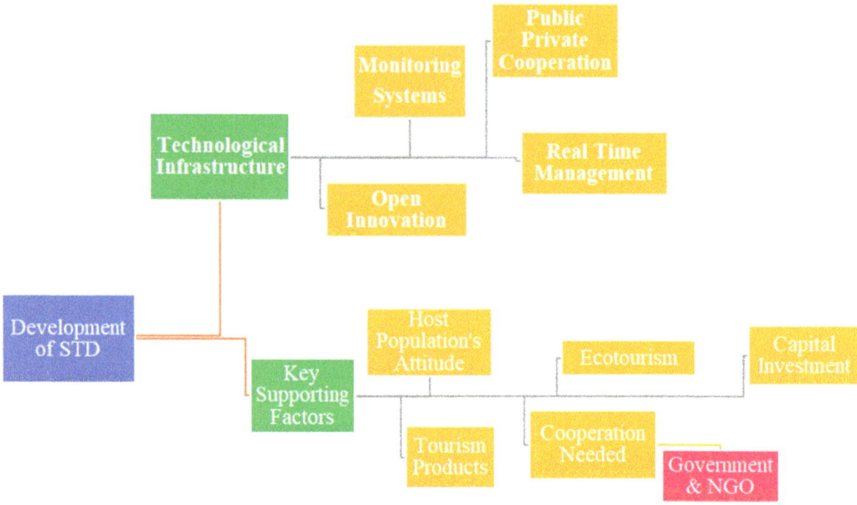

FIGURE 3.5 The process of sustainable tourism destination (STD) development.

Source: Author.

3.10.3 MITIGATING ENVIRONMENTAL IMPACT

3.10.3.1 SUSTAINABLE TOURISM PRACTICES

Smart tourism, a merger of technology integration and ecological considerations, appears as a powerful driver for positive changes within the tourist industry. The judicious deployment of smart systems to monitor and optimize energy and water consumption in hospitality establishments, exemplified by the implementation of intelligent heating and lighting settings in hotels, not only enhances operational efficiency but also contributes significantly to waste reduction (World Bank, 2018). In the world of sustainable transportation, the emergence of smart traffic control systems stands out as a crucial device. These systems not only improve garbage collection routes but also add to enhanced location beauty by easing crowding and cutting emissions (UNWTO, 2023). Cultural protection takes great importance, and technology plays a vital part in the protection of historical places. Innovations such as drone-assisted structure studies and virtual reality trips serve to minimize the effect of foot traffic on sensitive locations (European Commission, 2023). Engaging tourists with local communities and encouraging responsible behavior through smart platforms is important in ensuring that tourism becomes a force for community development while concurrently protecting traditional identities (UNWTO, 2023). Economically, the integration of smart platforms that connect tourists with local businesses fosters entrepreneurship and facilitates data-driven decision-making, thereby contributing to job creation and equitable distribution of tourism-generated income (European Commission, 2023; World Bank, 2018).

Exemplary efforts in places like Amsterdam, Barcelona, and Singapore highlight the concrete benefits of smart tourist strategies. These include flexible traffic control systems, green energy-powered infrastructure, and vast sensor networks tracking environmental factors (European Commission, 2023). These results illustrate the changing potential of smart tourism in improving tourist experiences while jointly achieving goals related to environmental conservation, cultural preservation, and economic growth. Efforts to address waste reduction are visible in the adoption of smart bins and recycling systems equipped with sensors, thereby raising the effectiveness of waste management and increasing recycling rates (Mingaleva et al., 2019). In protected areas, the introduction of environmental tracking systems involving drones and sensors adds to wildlife protection, aids in finding illegal activities, and refines land management practices (UNEP, 2019). Recognizing the importance of water conservation, especially in water-scarce areas, efforts such as low-flow taps, rainwater harvesting systems, and awareness programs play a key role in ensuring a sustainable water supply and protecting the natural environment. Drawing lessons from old Mediterranean water management practices, the tourism industry in Santorini, for instance, is set to accept sustainable measures such as the protection of traditional cisterns and the increase of storage capacity for eco-tourism (Enriquez et al., 2017). Furthermore, the growing growth in tourism offers difficulties for towns trying to keep their identity.

Addressing this requires a diverse approach involving governmental action, environmental measures, and the inclusion of triple or quadruple bottom-line strategies, often needing foreign partnerships and talks (Sutawa, 2011). The key work by Cessford and Dingwall (1998) highlights the deep environmental effects of the tourism sector, stressing the need for careful planning and sustainable tourism practices to create a subtle balance between extracting economic rewards and conserving the environment. Insights from experts such as Michalena and Tripanagnostopoulos (2010) and Petrevska et al. (2016) offer useful views on cost-effective energy solutions and the urge for greater knowledge and acceptance of green energy practices within the tourist industry. These research efforts support the critical role of educated decision-making in creating a sustainable and responsible future for the tourist sector.

3.10.3.2 TECHNOLOGY'S ROLE IN CONSERVATION

Smart tourism, based on data-driven technologies, appears as a potential answer to ease the environmental effects linked with traditional tourism practices (UNWTO, 2018). Utilizing smart grids and advanced building management systems, hotels and sites can effectively track and control energy consumption, resulting in significant reductions in power and water usage (Simpson et al., 2008). Complementary to this, intelligent transportation systems, driven by real-time traffic management leveraging sensors and artificial intelligence, not only ease congestion but also improve public transport routes, supporting eco-friendly options such as riding and walking (UNWTO, 2018). Remote sensing technologies play a vital part in tracking and assessing ecosystems. High-resolution image capabilities of satellites allow researchers to watch changes in land cover, deforestation, and habitat loss, as illustrated by the work of Pettorelli et al. (2014), showing the usefulness of remote sensing in evaluating the outcome of human actions on wildlife areas.

The merging of big data analytics greatly improves our knowledge of natural trends and processes. Anderson et al. (2015) stress the importance of predicting species spread, rating biodiversity, and projecting environmental changes on the successful use of big data. Conservation efforts gain from the ability to pick suitable places, spot trends in big databases, and create efficient management plans. Moreover, smart tourism adds to tourist education and involvement through digital platforms, virtual reality apps, and interactive displays. These tools not only raise knowledge about environmental problems but also support responsible behavior among tourists (Loureiro and Nascimento, 2021). The benefits of smart tourism stretch beyond environmental protection, covering an improved tourist experience defined by individual suggestions, efficient booking systems, and optimized mobility. This, in turn, leads to greater guest happiness and loyalty. From an economic viewpoint, the benefits include heightened resource efficiency, reduced running costs, and the potential to attract eco-conscious tourists, strengthening the changing potential of smart tourism (Mingaleva et al., 2019; Simpson et al., 2008). In essence, smart tourism appears as a complete and technology-driven method to create a more safe and enjoyable future for both locations and tourists (Figure 3.6).

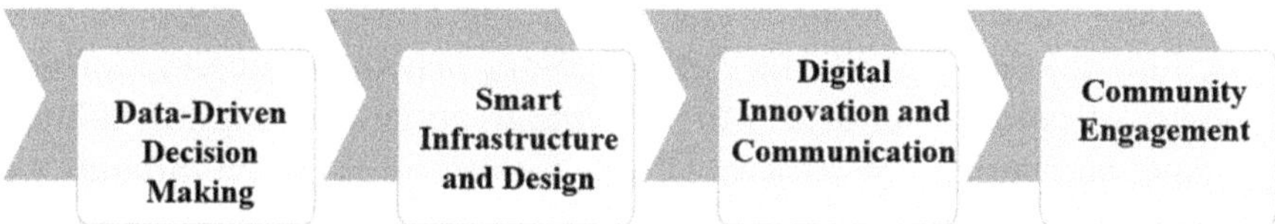

FIGURE 3.6 Key factors of smart tourism to mitigate environmental impact.
Source: Author.

3.10.4 BALANCING TECHNOLOGY WITH AUTHENTIC EXPERIENCES

3.10.4.1 MAINTAINING CULTURAL INTEGRITY AND STRATEGIES FOR PRESERVING LOCAL HERITAGE

In the changing environment of our current society, where technical breakthroughs and globalization have significant effects, a delicate balance must be maintained between the pursuit of innovation and the preservation of cultural legacy (Mowforth and Munt, 2015). This complicated balance is highlighted by a rising desire for real experiences among both visitors and locals, indicating a collective need for true relationships (Ning, 2017). As we negotiate this landscape, integrating the draw of face-to-face meetings with the practicality of technology emerges as a vital problem, requiring conscious management of screen time and a commitment to preserving real-life connections (Camilleri, 2018). At the core of the tension lies the essential authenticity of cultural interactions, with tourists attracted to the appeal of ancient knowledge, lively folklore, and the incorporation of traditional behaviors (Hall and Page, 2014). However, its authenticity faces the prospect of degradation via excessive technology intrusion. The growth of online booking platforms and social media advertisements has contributed to the commercialization of cultural expression, reducing communities into simply tourist attractions (Adie, 2017; World Heritage Committee, 2002).

Holden et al. (2011) believe that the turning of traditional traditions into staged performances for tourists jeopardizes their holy character and deep importance. A thorough assessment of the literature on the protection of cultural assets undertaken by Mendoza et al. (2023) uncovered 146 papers published between 2018 and 2022. Notably, 70% of these chapters put a considerable focus on technology, with augmented reality/virtual reality (15%) and 3D digital technologies (44%) appearing as key topics. This underlines the rising significance of technology in both the preservation and progress of cultural heritage. Efficient preservation of intangible cultural heritage, emphasized by Idris et al. (2016), can be achieved through the adoption of innovative digital technologies. The employment of virtual reality and human-computer interaction technologies in cultural heritage preservation not only assures conservation but also increases sharing and inheritance via immersive, multimedia platforms (Zhong et al., 2021). In this complicated dance between tradition and technology, a careful approach becomes necessary to minimize any harmful impacts on mental health and general well-being while protecting the authenticity of cultural experiences in our linked world (Figure 3.7).

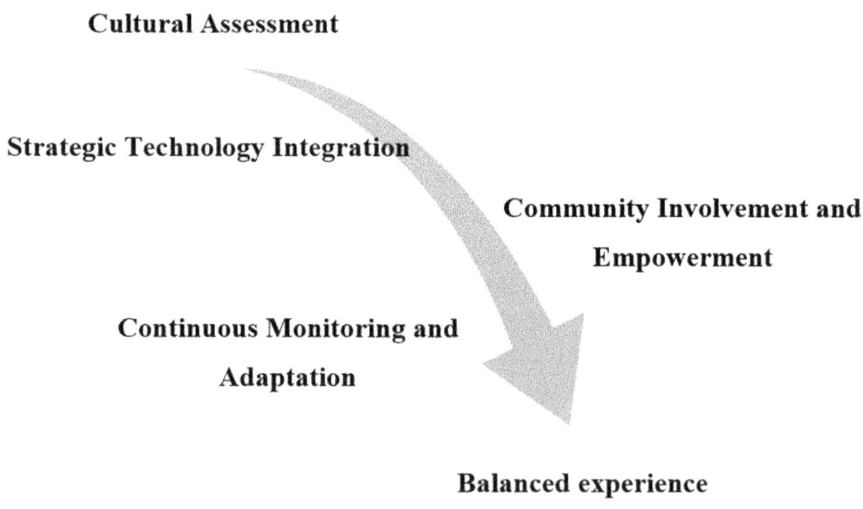

FIGURE 3.7 Process for a balanced experience.

Source: Author.

3.11 DISCUSSION

This chapter provides an in-depth analysis of how disruptive technologies such as the IoT, AI, AR, and VR are converting the tourism industry and transforming it into "smart tourism." The key discussion areas include the role of Artificial Intelligence in personalization, voice assistants, recommendation systems, and demand prediction. IoT applications to the hospitality sector, like inventory management, personalized hotel experiences, and real-time monitoring and controlling, are also addressed.

Offering immersive virtual tours, decision-making assistance, and educational experiences, the integration of AR and VR technologies is highlighted as a game-changer, while also taking part in the preservation of cultural heritage. For sustainable tourism development, strategies like monitoring resource consumption, implementing efficient transportation, and educating stakeholders are explored. Some case studies to showcase the social and economic benefits of sustainable tourism initiatives are also discussed.

3.12 CONCLUSION

The concept of smart tourism has been raised by the integration of cutting-edge technologies like AI, IoT, and AR/VR in the tourism industry. These modern technologies offer various benefits like personalized experiences, operational efficiency, resource optimization, and data-driven decision-making. The smart tourism concept gives an opportunity to balance innovation and the preservation of cultural authenticity and integrity. By using virtual reality technologies and computer-human interaction, cultural heritage can be preserved and shared in an accessible and immersive manner, minimizing the negative impact on sensitive sites.

Moreover, smart tourism holds the potential to promote sustainable tourism practices by properly monitoring and optimizing resource consumption, planting efficient transportation systems, and educating both local communities and tourists about responsible tourism behavior. Highlighting the significance of balancing economic improvement and environmental preservation, case studies were demonstrated, indicating the social and economic benefits of sustainable tourism.

3.13 PRACTICAL IMPLICATIONS

The integration of AI, IoT, and AR/VR technologies in the tourism industry can bring increased operational efficiency, personalized experiences, and improved resource management. If adopted, smart tourism technologies can lead to data-driven decision-making, real-time monitoring, predictive maintenance, and optimized tourism services. The implementation of augmented reality, virtual tours, and other related technologies can improve visitor experiences, foster cultural preservation, and provide accessible ways to explore a destination. The utilization of smart tourism technologies may contribute to sustainable tourism approaches by optimizing and monitoring resource consumption, providing an efficient transportation system, and educating tourists and other stakeholders. To enhance customer service, personalized recommendation systems, conversational AI, and voice assistants can be developed to cater to personal preferences, ultimately leading to more tourist satisfaction. In the hospitality sector, the integration of IoT can bring personalized hotel services, efficient management of inventory, and real-time monitoring and controlling of facilities.

3.14 THEORETICAL IMPLICATIONS

The "smart tourism" concept emerges as a theoretical framework that connects disruptive modern technologies with the tourism industry and its operations, services, and experiences. This chapter contributes to the reasoning of how cutting-edge technologies such as AI, AR/VR, and IoT are transforming the tourism industry and shaping the future of the travel experience. The chapter also provides insights about the potential implications of these modern technologies in numerous aspects of travel and tourism, like recommendation systems, personalized services, virtual tours, and preservation of cultural heritage. It also highlights the significance of striking a balance between technological innovation and the preservation of cultural authenticity and integrity, which contributes to the theoretical discourse on sustainable and responsible tourism development. By exploring the role of modern technology in mitigating the impacts on the environment, promoting sustainable tourism, and boosting the local economy, it contributes to the theoretical understanding of sustainable tourism management.

KEYWORDS

- **artificial intelligence**
- **augmented reality**
- **cybersecurity**
- **data privacy**
- **data-driven decisions**
- **Internet of Things**
- **operational efficiency**
- **smart tourism**
- **virtual reality**

REFERENCES

Adie, B. A. (2017). Franchising our heritage: The UNESCO World Heritage brand. *Tourism Management Perspectives*, *24*, 48–53.

Ahmad, K., Maabreh, M., Ghaly, M., Khan, K., Qadir, J., & Al-Fuqaha, A. (2022). Developing future human-centered smart cities: Critical analysis of smart city security, data management, and ethical challenges. *Computer Science Review*, *43*, 100452.

Akdu, U. (2020). Smart tourism: Issues, challenges and opportunities. In *The Emerald Handbook of ICT in Tourism and Hospitality* (pp. 291–308).

Arcuri, M. C., Gai, L., Ielasi, F., & Ventisette, E. (2020). Cyber-attacks on hospitality sector: Stock market reaction. *Journal of Hospitality and Tourism Technology*, *11*(2), 277–290.

Ardito, L., Cerchione, R., Del Vecchio, P., & Raguseo, E. (2019). Big data in smart tourism: Challenges, issues and opportunities. *Current Issues in Tourism*, *22*(15), 1805–1809.

BBC. (2021). *Civilizations AR* [Mobile app]. https://www.bbc.co.uk/taster/pilots/civilisations-ar (accessed on 8th July 2025).

Bekele, M. K., Pierdicca, R., Frontoni, E., Malinverni, E. S., & Gain, J. (2018). A survey of augmented, virtual, and mixed reality for cultural heritage. *Journal on Computing and Cultural Heritage*, *11*(2), 1–36.

Birenboim, A., Bulis, Y., & Omer, I. (2023). A typology of tourism mobility apps. *Tourism Management Perspectives*, *48*, 101161.

Bogicevic, V., Seo, S., Kandampully, J. A., Liu, S. Q., & Rudd, N. A. (2019). Virtual reality presence as a preamble of tourism experience: The role of mental imagery. *Tourism Management*, *74*, 55–64.

Bohdanowicz, P., Churie-Kallhauge, A., Martinac, I., & Rezachek, D. (2001). Energy-efficiency and conservation in hotels–Towards sustainable tourism. In *4th International Symposium on Asia Pacific Architecture* (Vol. 21).

Boothby, D., Dufour, A., & Tang, J. (2010). Technology adoption, training, and productivity performance. *Research Policy, 39*(5), 650–661. https://doi.org/10.1016/j.respol.2010.02.011 (accessed on 8th July 2025).

Buhalis, D., & Amaranggana, A. (2015). Smart tourism destinations enhancing tourism experience through personalization of services. In *Information and Communication Technologies in Tourism 2015: Proceedings of the International Conference in Lugano, Switzerland, February 3–6, 2015* (pp. 377–389).

Buhalis, D., Harwood, T., Bogicevic, V., Viglia, G., Beldona, S., & Hofacker, C. (2019). Technological disruptions in services: Lessons from tourism and hospitality. *Journal of Service Management, 30*(4), 484–506.

Bulchand-Gidumal, J. (2022). Impact of artificial intelligence in travel, tourism, and hospitality. In *Handbook of E-Tourism* (pp. 1943–1962). Cham: Springer International Publishing.

Buonincontri, P., & Micera, R. (2016). The experience co-creation in smart tourism destinations: A multiple case analysis of European destinations. *Information Technology & Tourism, 16*, 285–315.

Camilleri, M. A. (2018). The promotion of responsible tourism management through digital media. *Tourism Planning & Development, 15*(6), 653–671.

Car, T., Stifanich, L. P., & Šimunić, M. (2019). Internet of Things (IoT) in tourism and hospitality: Opportunities and challenges. *Tourism in South East Europe, 5*, 163–175.

Celik, Z. B., Babun, L., Sikder, A. K., Aksu, H., Tan, G., McDaniel, P., & Uluagac, A. S. (2018). Sensitive information tracking in commodity IoT. In *27th USENIX Security Symposium (USENIX Security 18)* (pp. 1687–1704).

Chiu, Y. B., & Yeh, L. T. (2017). The threshold effects of the tourism-led growth hypothesis: Evidence from a cross-sectional model. *Journal of Travel Research, 56*(5), 625–637. https://doi.org/10.1177/0047287516650938 (accessed on 8th July 2025).

Chouki, M., Talea, M., Okar, C., & Chroqui, R. (2020). Barriers to information technology adoption within small and medium enterprises: A systematic literature review. *International Journal of Innovation and Technology Management, 17*(01), 2050007.

Claveria, O., Monte, E., & Torra, S. (2015). A new forecasting approach for the hospitality industry. *International Journal of Contemporary Hospitality Management, 27*(7), 1520–1538.

Cohen, E. (2004). *Contemporary Tourism: Diversity and Change*.

Daly, P., Barenstein, J. D., Hollenbach, P., & Ninglekhu, S. (2017). *Post-Disaster Housing Reconstruction in Urban Areas in Nepal*. IIED, London.

Davide, M., Cristina, R., & Sabina, L. (2021). Accessible tourism and augmented and virtual reality: A review and analysis of current trends. In J. Abidin & L. Adu (Eds.), *Innovation and Transformation in Tourism Mobility* (pp. 89–105). IGI Global.

Dorcic, J., Komsic, J., & Markovic, S. (2019). Mobile technologies and applications towards smart tourism–State of the art. *Tourism Review, 74*(1), 82–103.

Dyer, P., Gursoy, D., Sharma, B., & Carter, J. (2007). Structural modeling of resident perceptions of tourism and associated development on the Sunshine Coast, Australia. *Tourism Management, 28*(2), 409–422.

Edwards, L. (2016). Privacy, security, and data protection in smart cities: A critical EU law perspective. *European Data Protection Law Review, 2*, 28.

Elshaer, A. M., & Marzouk, A. M. (2022). Memorable tourist experiences: The role of smart tourism technologies and hotel innovations. *Tourism Recreation Research*, 1–13.

Erjavec, K., Arsenijević, O., & Starc, J. (2018). Satisfaction with managers' use of communication channels and its effect on employee-organization relationships. *Journal of East European Management Studies, 23*(4), 559–578.

European Commission. (2023). *Leading Practices in Smart and Sustainable Tourism: 2023 Best Practices Booklet*. Brussels: European Commission. https://smart-tourism-capital.ec.europa.eu/leading-examples-smart-tourism-practices-europe_en (accessed on 8th July 2025).

Femenia-Serra, F., Ioannou, A., & Tussyadiah, I. P. (2022). Is smart scary? A mixed-methods study on privacy in smart tourism. *Current Issues in Tourism, 25*(14), 2212–2238.

Gani, A. A., Mahdzar, M., Mohamad, R., Abdullah, N., & Awang, K. W. (2017). Linking image and satisfaction of food tourism in Penang, Malaysia. *E-Proceeding of the 6th International Conference on Social Sciences Research 2017*.

Google. (2019). Live View with AR directions starts rolling out in Maps [Blog post]. https://blog.google/products/maps/live-view-ar-directions-maps/ (accessed on 8th July 2025).

Gretzel, U. (2011). Intelligent systems in tourism: A social science perspective. *Annals of Tourism Research, 38*(3), 757–779.

Gretzel, U., Sigala, M., Xiang, Z., & Koo, C. (2015). Smart tourism: Foundations and developments. *Electronic Markets, 25*, 179–188.

Gretzel, U., Werthner, H., Koo, C., & Lamsfus, C. (2015). Conceptual foundations for understanding smart tourism ecosystems. *Computers in Human Behavior, 50*, 558–563.

Griffin, T., Giberson, J., Lee, S. H. M., Guttentag, D., Kandaurova, M., Sergueeva, K., & Dimanche, F. (2017). Virtual reality and implications for destination marketing.

Gunter, U., & Önder, I. (2016). Forecasting city arrivals with Google Analytics. *Annals of Tourism Research, 61*, 199–212.

Gursoy, D., & Rutherford, D. G. (2004). Host attitudes toward tourism: An improved structural model. *Annals of Tourism Research, 31*(3), 495–516.

Guttentag, D. A. (2010). Virtual reality: Applications and implications for tourism. *Tourism Management, 31*(5), 637–651.

Hall, C. M., & Page, S. J. (2014). *The Geography of Tourism and Recreation: Environment, Place and Space*. Routledge.

Holden, A., Sonne, J., & Novelli, M. (2011). Tourism and poverty reduction: An interpretation by the poor of Elmina, Ghana. *Tourism Planning & Development, 8*(3), 317–334.

Huang, C. D., Goo, J., Nam, K., & Yoo, C. W. (2017). Smart tourism technologies in travel planning: The role of exploration and exploitation. *Information & Management, 54*(6), 757–770.

Huang, Y. C., Backman, K. F., Backman, S. J., & Chang, L. L. (2016). Exploring the implications of virtual reality technology in tourism marketing: An integrated research framework. *International Journal of Tourism Research, 18*(2), 116–128.

Huang, Y. C., Backman, S. J., Backman, K. F., & Moore, D. (2019). Virtual reality for enhancing the tourism experience: A perspective article. *Journal of Hospitality and Tourism Technology, 10*(2), 241–249.

Huang, Y. C., Backman, S. J., Backman, K. F., & Moore, D. (2020). Virtual reality: Emerging opportunities for tourism research and practice. *Journal of Travel Research, 60*(5), 1153–1170.

Idris, M. Z., Mustaffa, N. B., & Yusoff, S. O. S. (2016). Preservation of intangible cultural heritage using advanced digital technology: Issues and challenges. *Harmonia: Journal of Arts Research and Education, 16*(1), 1–13.

Jasrotia, A., & Gangotia, A. (2018). Smart cities to smart tourism destinations: A review paper. *Journal of Tourism Intelligence and Smartness, 1*(1), 47–56.

Jenny, S. (2017). *Enhancing tourism with augmented and virtual reality.* Bachelor's thesis. Visamäki–Hämeenlinna, Finland.

Jeong, M., & Shin, H. H. (2020). Tourists' experiences with smart tourism technology at smart destinations and their behavior intentions. *Journal of Travel Research, 59*(8), 1464–1477.

Jung, T. H., Lee, H., Chung, N., & Tom Dieck, M. C. (2018). Cross-cultural differences in adopting mobile augmented reality at cultural heritage tourism sites. *International Journal of Contemporary Hospitality Management, 30*(3), 1621–1645. https://doi.org/10.1108/IJCHM-02-2017-0084 (accessed on 8th July 2025).

Kapiki, S. (2021). Smart city and IoT technologies enabling smart tourism: The case of Greece. In *Higher Education in Smart City Technologies: European, Kazakh, Mongolian, Russian Universities Approach.*

Katkuri, P. K., Mantri, A., & Anireddy, S. (2019). Innovations in tourism industry & development using augmented reality (AR), virtual reality (VR). In *TENCON 2019–2019 IEEE Region 10 Conference (TENCON)* (pp. 2578–2581). IEEE.

Keane, A., Lund, J. F., Bluwstein, J., Burgess, N. D., Nielsen, M. R., & Homewood, K. (2020). Impact of Tanzania's wildlife management areas on household wealth. *Nature Sustainability, 3*(3), 226–233.

Khan, H. U., & Fournier-Bonilla, S. D. (2016). Technological infrastructure effects on export diversification: A case study of Qatar. *Northeast Decision Sciences Institute Conference, 31.*

Khan, H. U., & Smuts, R. G. (2019). A comparison of change management guidelines to address technology adoption barriers: A case study of higher educational institutions. *Journal of Theoretical and Applied Information Technology, 97*(7), 1999–2020.

Khan, M. S., Woo, M., Nam, K., & Chathoth, P. K. (2017). Smart city and smart tourism: A case of Dubai. *Sustainability, 9*(12), 2279.

Kontogianni, A., & Alepis, E. (2020). Smart tourism: State of the art and literature review for the last six years. *Array, 6*, 100020.

Kourouthanassis, P., Boletsis, C., Bardaki, C., & Chasanidou, D. (2015). Tourists' responses to mobile augmented reality travel guides: The role of emotions on adoption behavior. *Pervasive and Mobile Computing, 18*, 71–87.

Kumar, N., Madhuri, J., & Channegowda, M. (2017). Review on security and privacy concerns in Internet of Things. In *2017 International Conference on IoT and Application (ICIOT)* (pp. 1–5). IEEE.

Lai, W.C. & Hung, W.H. (2018). A framework of cloud and AI based intelligent hotel. In *Proceedings of The 18th International Conference on Electronic Business* (pp. 36–43). ICEB, Guilin, China, December 2–6.

Leung, R. (2022). Development of information and communication technology: From e-tourism to smart tourism. In *Handbook of E-Tourism* (pp. 1–33).

Liu, Z., & Chen, Y. (2016). The investment choices to deal with the slowdown in economic growth—Based on the analysis of the effect of human capital investment. *Open Journal of Business and Management, 4*(4), 558–570.

Loureiro, S. M. C., & Nascimento, J. (2021). Shaping a view on the influence of technologies on sustainable tourism. *Sustainability, 13*(22), 12691.

Loureiro, S. M. C., Guerreiro, J., & Ali, F. (2020). 20 years of research on virtual reality and augmented reality in tourism context: A text-mining approach. *Tourism Management, 77*, 104028.

Marasco, A., Buonincontri, P., van Niekerk, M., Orlowski, M., & Okumus, F. (2018). Exploring the role of next-generation virtual reality in hospitality education: A comparative study among teaching case scenarios. *International Journal of Hospitality Management, 74*, 65–75.

Masseno, M. D., & Santos, C. T. (2018). Assuring privacy and data protection within the framework of smart tourism destinations. *Medialaws–Rivista di Diritto dei Media, (2)*, 251–266.

McNaughton, M., Rao, L., & Verma, S. (2020). Building smart communities for sustainable development: Community tourism in Treasure Beach, Jamaica. *Worldwide Hospitality and Tourism Themes, 12*(3), 337–352.

Mendoza, M. A. D., De La Hoz Franco, E., & Gómez, J. E. G. (2023). Technologies for the preservation of cultural heritage—A systematic review of the literature. *Sustainability, 15*(2), 1059.

Mingaleva, Z., Vukovic, N., Volkova, I., & Salimova, T. (2019). Waste management in green and smart cities: A case study of Russia. *Sustainability, 12*(1), 94.

Mingjun, W., Zhen, Y., Wei, Z., Xishang, D., Xiaofei, Y., Chenggang, S., & Jinghai, H. (2012). Research on experimental system for Internet of Things major and application project. In *2012 3rd International Conference on System Science, Engineering Design and Manufacturing Informatization* (Vol. 1, pp. 261–263).

Miorandi, D., Sicari, S., De Pellegrini, F., & Chlamtac, I. (2012). Internet of Things: Vision, applications and research challenges. *Ad Hoc Networks, 10*(7), 1497–1516.

Mišúnová, E. (2015). Trends in the development of tourism marketing communications of the towns. In *Proceedings of The 5th Central European Conference in Regional Science – CERS.* Košice, Slovak Republic.

Mofokeng, N. E. M., & Matima, T. K. (2018). Future tourism trends: Virtual reality-based tourism utilizing distributed ledger technologies. *African Journal of Hospitality, Tourism and Leisure, 7*(3), 1–14.

Moiseeva, V. Y., & Zolotovskiy, V. A. (2020). Preconditions of development and perspectives of use of smart city technologies for regional market of tourism. *Institute of Scientific Communications Conference*, 85–91.

Mowforth, M., & Munt, I. (2015). *Tourism and Sustainability: Development, Globalization and New Tourism in the Third World.* Routledge.

NASA. (2018). Access Mars [Press release]. https://www.nasa.gov/feature/access-mars-explorers-wanted (accessed on 8th July 2025).

Nayyar, A., Mahapatra, B., Le, D., & Suseendran, G. (2018). Virtual reality (VR) & augmented reality (AR) technologies for tourism and hospitality industry. *International Journal of Engineering & Technology, 7*(2.21), 156–160.

Neumeyer, X., Santos, S. C., & Morris, M. H. (2020). Overcoming barriers to technology adoption when fostering entrepreneurship among the poor: The role of technology and digital literacy. *IEEE Transactions on Engineering Management*, 1–14. https://doi.org/10.1109/TEM.2020.2989740 (accessed on 8th July 2025).

Ning, W. (2017). Rethinking authenticity in tourism experience. In *The Political Nature of Cultural Heritage and Tourism* (pp. 469–490). Routledge.

Novera, C. N., Ahmed, Z., Kushol, R., Wanke, P., & Azad, M. A. K. (2022). Internet of Things (IoT) in smart tourism: A literature review. *Spanish Journal of Marketing–ESIC*, *26*(3), 325–344.

Nyaupane, G. P., Morais, D. B., & Dowler, L. (2006). The role of community involvement and number/type of visitors on tourism impacts: A controlled comparison of Annapurna, Nepal and Northwest Yunnan, China. *Tourism Management*, *27*(6), 1373–1385.

OECD. (2019). *Recommendation of the Council on Responsible Individual Usage of Artificial Intelligence in Education, Government, and Work.* Organization for Economic Co-operation and Development. https://legalinstruments.oecd.org/en/instruments/oecd-legal-0449 (accessed on 8th July 2025).

Ozdemir, M. A. (2021). Virtual reality (VR) and augmented reality (AR) technologies for accessibility and marketing in the tourism industry. In *ICT Tools and Applications for Accessible Tourism* (pp. 277–301). IGI Global.

Priyambodo, T. K., & Artianingsih, M. D. (2022). Strategy for sustainable smart tourism village development in Ponggok Village, Klaten, Central Java. *International Journal of Sustainable Competitiveness on Tourism*, *1*(02), 1–10.

Puah, C. H., Jong, M. C., Ayob, N., & Ismail, S. (2018). The impact of tourism on the local economy in Malaysia. *International Journal of Business and Management*, *13*(12), 151. https://doi.org/10.5539/ijbm.v13n12p151 (accessed on 8th July 2025).

Rainoldi, M., Driescher, V., Lisnevska, A., Zvereva, D., Stavinska, A., Relota, J., & Egger, R. (2018). Virtual reality: An innovative tool in destinations' marketing. *The Gaze: Journal of Tourism and Hospitality*, *9*(1), 53–68.

Rome Reborn. (2020). *Rome Reborn* [Virtual reality experience]. https://www.romereborn.org/ (accessed on 8th July 2025).

Rudas, I. J., & Fodor, J. (2008). Intelligent systems. *International Journal of Computers, Communications & Control*, *3*(3), 132–138.

Ruekeith, G. O. (2014). *EATOF Members Sign Declaration to Jointly Promote Tourism, Especially Eco-Tourism.* https://scholar.google.com/scholar?hl=en&as_sdt=0%2C5&q=Ruekeith%2C+G.+O.+%282014%2C+September+6%29.+EATOF+Members+Sign+Declaration+To+Jointly+Promote+Tourism%2C+Especially+Eco-Tourism.+Retrieved+From+The+Borneo+Post%3A+Http%3A%2F%2Fwww.Theborneopost.Com%2F2014%2F09%2F06%2Featof-Members+-Sign-Declaration-To-Jointly-Promote-Tourism-Especially-Eco-Tourism%2F (accessed on 8th July 2025).

Saif, I., Peasley, S., & Perinkolam, A. (2015). Safeguarding the Internet of Things. *Deloitte Review*.

Samaddar, K., & Mondal, S. (2023). AR and VR-based travel: A responsible practice towards sustainable tourism. *International Journal of Tourism Cities*.

Saoud, J., & Jung, T. (2018). An ethical perspective of the use of AR technology in the tourism industry. In *Augmented Reality and Virtual Reality: Empowering Human, Place and Business* (pp. 33–46).

Saranya, C. M., & Nitha, K. P. (2015). Analysis of security methods in the Internet of Things. *International Journal on Recent and Innovation Trends in Computing and Communication*, *3*(4), 1970–1974.

Seong, B. H., & Hong, C. Y. (2022). Corroborating the effect of positive technology readiness on the intention to use the virtual reality sports game "Screen Golf": Focusing on the Technology Readiness and Acceptance Model. *Information Processing & Management*, *59*(4), 102994.

Shen, S., Xu, K., Sotiriadis, M., & Wang, Y. (2022). Exploring the factors influencing the adoption and usage of augmented reality and virtual reality applications in tourism education within the context of COVID-19 pandemic. *Journal of Hospitality, Leisure, Sport & Tourism Education*, *30*, 100373.

Siddiqui, M. S., Syed, T. A., Nadeem, A., Nawaz, W., & Alkhodre, A. (2022). Virtual tourism and digital heritage: An analysis of VR/AR technologies and applications. *International Journal of Advanced Computer Science and Applications*, *13*(7).

Simpson, M. C., Gössling, S., Scott, D., Hall, C. M., & Gladin, E. (2008). *Climate Change Adaptation and Mitigation in the Tourism Sector: Frameworks, Tools and Practices.* UNEP, University of Oxford, UNWTO, WMO: Paris, France.

Transfr. (2021). *Virtual Reality Training for The Hospitality Industry*. https://transfrinc.com/products/hospitality/ (accessed on 8th July 2025).

Tripadvisor. (2022). *Tripadvisor AR* [Mobile app]. https://www.tripadvisor.com/AR (accessed on 8th July 2025).

Tsaih, R. H., & Hsu, C. C. (2018). Artificial intelligence in smart tourism: A conceptual framework. In *Proceedings of The 18th International Conference on Electronic Business* (pp. 124–133). ICEB, Guilin, China, December 2–6.

Tung, R. L. (2016). New perspectives on human resource management in a global context. *Journal of World Business*, *51*(1), 142–152.

Tussyadiah, I. P., Wang, D., Jung, T. H., & tom Dieck, M. C. (2018). Virtual reality, presence, and attitude change: Empirical evidence from tourism. *Tourism Management*, *66*, 140–154.

Tussyadiah, I., & Miller, G. (2019). Perceived impacts of artificial intelligence and responses to positive behavior change intervention. In *Information and Communication Technologies in Tourism 2019: Proceedings of the International Conference in Nicosia, Cyprus, January 30–February 1, 2019* (pp. 359–370).

UNEP. (2019). *Protected Areas and World Heritage Sites: Innovation in Monitoring and Management Using Drones and Sensors*. Nairobi: United Nations Environment Programmed. https://apps1.unep.org/resolutions/uploads/Inf_9.pdf (accessed on 8th July 2025).

UNESCO. (2019). *UNESCO Leads Ground-Breaking Initiative to Create First Complete Virtual Reality Model of an Iraqi World Heritage Site* [Press release]. https://en.unesco.org/news/unesco-leads-ground-breaking-initiative-create-first-complete-virtual-reality-model-iraqi-world (accessed on 8th July 2025).

UNWTO. (2018). *Smart Cities for Tourism: An Opportunity for Innovation and Inclusion.* https://www.unwto.org/Global/Press-Release/2018-10-16/Creating-Smart-Cities-Innovative-Tourism-Experiences (accessed on 8th July 2025).

UNWTO. (2023). *Sustainable Development Goals: Travel & Tourism's Contribution.* https://www.unwto.org/World-Tourism-Day-2023 (accessed on 8th July 2025).

UNWTO. (2020). *World Tourism Organization.* https://www.unwto.org/ (accessed on 8th July 2025).

Washington, M., & Ono, K. A. *Communication.* Oxford Research Encyclopedias.

World Bank. (2018). *Sustainable Tourism Development: Public & Private Perspectives.* https://documents1.worldbank.org/curated/en/558121506324624240/pdf/119954-WP-PUBLIC-Sustainabletourismdevelopment.pdf (accessed on 8th July 2025).

World Heritage Committee. (2002). *Operational Guidelines for the Implementation of the World Heritage Convention.*

Xiang, Z., & Fesenmaier, D. R. (2017). Big data analytics, tourism design, and smart tourism. In *Analytics in Smart Tourism Design: Concepts and Methods* (pp. 299–307).

Yavuz, M. C., Cavusoglu, M., & Corbaci, A. (2018). Reinventing tourism cities: Examining technologies, applications, and city branding in leading smart cities. *Journal of Global Business Insights*, *3*(1), 57–70.

Ye, B. H., Ye, H., & Law, R. (2020). Systematic review of smart tourism research. *Sustainability*, *12*(8), 3401.

YouVisit. (2023). *Virtual Tours & Virtual Reality.* https://www.youvisit.com/ (accessed on 8th July 2025).

Yovcheva, Z., Buhalis, D., & Gatzidis, C. (2013). Engineering augmented tourism experiences. In *Information and Communication Technologies in Tourism 2013* (pp. 24–35). Springer: Berlin/Heidelberg, Germany.

Zhong, H., Wang, L., & Zhang, H. (2021). The application of virtual reality technology in the digital preservation of cultural heritage. *Computer Science and Information Systems*, *18*(2), 535–551.

CHAPTER 4

BIG DATA AND SMART TOURISM: TRANSFORMING THE TRAVEL INDUSTRY

MD. ABDUL HALIM[1] and PRITOM KUNDU JOY[2*]

[1]*Department of Tourism and Hospitality Management, Rangamati Science and Technology University, Rangamati, Bangladesh*

[2]*School of Graduate Studies, Trent University, Peterborough, Ontario, Canada*

*Corresponding author

ABSTRACT

In this fourth industrial revolution, everything is getting eased by technology, where big data is the main driving force in providing data. Tourism enterprises are also adopting 4IR technologies such as big data, smart tourism, artificial intelligence, the internet of things, and so forth., which have transformed the entire business shape of the tourism system. Hotels use IoT services in their rooms to transform them into smart rooms; restaurants use robots to provide customer services. This chapter discusses big data and smart tourism and their components first. The components of 4IR have also been discussed, aligning with big data and smart tourism. Various tools and models have been used to show how big data can influence smart tourism, the challenges in data-sensitive tourism, and an evaluation of the importance of big data in the travel sector. Data for this chapter has been collected from secondary sources. The findings reveal that the tourism industry is adopting 4IR technologies rapidly to provide personalized tourist facilities and understand tourist behavior and their decision-making process. The key challenges and concerns about

Smart Tourism: Managing Technological Revolution in the Tourism Industry. Santus Kumar Deb (Ed.)

DOI: 10.1201/9781779641090-4

implementing big data analytics are high installation costs and security issues, which may negatively affect businesses and tourists.

4.1 INTRODUCTION

The tourism industry is directly and indirectly linked to different supply components of its system (Deb et al., 2023). Technology has been playing a central role in the tourism system in reshaping distribution and offering diverse services that enhance tourists' travel experiences. Evolving technologies like Artificial Intelligence, Internet of Things, Virtual Reality, Web 3.0, Blockchain, and Big Data rule the different sectors, including tourism. Smart cities are developing innovative services for their residents like IoT, where these technologies play an essential role in creating better facilities for their users. Data is used in these technologies to offer better, more customized, and user-oriented services. Data is also increasingly used in every business for different purposes, such as product development, product modifications, and customer targeting, which reinforce in archiving customer satisfaction. Social media sites like Threads, Facebook, Instagram, Twitter, and so forth, scrutinize billions of data of their users for a better understanding of users' choices through their interactions within and outside their platforms. According to Statista (2023), the global data production is projected to increase significantly, surpassing 180 zettabytes by 2025. Big Data is gathering data from different sources, analyzing them, and extracting information to make concise decisions. This does not imply instantaneous, inexpensive, and rapid triumph; rather, it requires significant financing (Espinet, 2019). Simultaneously, utilizing Big Data analysis is a straightforward and effective approach for businesses in the tourism industry to understand the preferences and requirements of visitors and tourists, regardless of their company type or scale (Li et al., 2018).

The amalgamation of Big Data and Smart Tourism is causing a paradigm shift in the travel sector. In an increasingly linked world, the travel industry is undergoing significant innovation. This is being propelled by data-driven insights and intelligent technology, transforming how we plan, experience, and contemplate our travels. The development of smart cities largely depends on the development of smart services offered to the residents, like IoT and other smart devices that enhance residents' life standards. IoT and other smart devices use users' data to make personalized services for them. A smart city utilizes data to provide information for individuals and to promote competitiveness, innovations, and quality of life via improved connections (Albino et al., 2015). Big Data helps devices to design and create personalized services, which contributes to developing smart cities and smart tourism. Smart tourism may leverage the concept of smart cities, since it focuses on improving urban living standards and boosting the value and experience of the destination via the creation, exchange, and consumption of services (Gretzel, 2011). Big Data refers to an extensive compilation of intricate and substantial data derived from diverse sources, such as healthcare, retail, sensors, social media, and mobile phone GPS signals. Conventional databases have some difficulties in processing and analyzing enormous amounts of data like big data. Still, it helps firms in targeting prospective clients and suggesting desired services. According to Gantz and Reinsel (2011), the methods or approaches used to describe the latest technologies and architectures aim to extract value from vast amounts of diverse data efficiently by facilitating rapid collection, discovery, and analysis.

Emmanuel and Stanier (2016) concluded that Big Data refers to the data center where architectures that are scalable are used to meet the needs of analyzing and other applications that handle large amounts of data with fast speed. This data may have many forms and may involve rapid data capture. In this era of "big data," different types of big data, along with new ideas and technologies, are used in many fields, such as science, engineering, healthcare, management, business, travel, and more (Hashem et al., 2015). The success of smart tourism relies on the capacity of tourist enterprises and destinations to efficiently gather vast quantities of data and effectively manage, manipulate, merge, analyze, and use big data for the purpose of developing tourism operations, services, and commercial advancements (Xiang and Fesenmaier, 2017).

The use of big data in tourist research is now in its nascent phase, and it has the potential to provide a significant edge over competitors (Li et al., 2018). This chapter assesses the connection between Big Data and Smart Tourism, emphasizing their transformative impact on the travel industry. The combination of Big Data and Smart Tourism is poised to revolutionize industry norms and satisfy changing customer demands by anticipating tourist preferences and using cutting-edge technology. This chapter will also identify the challenges, possibilities, and ethical questions that arise when the tourism sector adopts this technology revolution.

The following is how this study is organized: the next part is the literature review, which demonstrates the concept and evolution of big data and smart tourism with its different components; the following part describes the method of data used in the chapter; part four has demonstrated the relationship among 4IR, big data, and smart tourism with different aspects such as their application, tools, challenges, and prospects in the tourism and hospitality industry. In the end, theoretical consequences and possible extensions of big data and smart tourism are discussed.

4.2 LITERATURE REVIEW

The fast advancement of computer science and internet technology has led to the generation, recording, storage, and accumulation of large amounts of data in both organized and unstructured formats. This has given rise to the concept of big data and ushered in a new era (Kambatla et al., 2014). Tourism-related big data is generated from people, devices, and operations (Li et al., 2018). First, the internet has facilitated a fast increase in social media, providing a vast platform for disseminating user-generated content (UGC) data, including texts, photographs, and more (Xiang et al., 2017). Second, the Internet of Things (IoT) has revolutionized the monitoring of tourist movements and environmental conditions, generating vast amounts of spatial-temporal big data from GPS, mobile roaming, and Bluetooth (Shoval and Ahas, 2016). According to Mariani (2020), most tourism studies demonstrate that Big Data uses UGC data, such as social media posts and online travel reviews, and the use of location data, such as GPS and travel information search data, follows this. Kachniewska (2020) examined that the accuracy of tourism sales forecasts is strongly influenced by the kind of data it uses; specifically, UGC yields greater prediction accuracy compared to marketer-generated content (MGC). Furthermore, the study determined that augmenting the magnitude of big data can expedite the pace of sales expansion.

Furthermore, tourism is a multifaceted system encompassing activities such as doing searches on the internet, visiting web pages, and making reservations online. Analyzing large datasets from various sources may provide valuable insights about visitor behaviour and enable more effective marketing strategies, helping both academics and companies by deepening their expertise. From a positive viewpoint, the utilization of big data, tourism analytics, and smart tourism will enhance the tourism industry and benefit travelers by enhancing their ability to gather, assess, and comprehend data. At the same time, these innovative tools will stimulate the tourism industry's pursuit of adding value, creativity, and effective management of tourism destinations (Xiang and Fesenmaier, 2017).

Firican (2022) has classified the evolution of Big Data into three distinct parts. These are prehistoric, 20th-century, and internet-era. According to him, data storage and analysis have been utilized by humans since ancient times, with Paleolithic cultures using tools like sticks and bones to monitor transactions. The abacus, library, and military statistics were first used around 2400 BCE. Ancient civilizations like the Egyptians and Roman Empire used data storage for strategic purposes. In the 20th century, the first significant data effort monitored 26 million Americans and 3 million employers' contributions in 1937. To decipher Nazi codes, Colossus, the first data-processing computer, was constructed in 1943. Over 742 million tax returns and 175 million fingerprints were stored in the first US government data centre built in 1965. In the emergence of the internet era, records were systematically transferred to magnetic computer tape. Roger Mougalas identified the phrase "Big Data" in 2005 to describe the massive amounts of data that typical business intelligence tools failed to handle. It transformed the industry and human behaviour. Healthcare maps illness outbreaks, NASA examines the cosmos, music studies replace intuition, utilities research consumer behaviour, Nike utilizes health monitoring devices, and cybersecurity prevents crime using Big Data.

Xiang and Fesenmaier (2017) have classified tourism development into three stages, driven by Information Technology (IT): the initial stage, which took place between 1991 and 2000, involved tourism industry leaders recognizing their role as information intermediaries and utilizing the internet to communicate efficiently with both current and prospective customers. The internet was seen as an essential medium for commercial marketing during this era (Deb et al., 2024). E-catalogues and web pages took the role of printed brochures for all destinations and travel-related services globally. The second phase of development, occurring approximately from 2000–2010, marked a realization and appreciation among tourism industry executives that travel experiences may be packaged and marketed as goods with the use of information technology. The fundamental business model of several tourist firms has transformed because of IT (Pine and Gilmore, 2013). The tourist business encountered difficulties as a result of the need for customized experiences and intense competition in the worldwide market. Online organizations such as Expedia and

Travelocity have supplanted traditional travel agents since major suppliers have established direct connections with clients. Online search engines like Google and meta-search engines have enhanced the accessibility and transparency of product distribution.

The third phase of growth began around 2010 and beyond, during which progress in domains such as web browsers, social media, the IoT, and smartphone technology facilitated more change in the tourist business (Xiang et al., 2015). During this time, the emergence of Web 2.0 changed the business model again. The innovation of new gadgets like smartphones and portable computing devices embraces many technologies comprising communications, GPS, and camera work, which improve social settings in a way that gives people more power to choose how they travel. The tourist industry has adopted linked technologies like RFID networks, cloud computing, and Internet of Things, along with instrumented IT infrastructure like sensors, in this day and age. Since tourism is a data-centric industry with a strong reliance on IT, smart tourism has been developed to define this present stage of tourist growth (Gretzel et al., 2015). During this phase of advancement, there is a significant increase in the utilization of social media platforms (Sigala et al., 2012). Xiang and Fesenmaier (2017) also added that smart tourism, a natural extension of conventional and E-tourism, is rooted in technological advancements and innovation through the widespread implementation of information technology in the industry and consumers. During this phase of development, smart systems can be implemented to assist passengers in three ways: first, forecasting tourist demands and offering suggestions; second, improving the tourists' satisfaction during their travel based on data; and lastly, encouraging them to exchange travel memories which will influence travel decision of others and emphasizing positive feedback.

In 2019, Wise and Heidari presented a Model describing the flow of data from the Internet of Things to apps that facilitate tourism. The procedure starts with the sensing devices establishing secure communication with an IoT platform. The data is sent across several devices and analytics systems to facilitate the delivery of pertinent information to apps that may effectively cater to industry and individual requirements. Online Travel Agencies (OTAs) are crucial in providing valuable data for tourism research, primarily by using online reviews. The data was mostly sourced from TripAdvisor, the leading platform for tourism-related social media and now a significant OTA (Guo et al. 2017). Espinet (2019) has concluded that OTAs may immediately use the potential of big data to create innovative and tailored marketing tactics as it can trace and monitor every digital activity of consumers, starting from their first interaction with the platform until they complete the customer satisfaction survey, which ultimately will benefit OTAs.

4.2.1 COMPONENTS OF BIG DATA

Emmanuel and Stanier (2016) have discussed three components of big data: (i) data characteristics; (ii) application of big data; and (iii) architecture and processing for big data. The three key characteristics, namely Volume, Velocity, and Variety, are universally acknowledged as the fundamental elements that define Big Data (Bedi et al., 2014; Demchenko et al., 2013; Demchenko et al., 2014; Uddin and Gupta, 2014). Doug Laney provided a more detailed explanation of the concept of "Big Data" (Laney, 2001), defining it based on three key attributes: Volume, Variety, and Velocity (Gantz and Reinsel, 2011; Manyika et al., 2011). In 2012, IBM added two additional components to the concept of Big Data, namely Value and Veracity, bringing the total number of components to five (Matsudaira, 2013; Zikopoulos et al., 2012). Additionally, in 2013, another V was suggested to be included as "Variability" in order to expand the existing five Vs of Big Data to a total of six Vs (Demchenko et al., 2013; Zikopoulos et al., 2013). Table 4.1 demonstrates the seven Vs of Big Data, where Bedi et al. (2014) have proposed seven Vs of Big Data.

4.2.2 OVERVIEW OF SMART TOURISM

Smart tourism, an evolved phase of digital tourism (Jia et al., 2022), is a significant advancement in the amalgamation of ICT in the tourism industry. It involves the digitalization of both the physical and governance aspects of tourism, resulting in enhanced levels of intelligence in tourism systems (Gretzel, 2011). Smart cities employ data to tailor information, promote competitiveness and innovation, and improve quality of life via connection. Tourists are also affected by this phenomenon as they get advantages from digital infrastructures, combined urban data, platform services, regulations that facilitate procedures, and city residents who provide information about the economy, community, culture and entertainment, transportation and mobility, as well as urban locations and areas (Koo et al., 2017). The use of online connectivity and intelligent mobile devices is becoming more advantageous for those involved in company ownership, management, planning, as well as for local residents and visitors (Wise and Heidari, 2019). Via extensive, well-coordinated efforts and prudent financing in technology infrastructure, smart tourism ultimately aims to encourage mobility, innovation, resource availability and allocation, long-term viability, and a high standard of life and visitors (Xiang and Fesenmaier, 2017). They also add that to accomplish this objective, smart destinations must construct an information infrastructure that promotes customers' active or implied data exchange.

TABLE 4.1 Seven Vs of Big Data

Characteristics	Details
Volume	The volume of data is being produced in real-time by many data sources that contribute to big data.
Variety	Data comes from several sources with different types. Blogs, emails, social media data, audio and video chats, XML files, logs, database data, and data warehouse data are all included.
Velocity	Velocity pertains to the rapid pace at which the data is being created. It encompasses not only the data creation rate, the processing time required to input the data into the system, and the frequency at which the data is provided.
Value	A business needs to analyse and forecast the significance of data to anticipate its consumers' behaviour, enabling the formulation of policies or decisions aimed at maximizing revenues.
Veracity	Big Data veracity depends on the accuracy and trustworthiness of data collection and processing procedures, infrastructure, and data sources. This protects the data against unauthorized access and changes throughout its life.
Variability	Fluctuations in data may result in sudden increases in data load; hence, it is crucial to process and assimilate the data efficiently. As a result of a wave-like phenomenon, it is necessary to modify the data model and relationships to deduce information from the growing data pool.
Viability	The multidimensional nature of data leads to inefficiencies in processing. Efficient processing and space usage in enterprises need the precise selection of dimensions and parameters to forecast critical results in the real-time analysis of enormous datasets.

Source: Bedi et al. (2014).

Figure 4.1 illustrates that smart tourism development relies on the gathering, sharing, and analysis of data produced by many elements within the system, including the customer, the company, and the destination as a whole. Gretzel et al. (2015) have also extended that smart tourism employs information and communication technologies (ICTs) to streamline many components, with a particular emphasis on the needs of both local inhabitants and visitors, and smart destinations use smart city ideas in both urban and rural locations, enhancing mobility, resource accessibility, sustainability, and overall quality of life. In the context of smart tourism, this technology serves as the fundamental element of information systems that aim to give tourist customers and service providers more pertinent information, improved decision-making assistance, enhanced mobility, and ultimately, more pleasurable tourism experiences (Gretzel, 2011; Sigala and Chalkiti, 2014; Werthner, 2003). Within the context of smart tourism, technology is considered an essential framework rather than a distinct information system, and it comprises a variety of advanced computing technologies that combine hardware, software, and network technologies to enable immediate understanding of the physical environment and use sophisticated data analysis to aid people in making well-informed choices and improving corporate operations and efficiency (Washburn et al. 2010).

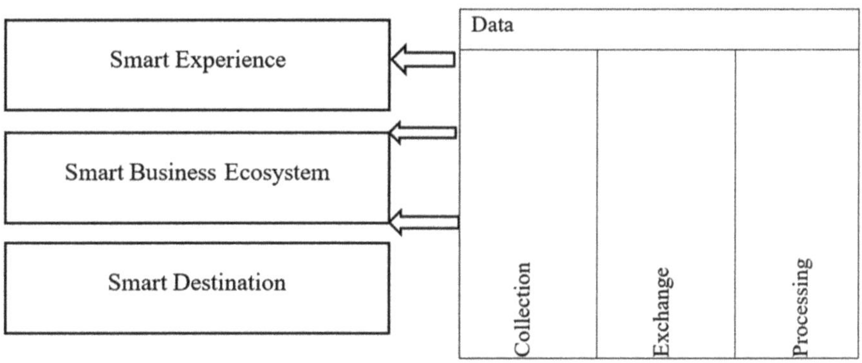

FIGURE 4.1 Components and layers of smart tourism.

Source: Reprinted from Gretzel et al. (2015). https://creativecommons.org/licenses/by/4.0/

4.3 METHODOLOGY

The prime purpose of this chapter is to assess the concept of big data and smart tourism and their implication in the travel industry. For this purpose, various techniques and models have been used to demonstrate the justification of the concept. The data for this chapter was gathered from secondary sources. A conceptual framework has been proposed on how big data contributes to developing smart cities and smart tourism through the innovation of industry 4.0.

The authors have used reliable online sites such as Scholar, Google, and Scopus in order to perform an exhaustive search for further material that is pertinent to the topic at hand. Relevant keywords, such as "big data," "big data in tourism," and "smart tourism," were employed by the researchers in order to get access to the publications that were utilized in this study. When it comes to the selection of publications for this research, the authors have reviewed the papers that were selected based on their relevance to the subject that is being investigated, the dependability of the publishers, and the quality of the data and information that they give.

4.4 INDUSTRY 4.0, BIG DATA AND SMART TOURISM

The digital revolution has transformed the tourist industry through the sophisticated development of Industry 4.0, using mobile devices, kiosks, social networking channels, and artificially intelligent technology to create content, monitor travelers' online activities, and deliver customized services, eventually improving the tourism sector (Dalkiran, 2022). The Fourth Industrial Revolution (4IR) has its significance in providing the main driving force for technologies like Big Data, Smart Tourism, Artificial Intelligence, The Internet of Things, and so forth, and utilizing these technologies have the capacity to revolutionize the tourist sector (Al-Amin et al., 2019; Jankova et al., 2023). Utilizing big data in smart tourist destinations may aid in commercial decision-making, efficient resource distribution, and uncovering novel insights that impact markets and companies (Mariné-Roig and Clavé, 2015). Nevertheless, it should be emphasized in smart tourism that managing big data generated by social media may support tourism attractions and companies for developing the business ecosystem while also encouraging a pre-competitive development approach (Vecchio et al., 2018). This suggests that big data might be essential for the tourist sector to encourage innovation and boost value generation. Smart cities use advanced technologies such as Artificial Intelligence (AI), Big Data, Cloud Computing, IoT, Robots, and smart automation to adjust to the 4IR. In contrast, innovative technologies assist cities in engaging with many stakeholders, coordinating with businesses, states, and foreigners, and providing an adaptable form and functional strategy (Williams and Charles, 2023).

According to Dalkiran (2022), technology is essential for several aspects of operations in tourist firms, such as capacity evaluation, product creation, event organizing, and input procurement. With the advancement of technology, the tourism industry is adopting "smart" ideas such as "smart tourism," "smart destination," and "smart hotel" to enhance consumer happiness and preserve social, cultural, and natural resources for its sustainability. Jia et al. (2022) delve into the connection between smart tourism and big data, examining the development of smart tourism technologies within the realm of big data and prioritizing the cutting-edge technology to enhance the development of smart tourism application models. Big data is seen as a key factor in the success of smart tourism because it helps businesses learn more about their consumers and the marketplace, which is important for staying strong competitors and making good decisions (Vecchio et al., 2021). Thus, big data may improve the comprehension of consumer buying habits and market trends, boosting tourist enterprises' competitiveness. The success of smart tourism relies on the capacity to efficiently gather and manage vast quantities of data, as well as to effectively use big data for commercial advancements, operational improvements, and service enhancements (Gretzel et al., 2015). 4IR may have some disadvantages, especially in the job market. Industry 4.0's digitization in labor-intensive tourism manufacturing poses concerns. In the future, there may be a drop in demand for people to work, leading to the potential disappearance of some occupations (Dalkiran, 2022).

Industry 4.0 technological tools shown in Figure 4.2 considerably influence the tourist industry in the digital revolution; these tools facilitate service distinction, cost minimization, and monitoring measures, whereas big data facilitates the discovery of particular consumers, enabling the establishment of new markets in various tourist categories and seasons (Dalkiran, 2022). Abdurakhmanova et al. (2022) have discussed the concept of Tourism 4.0 to improve the value of tourism by using Industry 4.0 technology shown in Figure 4.2. Tourism 4.0 strives to provide customized tourist encounters, develop smart destinations, optimize performance, and promote ethical behavior among worldwide players (Goriup and Ratkajec, 2021). Korže (2019) has discussed the adoption of Industry 4.0 in tourism as Tourism 4.0, stating that big data is considered a crucial component of Tourism 4.0. According to Korže (2019), only 10 nations have adopted the Tourism 4.0 framework. This number is skewed toward European nations like Slovenia, Portugal, Italy, Turkey, Spain, and Finland, as well as Southeast Asian nations like Thailand, Bali, and Malaysia.

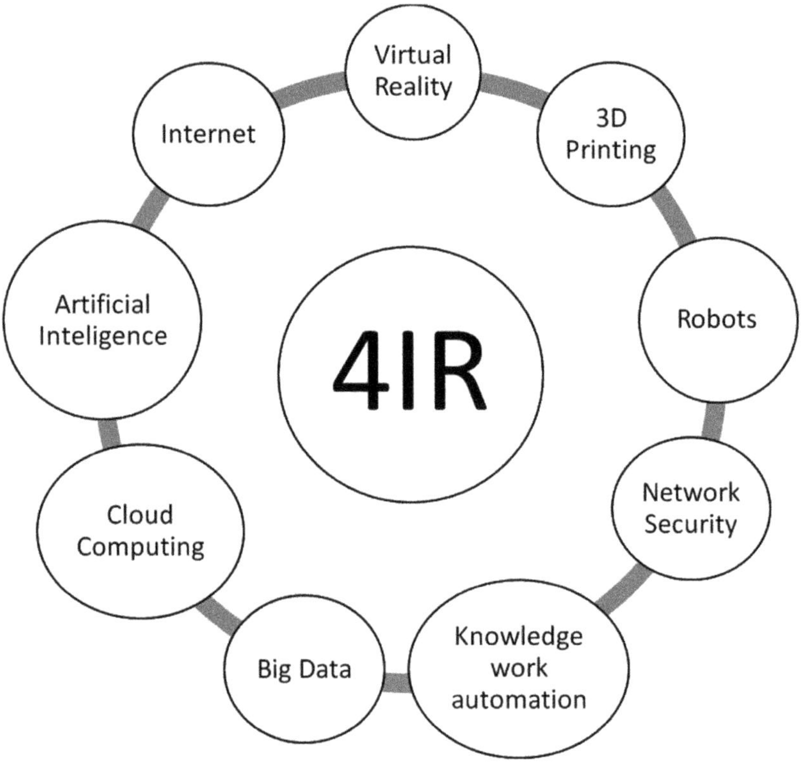

FIGURE 4.2 Components of industry 4.0.

Source: Adapted from Yusuf et al. (2020).

According to Stankov and Gretzel (2020), Tourism 4.0, inspired by Industry 4.0 advancements, seeks to evolve the tourist sector into intricate and tangible systems. There is an underlying belief that the technical advancements of Tourism 4.0 are more effective in providing advantages to visitors, as the fast advancement and integration of modern technology is testing traditional methods of creating user experiences. They added that Tourism 4.0 technologies provide advanced features that improve engagement, enrich the visitor experience, and support behavior modification and long-term user transformation. Technological improvements influence the tourist industry and other associated industries, increasing anticipated advantages; automation in the agricultural industry may enhance productivity, and innovative technology can be used for storage and control of inventories and identifying depleting supplies in fridges, kitchens, and hotels (Dalkiran, 2022).

4.5 CHALLENGES IN IMPLEMENTING BIG DATA IN THE TRAVEL INDUSTRY/DATA-SENSITIVE CHALLENGES IN TOURISM

The expansion of social media has generated curiosity in using online textual big data for the purpose of researching tourism. Nevertheless, previous research relies on limited datasets, which may introduce biases in the selection and estimating process, possibly yielding inaccurate outcomes (Li et al., 2018). They also extended that when it comes to drawbacks, the use of big data in tourist research still faces some complex challenges related to data quality, data cost, and privacy issues (Table 4.2).

According to Anuar and Gretzel (2011), numerous obstacles and difficulties lie ahead in using big data, and privacy concerns are one of them, particularly in geolocation services. The security issues have been scrutinized by Vu et al. (2019), which are also accessed by location-based social media (LBSM). They stated that LBSM users' knowledge gap about possible risks is the reason behind it and encouraged management and researchers to prioritize privacy-protection methods in their mechanism. Though initially, it has been observed that travelers' express privacy worries while sharing these data on LBSM platforms (Ahas et al., 2008; Anuar and Gretzel, 2011; Clarke, 2001). In their study, Sivarajah et al. (2017) have put up a comprehensive model that outlines the major issues faced by OTAs in dealing with big data. These challenges have been classified into three distinct categories (Figure 4.3).

Jia et al. (2022) have discussed one big disadvantage with smart tourism big data tools: they can't be monitored and haven't been fully evolved yet to address the demands of smart tourism. Furthermore, according to Afolabi et al. (2021), sustainability and financial advantages may be affected by privacy issues of utilizing big data in tourism destinations, and they also addressed the moral and responsible data integration in dealing

TABLE 4.2 Comparison Between Advantages and Disadvantages of Big Data in Tourism

Data Type	Advantages	Disadvantages
User Generated Content (UGC)	• Inexpensive; • Expressing tourist feelings; • Inexpensive; • Comprising multiple metadata.	• Issues about reliability; • Inaccurate location accuracy.
Devices	• Global; • High accuracy; • Availability in remote locations; • Permitting the monitoring of tourists' points of origin; • Availability in crowded indoor environments; • Permitting surreptitious surveillance; • Inexpensive; • High accuracy; • Presence in congested indoor environments • Presence in crowded indoor environments; • Permitting surreptitious monitoring; • Weather variables.	• Expensive; • Privacy issues; • Privacy issues; • Inaccurate location accuracy; • Limited scope; Privacy issues; • Limited coverage; Privacy issues; • Limited coverage area; • Concerns about privacy; • Challenges in handling data of various formats.
Operations	• Inexpensive; • Capturing public interest; • Documenting visitors' activities in tourist markets.	• Potential estimation biases; • Privacy issues.

Source: Li et al. (2018).

with big data. Gretzel et al. (2015) have also addressed privacy issues by identifying data as prime of all smart tourism operations.

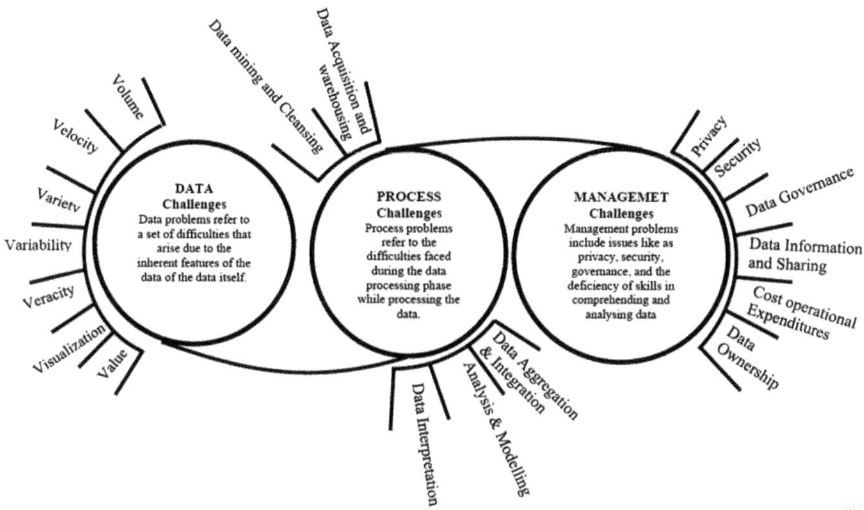

FIGURE 4.3 Challenges of big data in online travel agencies (OTAs).
Source: Reprinted with permission from Sivarajah et al. (2017). Copyright © 2017 Elsevier.

4.6 DATA AND BIG DATA IN TOURISM AND HOSPITALITY

Data is the most influential element for businesses, policymakers, and planners to address issues and make more accurate plans or strategies for the enterprise. The significance of tourist data has also been broadly acknowledged by scholars (Latham and Edwards, 2003; Smith, 1988; Volo, 2018). The tourist sector generates a significant amount of data via routine and legal processes, such as data related to airline or hotel reservations (Guan and Du, 2016). According to Volo (2018), the tourist statistics system is categorized into data sources for supply and demand, as shown in Figure 4.4. The supply includes the categorization, quantity, and geographical distribution of lodging facilities, the number of visitors, and the duration of their stays. Additionally, temporary surveys are conducted during periods of high demand. Domestic or international tourist demand in a country is assessed using several methods, including household sample surveys, counts of foreigners at borders, and surveys of both natives and foreigners at borders.

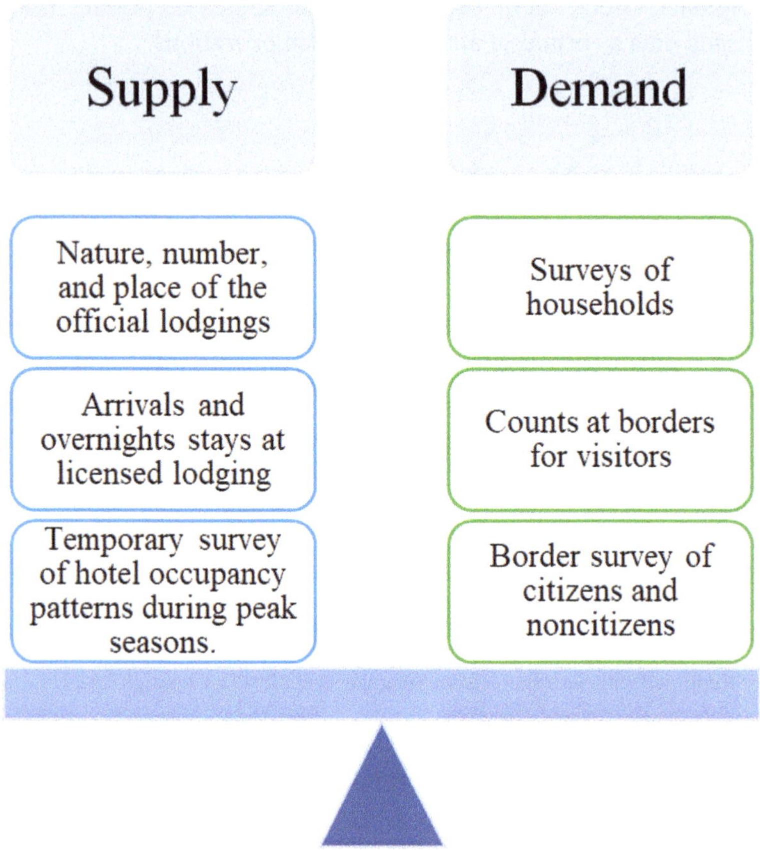

FIGURE 4.4 Tourism data sources.

Source: Volo (2018), P-194.

Jia et al. (2022) suggested a smart tourism concept that relies on big data technology that gathers extensive data from three primary sources: visitors, their gadgets, and operational procedures. Big data from these origins has enhanced the tourism business by offering valuable information on visitor preference and market trends, leading to increased success and efficiency. Furthermore, Volo (2020) has addressed that big data, which refers to vast and intricate data sets, has become a significant source of information that can be tracked digitally for different human activities. Unstructured and unregistered big data is significant for researchers and companies since it enables monitoring of individual tourist choices, attitudes, sites of interest, actions, and spending (Volo, 2018). It is significant for developing nations to incorporate new data sources into tourist statistics, and more integration and reinforcement of conventional statistics with big data is required as these nations see an increase in international tourism (Volo, 2020).

The utilization of big data is implemented and may be internally employed in several manners, such as identifying user trends, habits, expectations, and wishes (Banerjee and Chua, 2016), as well as establishing more effective pricing strategies (Steppe, 2017), developing profit or return management methods (Stringam and Gerdes, 2010), developing notifications and indications, as well as constructing customer journey maps (Espinet, 2019). Marketing in tourist enterprises uses Big Data analytics to enhance standing, foster client loyalty, augment earnings, and strengthen corporate position relative to competitors (Belias et al., 2021) and to improve the accuracy of vacation package sales predictions (Kachniewska, 2020).

Technology and the fast progress of the internet have transformed how tourists choose travel products (Halim, 2022; Middleton et al., 2009). The use of big data has significantly improved the capacity to engage directly with customers and visitors, allowing the highest level of customization in all communication endeavors (Fan et al., 2015). They also discussed the application of the 5Ps (product, price, place, promotion, and people) of the marketing mix in Big Data; hence, big data facilitates innovations that include substantial modifications in product design, advertisement, and price adjustments. According to Mariani (2020), the tourism sector is progressively embracing Big Data. Fuchs et al. (2014) have examined big data in analyzing tourist destinations in Sweden (Marine-Roig and Clavé, 2015), in Barcelona (Irudeen and Samaraweera, 2013), and in Sri Lanka.

Big Data may serve as a strategic instrument for acquiring insights and enhancing comprehension of visitors' behavior in a particular location (Fuchs et al., 2014). Aguiar and Szekut (2019) state that tourism data gathering is complicated due to the industry's mobility, movement, and intangibility, and Big Data provides advantages over traditional research methods because it analyzes data in real-time and quickly captures visitor activities. Big Data significantly improves the tourist industry via a variety of contributions. In their study, they discussed the different contributions of big data, as demonstrated in Figure 4.5.

Trends — Identifying patterns and actions

- Big Data allows managers to uncover behavioral patterns and customer trends that are undetectable by conventional approaches.

Demand — Understanding demand and tailoring goods and services

- Big data helps in assessing market demand and customizing products and services by investigating customer interactions in diffrent channels.

Development — Innovation of new products and services

- Using statistical analysis, Big Data provides competitive goods and services, allowing organizations to capitalize on new possibilities promptly.

Decision — Agility in decision-making

- Big Data allows for immediate investigation and automated pricing and services depending on client demand and engagement across multiple channels of distribution.

Function — Enhancing procedures and internal operations

- Enterprise leverages Big Data to develop customer-centered methods to optimize operations and enhance their operations based on skills and expertiese they have.

Revenue — Managing Revenue

- Businesses may enhance pricing strategies by using big data, employing empirical methods and algorithms to monitor competition prices, and improving their profitability and competitiveness.

FIGURE 4.5 Contribution of big data in tourism.

Source: Aguiar and Szekut (2019).

4.7 INTRODUCTION TO BIG DATA TOOLS

In today's digital age, the tourism industry is experiencing a profound transformation driven by the proliferation of big data and advanced analytics tools. Big data, characterized by its vast volume, velocity, and variety, is pivotal in shaping insights and decision-making processes across various sectors, including tourism. Many tools and technologies have emerged to effectively harness the power of big data within the realm of smart tourism, facilitating data collection, storage, processing, analysis, and visualization. One of the foundational tools in the realm of big data is Hadoop. Hadoop, an open-source framework, enables the distributed processing of datasets across clusters of computers using simple programming models. It comprises two core components: the Hadoop Distributed

File System (HDFS) for reliable storage and MapReduce for parallel data processing. Hadoop empowers organizations within the tourism industry to efficiently store and analyse massive volumes of structured and unstructured data, encompassing everything from customer transactions to social media interactions (White, 2015).

Another indispensable tool for big data analytics in smart tourism is Apache Spark, a lightning-fast cluster computing framework renowned for its in-memory processing capabilities, facilitating real-time data analytics and iterative algorithms. Spark's versatility allows tourism enterprises to execute a wide range of tasks, including machine learning, graph processing, and stream processing, with unprecedented speed and scalability. By leveraging Spark, businesses can extract valuable insights from diverse data sources, such as customer reviews, sensor data, and geospatial information, to optimize operations and enhance the visitor experience (Zaharia et al., 2016). Furthermore, the advent of cloud computing platforms, such as Amazon Web Services (AWS) and Microsoft Azure, has democratized access to big data tools and resources for tourism organizations of all sizes. These cloud-based services offer scalable storage and computing capabilities, enabling businesses to deploy advanced analytics solutions without significant upfront investments in infrastructure. Additionally, cloud platforms provide various managed services, including data lakes, data warehouses, and machine learning tools, streamlining the big data lifecycle and accelerating time-to-insight.

In addition to Hadoop and Spark, numerous specialized tools and frameworks cater to specific use cases in smart tourism analytics. For instance, Apache Kafka facilitates real-time data streaming and event-driven architectures, allowing businesses to respond promptly to changing market trends and customer preferences (Buhalis et al., 2023). Similarly, Apache Flink offers high-throughput and low-latency processing for continuous data streams, making it ideal for applications such as dynamic pricing and personalized recommendations in the travel industry (Carbone et al., 2015). As big data continues to reshape the landscape of smart tourism, organizations must invest in robust infrastructure and talent to harness its full potential. By leveraging diverse tools and technologies, tourism enterprises can unlock actionable insights, drive innovation, and deliver unparalleled experiences to travelers in the digital age.

4.8 APPLICATION OF BIG DATA IN SMART TOURISM

Integrating big data analytics has significantly impacted the tourism industry, revolutionizing destination management, traveler engagement, and overall visitor experiences. Big data analytics facilitates processing extensive datasets from various sources, leading to data-driven insights that inform decision-making processes, personalize services, and enable predictive analytics, thereby advancing the tourism sector. A prominent application of big data analytics in smart tourism lies in destination management and marketing. By analyzing comprehensive datasets encompassing tourist preferences, behaviors, and demographics, destination management organizations (DMOs) can adeptly tailor marketing strategies and promotional campaigns to target specific audience segments. Utilizing sentiment analysis of social media data and online reviews provides destinations with invaluable insights into visitor satisfaction levels and areas for improvement, facilitating the refinement of marketing initiatives and the efficient allocation of resources (Sarker et al., 2020).

Big data analytics enhances operational efficiency and resource allocation within tourism destinations. Destination managers can optimize infrastructure development, traffic management, and crowd control measures by analyzing real-time data on visitor flows, crowd densities, and resource utilization. Leveraging predictive analytics models based on historical data enables destinations to forecast visitor demand patterns and adjust operational strategies accordingly, ensuring seamless and enjoyable experiences for travelers while maximizing resource utilization (Alarcón-del-Amo et al., 2024). In addition to destination management, big data analytics enables personalized experiences and services for individual travelers. By analyzing user-generated content, travel preferences, and past behaviors, travel companies can offer tailored recommendations, itinerary suggestions, and personalized offers aligned with each traveler's preferences and interests. This personalized approach enhances customer satisfaction and fosters loyalty and repeat visitation, ultimately driving long-term value for tourism businesses (Gretzel et al., 2015).

Moreover, big data analytics enables predictive analytics for risk management and crisis response in the tourism sector. Destinations can develop predictive models to anticipate potential risks and mitigate their impact on travelers by analyzing historical data on weather patterns, geopolitical events, and public health crises. Real-time monitoring of social media and news feeds allows destinations to promptly detect emerging crises and implement timely response measures, ensuring traveler safety and maintaining destination resilience (Fuchs et al., 2014).

According to Lee et al. (2020), smart tourism is the merging of ICT with the tourist industry, resulting in the modernization of tourism via technological advancements. The development of Industry 4.0 has contributed to the business by providing key technologies such as big data, AR, IoT, and VR. The fourth industrial revolution has helped tourism grow, which is why cities are building smart tourism environments using their current digital technology networks. This supports the idea of smart tourism cities (Lee et al., 2020). Data played a significant role in these innovations; for example, different IoT devices analyze billions of data collected from different data sources, which acted as big data. Figure 4.6 illustrates how big data contributes to the development of smart tourism. According to Lee et al. (2020), in smart tourism, tourists not only use data from their travels but also make it possible for future data to be created and shown in different ways.

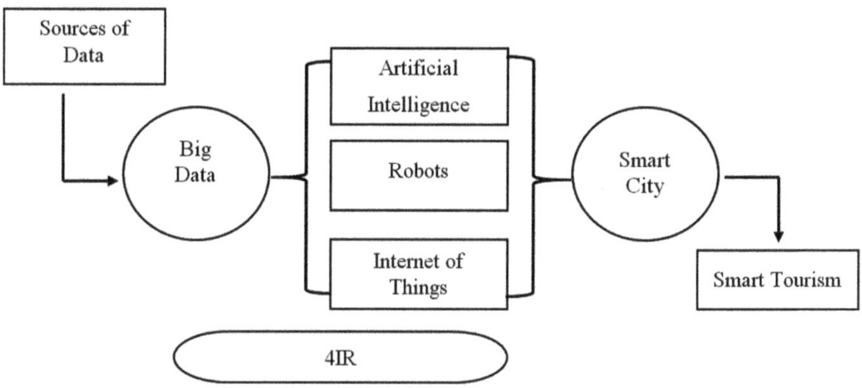

FIGURE 4.6 Big data in developing smart tourism.
Source: Authors.

Big data analytics applications in smart tourism are diverse and multifaceted, spanning destination management, marketing, operational efficiency, personalized services, and risk management. By harnessing the power of big data analytics, tourism stakeholders can unlock actionable insights, optimize resource allocation, and enhance traveler experiences, ultimately driving sustainable growth and competitiveness in the global tourism landscape.

4.9 EVALUATING THE IMPORTANCE OF BIG DATA IN THE TRAVEL SECTOR

Big data has emerged as a transformative force in the continually evolving travel sector, reshaping strategic decision-making, operational efficiency, and customer satisfaction. Its pervasive influence spans various industry dimensions, from acquiring deep customer insights and revenue maximization to optimizing operational processes and risk management, leading to groundbreaking advancements and heightened competitiveness. At the forefront of big data's impact is its capacity to unearth profound customer insights (Buhalis, 2023; McKinsey and Company, 2018). Companies can meticulously construct detailed customer profiles by leveraging extensive datasets from diverse channels such as online reservations, social media engagements, and historical travel patterns. These profiles serve as the cornerstone for tailoring personalized offerings to meet individual preferences and elevate the overall travel experience. Through data-driven segmentation and targeting strategies, businesses enhance customer satisfaction and cultivate brand loyalty and advocacy, fostering sustainable growth amidst an increasingly discerning market landscape.

Big data catalyzes revenue optimization, revolutionizing pricing strategies and marketing endeavors (Li et al., 2018; McKinsey and Company, 2018). Utilizing sophisticated analytics and machine learning algorithms, travel companies can dynamically adjust pricing in response to fluctuating demand dynamics, competitive actions, and market trends. This dynamic pricing approach maximizes revenue generation and ensures pricing consistency across distribution channels, fostering transparency and bolstering consumer trust. Additionally, data-driven insights facilitate targeted marketing initiatives, enabling companies to deploy personalized promotions and offers that optimize marketing expenditure and drive incremental revenue growth. Beyond customer-centric initiatives, big data permeates operational excellence across the entire travel value chain (Li et al., 2018; McKinsey and Company, 2018). Real-time analysis of operational data empowers companies to identify inefficiencies, optimize resource allocation, and automate processes. For instance, airlines employ big data analytics to optimize flight schedules, minimize fuel consumption, and optimize crew assignments, improving punctuality and cost-effectiveness. Similarly, hotels leverage data analytics to optimize room inventory management, enhance guest satisfaction, and capitalize on ancillary revenue opportunities through targeted upselling and cross-selling.

Big data serves as a linchpin for proactive risk management and decision support within the travel sector (McKinsey and Company, 2018; Wang et al., 2012). Through predictive analytics and scenario modeling, companies can anticipate and mitigate potential risks, encompassing natural disasters, geopolitical upheavals, economic downturns, and public health crises. Real-time monitoring of external variables and market indicators empowers companies to adapt swiftly to evolving circumstances, safeguarding operational continuity and traveler well-being. The profound impact of big data in the travel sector transcends conventional paradigms, revolutionizing business operations and redefining customer engagement strategies. As companies continue to harness the power of big data to drive innovation and agility, new opportunities for expansion and distinctiveness emerge, positioning them to thrive in a rapidly evolving and fiercely competitive market landscape.

4.10 FUTURE PROSPECTS OF BIG DATA AND SMART TOURISM

Integrating big data analytics into smart tourism heralds transformative possibilities for the industry's future landscape, promising advancements in destination management, traveler engagement, and operational efficiencies. By strategically utilizing big data, tourism stakeholders can anticipate emerging trends, personalize experiences, and optimize processes, fostering sustainable growth and maintaining competitiveness in an evolving market milieu.

Predictive analytics and forecasting within smart tourism are significant avenues for future development. By leveraging historical data, real-time information, and sophisticated analytics methodologies, tourism entities can construct predictive models to anticipate traveler preferences, demand fluctuations, and market dynamics (Xiang et al., 2017). Machine learning and AI promise to reshape smart tourism practices. AI-driven chatbots, virtual assistants, and recommendation systems have the potential to offer personalized support to travelers across various stages of their journey, leveraging significant data insights to understand individual preferences and behaviors (Almeida-Santana et al., 2020).

The expansion of IoT technologies offers innovative avenues for leveraging significant data in smart tourism. By deploying IoT devices and sensor networks, establishments can gather real-time data on visitor movements, environmental conditions, and infrastructure usage. This wealth of data empowers data-driven decision-making and facilitates personalized service delivery within the tourism sector (Gretzel et al., 2015). Ethical considerations and data privacy concerns will increasingly shape the future trajectory of big data in smart tourism. As the volume and granularity of data collected continue to expand, tourism organizations must prioritize data security, transparency, and compliance with regulatory frameworks (Gretzel et al., 2015). The prospects of big data in smart tourism are extensive and multifaceted, offering avenues for predictive analytics, AI-driven innovation, IoT integration, and ethical data governance. By strategically harnessing big data analytics, tourism stakeholders can unlock new insights, enhance traveler experiences, and foster sustainable growth in an increasingly digitalized tourism landscape.

4.11 CONCLUSION

Tourism, a multidimensional industry, depends on different supply components to offer quality services to tourists. Technology is one of the components of the tourism system that provides easy, accessible, and customized services. The industrial revolution has contributed to the development of different innovations, mainly changing the travel market. After e-tourism, big data plays a decisive role in shaping this industry based on market data, which contributes to creating the market demand and supply balance. The study broadly discussed the concepts of big data and smart tourism and their application in the travel industry and found that the tourism industry uses two sources of data: supply and demand. The critical challenges of travel market data are mobility, movement, and intangibility, which is solved by the development of big data with its real-time and fast data analysis capabilities. On the other hand, smart tourism depends on big data for data gathering from visitors, their gadgets, and operational procedures. Big data contributes to the tourism and hospitality industry by identifying emerging trends and patterns, forecasting market demand, developing customer-centred products and services, improving business operations, precise decision-making, and managing revenues. The study can be extended in the future to identify the cost-effective method of implementing big data and ensuring data privacy.

KEYWORDS

- **artificial intelligence**
- **big data**
- **Internet of Things (IoT)**
- **robots**
- **smart rooms**
- **smart tourism**
- **social media**

REFERENCES

Abdurakhmanova, G. K., Astanakulov, O. T., Goyipnazarov, S. B., & Irmatova, A. B. (2022). Tourism 4.0: Opportunities for applying Industry 4.0 technologies in tourism. In *Proceedings of the 6th International Conference on Future Networks & Distributed Systems* (pp. 33–38).

Afolabi, O., Ozturen, A., & İlkan, M. (2021). Effects of privacy concern, risk, and information control in a smart tourism destination. *Economic Research-Ekonomska Istraživanja, 34*(1), 3119–3138. https://doi.org/10.1080/1331677x.2020.1867215 (accessed on 8th July 2025).

Aguiar, A. B., & Szekut, A. (2019). Big data and tourism: Opportunities and applications in tourism destination management. *Applied Tourism, 4*(2), 36–47. http://dx.doi.org/10.14210/at.v4n2.p36-47 (accessed on 8th July 2025).

Ahas, R., Aasa, A., Roose, A., Mark, Ü., & Silm, S. (2008). Evaluating passive mobile positioning data for tourism surveys: An Estonian case study. *Tourism Management, 29*(3), 469–486.

Al-Amin, M., Sultana, N., Nafi, S. M., & Islam, S. M. (2019). Customers' perceptions about plastic money towards sustainable banking in Bangladesh: A technological adoption. *International Business Research, 12*(6).

Alarcón-del-Amo, M. D. C., Lorenzo-Romero, C., & Gómez-Borja, M. Á. (2024). Do sustainability motivations drive satisfaction and engagement in the sharing economy, or is it just price and convenience? Some answers in a collaborative housing context. *Journal of Hospitality and Tourism Insights, 7*(1), 291–311.

Albino, V., Berardi, U., & Dangelico, R. M. (2015). Smart cities: Definitions, dimensions, performance, and initiatives. *Journal of Urban Technology*, *22*(1), 3–21.
Almeida-Santana, A., David-Negre, T., & Moreno-Gil, S. (2020). New digital tourism ecosystem: Understanding the relationship between information sources and sharing economy platforms. *International Journal of Tourism Cities*, *6*(2), 335–345.
Anuar, F., & Gretzel, U. (2011). Privacy concerns in the context of location-based services for tourism. In *ENTER 2011 Conference*.
Banerjee, S., & Chua, A. Y. (2016). In search of patterns among travelers' hotel ratings in TripAdvisor. *Tourism Management*, *53*, 125–131.
Bedi, P., Jindal, V., & Gautam, A. (2014). Beginning with big data simplified. In *International Conference on Data Mining and Intelligent Computing (ICDMIC)*. IEEE, pp. 1–7.
Belias, D., Malik, S., Rossidis, I., & Mantas, C. (2021). The use of big data in tourism: Current trends and directions for future research. *Academic Journal of Interdisciplinary Studies*, *10*(5), 357–364.
Buhalis, D., O'Connor, P., & Leung, R. (2023). Smart hospitality: From smart cities and smart tourism towards agile business ecosystems in networked destinations. *International Journal of Contemporary Hospitality Management*, *35*(1), 369–393.
Carbone, P., Katsifodimos, A., Ewen, S., Markl, V., Haridi, S., & Tzoumas, K. (2015). Apache Flink: Stream and batch processing in a single engine. *IEEE Data Engineering Bulletin*, *38*(4), 28–38.
Clarke, R. (2001). Person location and person tracking-technologies, risks and policy implications. *Information Technology & People*, *14*(2), 206–231.
Dalkiran, G. B. (2022). The effects of Industry 4.0 components on the tourism sector. In *Logistics 4.0 and Future of Supply Chains* (pp. 235–250).
Deb, S. K., Kuri, B. C., Roy, H., Roy, B., & Nafi, S. M. (2023). The adaptability of tourism and hospitality supply chain management to manage COVID-19 crisis: Bangladesh perspective. *International Journal of Agile Systems and Management*, *16*(4), 484–511.
Deb, S. K., Nafi, S. M., & Valeri, M. (2024). Promoting tourism business through digital marketing in the new normal era: A sustainable approach. *European Journal of Innovation Management*, *27*(3), 775–799. https://doi.org/10.1108/EJIM-04-2022-0218 (accessed on 8th July 2025).
Demchenko, Y., Grosso, P., De Laat, C., & Membrey, P. (2013). Addressing big data issues in scientific data infrastructure. In *International Conference on Collaboration Technologies and Systems (CTS)*. IEEE, pp. 48–55.
Demchenko, Y., Gruengard, E., & Klous, S. (2014). Instructional model for building effective big data curricula for online and campus education. In *2014 IEEE 6th International Conference on Cloud Computing Technology and Science*. IEEE, pp. 935–941.
Emmanuel, I., & Stanier, C. (2016). Defining big data. In *Proceedings of the International Conference on Big Data and Advanced Wireless Technologies* (pp. 1–6).
Espinet, J. M. (2019). Big data in online travel agencies and its application through electronic devices. In *Big Data and Innovation in Tourism, Travel, and Hospitality: Managerial Approaches, Techniques, and Applications* (pp. 31–55).
Fan, S., Lau, R. Y., & Zhao, J. L. (2015). Demystifying big data analytics for business intelligence through the lens of marketing mix. *Big Data Research*, *2*(1), 28–32.
Firican, G. (2022). *The History of Big Data*. LightsOnData. https://lightsondata.com/the-history-of-big-data (accessed on 8th July 2025).

Fuchs, M., Höpken, W., & Lexhagen, M. (2014). Big data analytics for knowledge generation in tourism destinations–A case from Sweden. *Journal of Destination Marketing & Management*, *3*(4), 198–209.

Gantz, J., & Reinsel, D. (2011). Extracting value from chaos. *IDC iView*, *1142*, 1–12.

Goriup, P. D., & Ratkajec, H. (2021). Preliminary application of Tourism 4.0 data analytics in Odessa City reveals challenges and opportunities for sustainable tourism development. *Economic Innovations*, *23*(4 (81)), 36–43.

Gretzel, U. (2011). Intelligent systems in tourism: A social science perspective. *Annals of Tourism Research*, *38*(3), 757–779.

Gretzel, U., Sigala, M., Xiang, Z., & Koo, C. (2015). Smart tourism: Foundations and developments. *Electronic Markets*, *25*, 179–188.

Guan, D., & Du, J. (2016). Cross-media big data tourism perception research based on multi-agent. In *Proceedings of the 2015 Chinese Intelligent Systems Conference*, *1*, 353–360. Springer Berlin Heidelberg.

Guo, Y., Barnes, S. J., & Jia, Q. (2017). Mining meaning from online ratings and reviews: Tourist satisfaction analysis using latent Dirichlet allocation. *Tourism Management*, *59*, 467–483.

Halim, M. A. (2022). Measuring tourists' intention to use digital platform in selecting travel products: A study on Bangladesh. *Asian Journal of Economics, Business and Accounting*, *22*(13), 34–46.

Hashem, I. A. T., Yaqoob, I., Anuar, N. B., Mokhtar, S., Gani, A., & Khan, S. U. (2015). The rise of "big data" on cloud computing: Review and open research issues. *Information Systems*, *47*, 98–115.

Irudeen, R., & Samaraweera, S. (2013). Big data solution for Sri Lankan development: A case study from travel and tourism. In *2013 International Conference on Advances in ICT for Emerging Regions (ICTer)* (pp. 207–216). IEEE.

Jankova, L., Auzina, A., & Zvirbule, A. (2023). Regional smart cultural tourism destinations in a region of Latvia. *Worldwide Hospitality and Tourism Themes*, *15*(5), 507–516. https://doi.org/10.1108/whatt-06-2023-0082 (accessed on 8th July 2025).

Jia, Q., Cui, Y., Liu, E. N. X. U. A. N., Young, J., Polly, Y., Sun, W., & Shen, H. (2022). Construction and design of a smart tourism model based on big data technologies. *Mobile Information Systems*, 1–7. https://doi.org/10.1155/2022/1120541 (accessed on 8th July 2025).

Kachniewska, M. (2020). The use of big data in tourism sales forecasting. *International Journal of Contemporary Management*, *19*(2), 7–35.

Kambatla, K., Kollias, G., Kumar, V., & Grama, A. (2014). Trends in big data analytics. *Journal of Parallel and Distributed Computing*, *74*(7), 2561–2573.

Koo, C., Park, J., & Lee, J. N. (2017). Smart tourism: Traveler, business, and organizational perspectives. *Information and Management*, *54*(6), 683–686.

Korže, S. Z. (2019). From Industry 4.0 to Tourism 4.0. *Innovative Issues and Approaches in Social Sciences*, *12*(3), 29–52.

Laney, D. (2001). 3D data management: Controlling data volume, velocity and variety. *META Group Research Note*, *6*(70), 1.

Latham, J., & Edwards, C. (2003). The statistical measurement of tourism. *Progress in Tourism, Recreation and Hospitality Management*, *1*, 55–76.

Lee, P., Hunter, W. C., & Chung, N. (2020). Smart tourism city: Developments and transformations. *Sustainability, 12*(10), 3958. https://doi.org/10.3390/su12103958 (accessed on 8th July 2025).

Li, J., Xu, L., Tang, L., Wang, S., & Li, L. (2018). Big data in tourism research: A literature review. *Tourism Management*, *68*, 301–323. https://doi.org/10.1016/j.is.2014.07.006 (accessed on 8th July 2025).

Manyika, J., Chui, M., Brown, B., Bughin, J., Dobbs, R., Roxburgh, C., & Hung Byers, A. (2011). *Big Data: The Next Frontier for Innovation, Competition, And Productivity*. McKinsey Global Institute.

Mariani, M. (2020). Big data and analytics in tourism and hospitality: A perspective article. *Tourism Review*, *75*(1), 299–303.

Marine-Roig, E., & Clavé, S. A. (2015). Tourism analytics with massive user-generated content: A case study of Barcelona. *Journal of Destination Marketing & Management*, *4*(3), 162–172.

Matsudaira, K. (2013). ACM Learning Center webcast, big data without big database—Extreme in-memory caching for better performance. Available at: https://www.learning.acm.org/webinar/ (accessed on 8th July 2025).

McKinsey & Company. (2018). *Travel and Logistics: Data Drives the Race for Customers*. Available at: https://www.mckinsey.com/~/media/mckinsey/industries/travel%20logistics%20and%20infrastructure/our%20insights/travel%20and%20logistics%20data%20drives%20the%20race%20for%20customers/mck-travel-and-logistics-data-drives-the-race-for-customers.pdf (accessed on 8th July 2025).

Middleton, V. T., Fyall, A., Morgan, M., & Ranchhod, A. (2009). *Marketing in Travel and Tourism*. Routledge.

Pine, B. J., & Gilmore, J. H. (2013). The experience economy: Past, present and future. In *Handbook on the Experience Economy* (Vol. 1, pp. 21–44). Edward Elgar Publishing.

Sarker, M. N. I., Khatun, M. N., Alam, G. M., & Islam, M. S. (2020). Big data driven smart city: Way to smart city governance. In *2020 International Conference on Computing and Information Technology (ICCIT-1441)* (pp. 1–8). IEEE.

Shoval, N., & Ahas, R. (2016). The use of tracking technologies in tourism research: The first decade. *Tourism Geographies*, *18*(5), 587–606.

Sigala, M., & Chalkiti, K. (2014). Investigating the exploitation of Web 2.0 for knowledge management in the Greek tourism industry: A utilization–importance analysis. *Computers in Human Behavior*, *30*, 800–812.

Sigala, M., Christou, E., & Gretzel, U. (2012). *Social Media in Travel, Tourism and Hospitality: Theory, Practice and Cases*. Ashgate Publishing, Ltd.

Sivarajah, U., Kamal, M. M., Irani, Z., & Weerakkody, V. (2017). Critical analysis of big data challenges and analytical methods. *Journal of Business Research*, *70*, 263–286.

Smith, S. L. (1988). Defining tourism: A supply-side view. *Annals of Tourism Research*, *15*(2), 179–190.

Stankov, U., & Gretzel, U. (2020). Tourism 4.0 technologies and tourist experiences: A human-centered design perspective. *Information Technology & Tourism*, *22*(3), 477–488.

Statista. (2023). Data growth worldwide 2010–2025. Available at: https://www.statista.com/statistics/871513/worldwide-data-created/ (accessed on 8th July 2025).

Steppe, R. (2017). Online price discrimination and personal data: A General Data Protection Regulation perspective. *Computer Law & Security Review, 33*(6), 768–785.

Stringam, B. B., & Gerdes Jr, J. (2010). An analysis of word-of-mouth ratings and guest comments of online hotel distribution sites. *Journal of Hospitality Marketing & Management, 19*(7), 773–796.

Uddin, M. F., & Gupta, N. (2014). Seven Vs of big data: Understanding big data to extract value. In *Proceedings of the 2014 Zone 1 Conference of the American Society for Engineering Education* (pp. 1–5). IEEE.

Vecchio, P., Malandugno, C., Passiante, G., & Sakka, G. (2021). Circular economy business model for smart tourism: The case of Ecobnb. *Euromed Journal of Business, 17*(1), 88–104. https://doi.org/10.1108/emjb-09-2020-0098 (accessed on 8th July 2025).

Vecchio, P., Mele, G., Ndou, V., & Secundo, G. (2018). Open innovation and social big data for sustainability: Evidence from the tourism industry. *Sustainability, 10*(9), 3215. https://doi.org/10.3390/su10093215 (accessed on 8th July 2025).

Volo, S. (2018). Tourism data sources: From official statistics to big data. In *The SAGE Handbook of Tourism Management* (pp. 193–201). SAGE, London.

Volo, S. (2020). Tourism statistics, indicators and big data: A perspective article. *Tourism Review, 75*(1), 304–309. https://doi.org/10.1108/TR-06-2019-0262 (accessed on 8th July 2025).

Vu, H. Q., Law, R., & Li, G. (2019). Breach of traveler privacy in location-based social media. *Current Issues in Tourism, 22*(15), 1825–1840.

Wang, D., Park, S., & Fesenmaier, D. R. (2012). The role of smartphones in mediating the touristic experience. *Journal of Travel Research, 51*(4), 371–387.

Washburn, D., Sindhu, U., Balaouras, S., Dines, R. A., Hayes, N., & Nelson, L. E. (2010). Helping CIOs understand "smart city" initiatives. *Growth, 17*(2), 1–17.

Werthner, H. (2003). Intelligent systems in travel and tourism. In *IJCAI, 3*, 1620–1625.

White, T. (2015). *Hadoop: The Definitive Guide*. O'Reilly Media.

Williams, D. A., & Charles, C. B. (2023). 4IR technologies to facilitate planning in smart cities of the future. In *Smart Cities: Social and Environmental Challenges and Opportunities for Local Authorities* (pp. 17–31). Cham: Springer International Publishing.

Wise, N., & Heidari, H. (2019). Developing smart tourism destinations with the Internet of Things. In *Big Data and Innovation in Tourism, Travel, and Hospitality: Managerial Approaches, Techniques, and Applications* (pp. 21–29).

Xiang, Z., & Fesenmaier, D. R. (2017). Big data analytics, tourism design and smart tourism. In *Analytics in Smart Tourism Design: Concepts and Methods* (pp. 299–307).

Xiang, Z., Du, Q., Ma, Y., & Fan, W. (2017). A comparative analysis of major online review platforms: Implications for social media analytics in hospitality and tourism. *Tourism Management, 58*, 51–65.

Xiang, Z., Wang, D., O'Leary, J. T., & Fesenmaier, D. R. (2015). Adapting to the internet: Trends in travelers' use of the web for trip planning. *Journal of Travel Research, 54*(4), 511–527.

Yusuf, B., Walters, L. M., & Sailin, S. N. (2020). Restructuring educational institutions for growth in the Fourth Industrial Revolution (4IR): A systematic review. *International Journal of Emerging Technology Learning, 15*(3), 93–109.

Zaharia, M., Xin, R. S., Wendell, P., Das, T., Armbrust, M., Dave, A., Ming, X., Rosen, J., Venkataraman, X., Franklin, M. J., Ghodis, A., Gonzalez, J., Shenker, S., & Stoica, I. (2016). Apache Spark: A unified analytics engine for big data processing. *Communications of the ACM*, *59*(11), 56–65.

Zikopoulos, P. C., Eaton, C., deRoos, D., Deutsch, T., & Lapis, G. (2012). *Understanding Big Data: Analytics for Enterprise Class Hadoop and Streaming Data*. McGraw-Hill.

Zikopoulos, P., deRoos, D., & Corrigan, K. P. (2013). *Harness the Power of Big Data: The IBM Big Data Platform*. US: McGraw-Hill.

CHAPTER 5

EMERGENCE OF AI CHATBOTS IN SMART TOURISM: A 24/7 TRAVEL AGENT

MD. TANVIR ALAM HIMEL* and SHAHRIN ASHRAF

Department of Marketing, University of Dhaka, Dhaka, Bangladesh

*Corresponding author

ABSTRACT

Artificial intelligence is heralding the rapid shift of frontline employees to chatbots (intelligent conversational agents) for quick, strategic, and personalized textual conversational service with customers. The purpose of this chapter is to explore the avenues of chatbot services that can be leveraged by the hospitality and tourism (H&T) industry in Bangladesh. The chapter perused the chatbot literature to find a novel solution, also known as AI chatbots, for the H&T industry, which has a growing ubiquitous acceptance in traveler engagement. 24/7 seamless support, personalized interaction, empathy, and privacy all lead to this service transition from human-centered support to this travel-bot support. In order to pursue the objective, a qualitative study was conducted by scrutinizing reliable and valid secondary sources. As smart tourism evolves, intelligent travel agents can interact in the natural language, manifest empathic tone in their textual conversations, and mirror the human conversation to humanize the bot presence as well as aptly satisfy the curious minds of human users. The findings suggest some existing and proposed usage and application of AI chatbots in different spheres of smart tourism in the emerging country of Bangladesh.

Smart Tourism: Managing Technological Revolution in the Tourism Industry. Santus Kumar Deb (Ed.)

DOI: 10.1201/9781779641090-5

5.1 INTRODUCTION

The rapid transformation of the economy of Bangladesh into a digital economy has been triggered and boosted by the swift expansion of digital technology. The current Prime Minister of Bangladesh, Sheikh Hasina, has envisioned a futuristic economy where the universal adoption of smartphones, the availability of 5G connectivity, and cashless payment systems are predominant features. The prophesied economy is termed “Smart Bangladesh.” Smart citizens, smart governments, smart societies, and smart economies are the four founding pillars of this concept (UNB, 2022). The implementation of this economy will empower and include all citizens of Bangladesh to contribute in an efficient and smart way. Well planned alignment, preparation, coordination, and execution of these four tenets of smart Bangladesh will help the people of Bangladesh realize this vision (Chowdhury, 2023).

With countries’ scenic beauty, delicious cuisines, hospitality, and after all the ease of access, connectivity, internet technology, and digitalization, tourism has become one of the biggest contributors to the world economy. Around $100 trillion world’s GDP, travel and tourism has the contribution of $7.7 trillion, which is around 7.7 per cent of the GDP (Statista, 2023). In the USD 454 billion economy of Bangladesh, the tourism sector has a margin of slightly greater than 3% of the GDP (Uddin et al., 2023). In order to increase this margin, the government of Bangladesh has initiated diverse private and public initiatives to contribute to this smart economy through smart tourism.

The recent trend in customer service is dominated by artificial agents, also known as chatbots (Belanche et al. 2021; Kautish and Sharma 2021; Okuda and Shoda 2018). Chatbots are not new in services for how highlighted they are due to their agility in providing services. The tourism sector has been celebrating the services provided by chatbots for personalizing and structuring the conversation at customer contact points, moving the bots to the frontlines (Belanche et al., 2021). The novel and diverse nature of contemporary travel and tourism services requires universal and parallel customer support and solutions. The upsurge trend of artificial intelligence is impelling tourism sectors to adopt intelligent technology to uplift tourists’ travel experiences.

“Chatbot” is the offspring of artificial intelligence, which is going to transform the current scenario of customer support services. Chatbot is a portmanteau word that uses “chat” from “chatting” and “bot” from "robot" to describe a new form of AI conception. Okuda and Shoda (2018) defined a chatbot as a program that mechanically interacts with people by replacing an entire human-based service center. Personalized communication (Zumstein and Hundertmark, 2017), 24/7 uninterrupted services (Okuda and Shoda, 2018), convenience, and personal privacy (Lai et al., 2019) are some of the defining characteristics that are promoting AI-based chatbots to replace human-centered customer care services. While AI-based chatbots come with numerous pros, they are not without cons. Breach of digital privacy and trust (Lappeman et al., 2022), security risks (Lia et al., 2019), and issues of security authentication, data integrity, system availability, and transparency (Bhuiyan et al., 2020) are the major stumbling blocks for the swift adoption of chatbots in tourism services.

The chapter is designed as follows: initially, the review of the literature regarding smart tourism, technologies of fourth industrial revolution, AI chatbots, the theories covering adoption of AI chatbots in tourism sector is explained, following that the method used to collect the data and contents of the chapter is discussed, in the later phases, the advantages and applications of AI chatbots is presented, and practical implications, biases, and challenges of using AI chatbots are particularized to end the chapter.

5.2 LITERATURE REVIEW

Chatbots are considered the most rudimentary and extensive illustration of intelligent Human-Computer interaction. Because of their uncomplicated nature and wide acceptance, corporations as well as governments are adopting this "intelligent agent" to render diverse front-line jobs. The data-driven and knowledge-intensive tourism industry demands constant communication among the stakeholders. Because of personal as well as organizational inefficiency (Bienstock and DeMoranville, 2003), service employees cannot provide consistent service. Whereas, to make business more cost-efficient and scalable, organizations are pursuing chatbot services for their service contact points over traditional helpline centers. In order to implement the smart city concept, the communication and user service experience should be instant, streamlined, and systematic, which can only be ensured using artificial intelligent chatbots at every touchpoint of a service. Moreover, in the dynamic and fast-paced conversational service era, AI chatbots are the next big invention, which aid service users to converse with messaging systems in natural language.

5.2.1 SMART TOURISM AND ITS CHARACTERISTICS

The nucleus of smart tourism is the smart destination. Smart destinations are mainly characterized by innovation, free access to internet facilities, well-planned connectivity, and a well-organized infrastructure. All the facilities are ensured to provide an unparalleled travel experience to the tourists. Barrionuevo et al. (2012) termed a smart city as a city where all available resources and technologies are integrated in a coordinated and intelligent way to establish urban centers which are altogether habitable, integrated, and sustainable. The main theme of a smart city is portrayed as a city which is instrumented, interconnected, and intelligent (Harrison et al., 2010). A smart city is equipped with all the state-of-the-art technologies and modern facilities. All the cities and facilities are connected through streamlined infrastructure. Whereas, intelligence is a "strip element" that binds other components of a smart city. An intelligent city is devoted to nurturing diversity, innovative culture, digital innovation, and social intelligence, thus turning it into a cultural capital of the world (Honeywill, 2016). Smart cities are bringing Taylor Swift or Beyoncé, music superstars, to make cities like Dubai, Vancouver, or Stockholm a cultural hub for the people around those cities and country. Thus, it makes the cultural capital the foundation of a new economy. Rudoft et al. (2007) delineated the conversion of six aspects of a city to convert it into a smart city: economy, mobility, living, environment, people, and governance. These dimensions are executed in an efficient, seamless, safe, and impactful way in an economy. The optimized resource allocations, smooth delivery of citizen services, protecting individuals from physical and cyberspace risks, and providing enjoyable and satisfying living and travel experiences are the core features of smart cities (Khan et al., 2017) (Figure 5.1).

There are several specific variables that should be taken into consideration to convert a tourist destination into a smart destination. Attributes of smart destinations are discussed below:

1. **Efficient Information and Communication Technology (ICT):** Smart destinations can incorporate information and communication technology to serve the diverse needs of tourists. In order to strengthen travel experiences at hotels, airplanes, attraction sites, museums, and shopping malls, accessibility to modern communication technologies and the availability of information are indispensable requirements of a smart city. Estonia, a country

by the Baltic Sea, has deployed ICT to offer services like information portals, online flight and hotel booking apps, and interactive maps (Ahas et al., 2019).

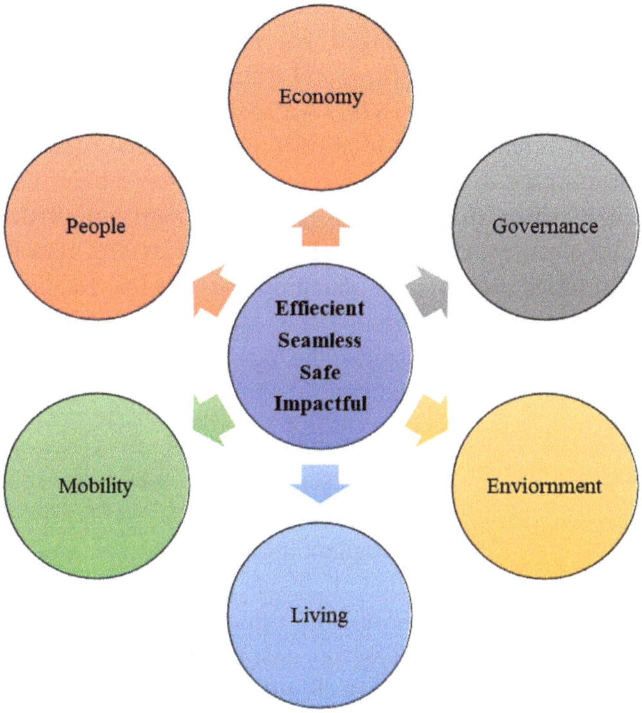

FIGURE 5.1 Pillars and dimensions of a smart destination.

Source: Adapted from Salem (2016) and Khan et al. (2017).

2. **Free Access to the Internet:** Nowadays, the general public, as well as tourists, are extensively dependent on free, public Wi-Fi hotspots (Bellini et al., 2017). In order to get the complete benefit of government and private services online, one needs to have access to the internet. This can only be feasible when citizens have access to the internet through public Wi-Fi hotspots. Seoul, Barcelona, Singapore, Dubai, and New York are cities offering 5G internet facilities to their citizens.
3. **Availability of Big Data:** From urban planning to public safety, tourism management to traffic management, and energy

optimization to water management, big data is playing a remarkable role in the efficient execution of these activities. The computational power of devices and the capacity to store large volumes of data have been developed substantially (Li et al., 2020). Dubai, Tokyo, Amsterdam, and Barcelona are analyzing tons of tourist data to offer personalized recommendations, optimize traffic routes, control visitor flow, and stimulate off-peak tourism.

4. **Internet of Things (IoTs):** An abundance of digital and online services can be provided through the incorporation of IoT technologies in tourism services. IoTs have been used by city security authorities and waste management authorities to make a city secure and clean. IoT technologies are also being used to optimize the energy consumption of city dwellers. A smart destination like San Francisco installed IoT-enabled garbage bins which automatically signal the Waste and Garbage Management authority to empty the bins.

If a country wants to implement the sustainable development goals by 2030, it should emphasize bringing sustainability into every sphere of life. Sustainability is also at the core of the smart city as well as the destination concept (Khan, et al., 2017). The planning, development, and execution of a smart city are based on the idea of cleaner production, energy efficiency, extensive use of technologies and information, and the ubiquity of the internet and the IoTs. All these factors ensure a destination's sustainability and the nurturing of natural beauty (Figure 5.2).

5.2.2 SMART TOURISM AND 4IR

The adoption of the internet and diverse technologies has facilitated the concept of smart tourism. With the emergence of broad technologies like Big Data and Artificial Intelligence (BDAI), machine learning, and 5G internet, and specific technologies like the IoTs, Near Field Communication (NFC), chatbots, and Radio Frequency Identification (RFID), conventional travel experiences have turned into smart ones. The concept of smart tourism is rooted in the idea of smart city principles, known as smart destinations (Gretzel et al., 2015). Thus, smart tourism is the process of bridging the physical and digital domains of travel experiences through digital technology. Several technologies of the 4th Industrial Revolution

(4IR) are currently being used by the travel and tourism authority of countries to implement smart tourism in their economy (Table 5.1).

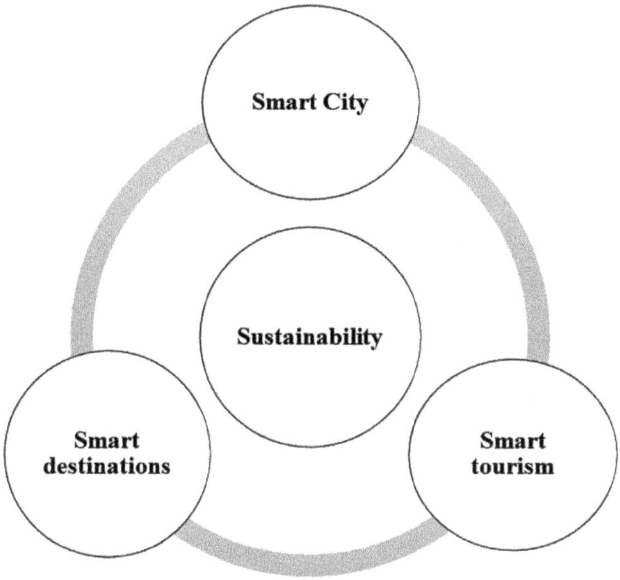

FIGURE 5.2 Relationship among sustainability, smart city, smart destinations, and smart tourism.

Source: Authors' Constructed.

5.2.3 TYPE OF CHATBOTS AND SMART TOURISM

A chatbot is defined as software that simulates human-like conversation and converses with individuals in a textual or dialogue format (Brandtzaeg and Følstad, 2018). Generally, chatbots are of two types: rule-based and neural-based (Agarwal and Wadhwa, 2020). Rule-based chatbots are primitive and respond to the user's questions following a pre-determined programmed approach. On the contrary, neural-based chatbots, also known as AI chatbots, use natural language processing (NLP) technology to comprehend written and spoken text, conduct sentiment analysis, and perform automatic text summarization duties. Computers use NLP to apprehend text, languages, and speech as humans do.

The AI Chatbot spheres are dynamic, as their classifications can be based on various criteria. For instance, chatbots can be cataloged in line with goal

TABLE 5.1 4th Industrial Revolution (4IR) Technologies and the Implications in Tourism Sector

Technologies	Implications for Travelers	Implications for Organizers
IoT	Customization and personalization of travel services; tracking luggage in the airport; smart parking at airport; real time information.	Offer individualized services to travelers; optimize business operations; energy saving operations; preventive maintenance.
Cloud Computing	Ease of accessibility to travelers for all travel experiences from anywhere, anytime; provide security through robust encryption system; help in collaborative planning with peers and family members.	Cost saving through reducing investment in servers; achieve scalability through scaling up/down the fluctuations of demand; provide flexibility to all stakeholders of tourism industry; offer centralized data management.
Big data and analytics	Diverse options for booking; personalized recommendations; real time updates; Offering competitive pricing; in-depth overview of travel sites.	Offer location based advertising and promotional campaigns; Revenue management; reputation management; personalized traveler experience; efficient ways segmenting travelers.'
Virtual Reality	Reduce travel cost for travelers; intensifying travelers' experience through virtual trips and immersive experience; enable travelers' to virtually experiencing the hotels or destinations before booking (trying before buying).	Marketing a destination through immersive services; promoting sustainable tourism.
Robots	Handle baggage; use chatbots for hotels and flight bookings.	Use robots in different customer service touch points at hotels and restaurants.

Source: Developed by Authors.

orientation, domain openness, interaction mode, or design technique. The orientation of chatbots can be broadly classified into task and non-task orientation (Lidén and Nilros, 2020). Task-oriented chatbots are usually designed for specific goal accomplishment, For example, translation bots, travel bots, and informational bots for accumulating information for users. The tasks are accomplished in closed domains. Whereas non-task bots are not restricted to any specificity and use their natural language processing to carry on human-like conversations with the user. They normally operate in open domains and are denoted as generalist bots, For example, Siri by Apple, Alexa by Amazon. These bots can manage social interaction or mimic the user for personalized interaction (Lidén and Nilros, 2020) (Figure 5.3).

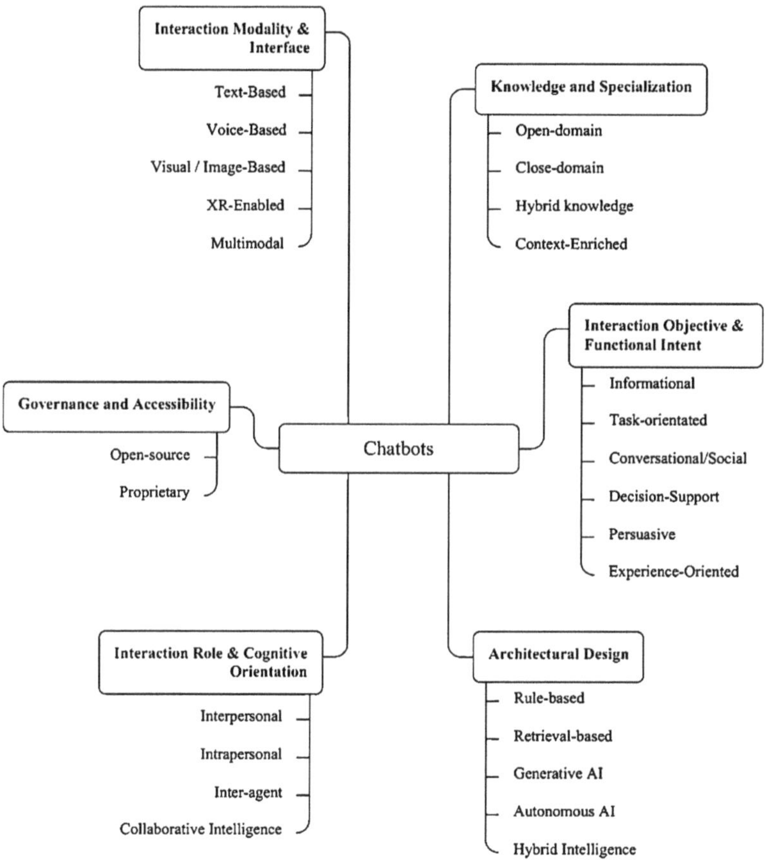

FIGURE 5.3 Broad categorization of chatbots.

Source: Modified from Adamopoulou & Moussiades, 2020a.

5.2.4 FACTORS INFLUENCING THE ADOPTION AND USAGE OF CHATBOTS

In recent years, the adoption and usage of chatbots has increased in marketing, customer support service, education, healthcare, and entertainment sectors (Adamopoulou and Moussiades, 2020b). The birth of chatbots is due to convenience, consistency, innovativeness, and low-pressure self-service (Terpening and Littleton, 2016), whereas the extension and sustainability are subject to the advancement of artificial intelligence, machine learning, NLP (Rahman et al., 2017), natural language understanding (NLU), and recurrent neural networks (Adamopoulou and Moussiades, 2020b). In studying the dimensions affecting users' intention to adopt, accept, or continue, a few well-established and well-explained theories were used to explain users' behavior. Technology Acceptance Model (TAM) (Pereira et al., 2022; Salloum et al., 2023), Extended TAM (Cicco et al., 2021; Pillai and Sivathanu, 2020), Unified Theory of Acceptance and Use of Technology (UTAUT), UTAUT2 (Hanji et al., 2023; Melián-González et al., 2021; Zhang et al., 2023), and the diffusion of innovation model (Hanji et al., 2023) were widely used by researchers to study the dimensions that users consider and emphasize in adopting or using a chatbot service (Table 5.2).

5.3 METHODOLOGY

The major objective of this chapter is to portray the current practice, prospects, and challenges of using AI chatbots in providing constant and consistent services to the travelers of smart destinations from the emerging country's point of view. For this objective, numerous models and frameworks have been studied and presented to justify the concept. Secondary sources are searched and investigated to gather data for this chapter. A systematic analysis and delineation of the information are provided in this study to make aware of the potential usability and challenges of using chatbots in the tourism sector in emerging economies.

The study has utilized authentic and reliable online sources such as Google Scholar, Scopus, and Web of Science database to delve into the core of this topic. As this chapter focused on the qualitative aspects of the chatbots adoption, applications, biases, and challenges, several publishers' journals are scrutinized to apprehend the objectives of the study, namely Taylor and Francis, Sage, Wiley, MDPI, Elsevier, and Emerald Group Publishing. Specific keywords, like "Chatbots," "Artificial Intelligence,"

TABLE 5.2 Factors of Adopting, Acceptance, and Usage Continuance of Chatbots

Authors	Theory	Factors
Pillai and Sivathanu (2020)	Extended TAM Approach with Context-specific variables.	Perceived ease of use, perceived usefulness, technological anxiety, perceived trust, perceived intelligence, anthropomorphism.
Cicco et al. (2021)	Extended TAM Approach.	Perceive ease of use, perceived usefulness, perceived enjoyment, trust, compatibility.
Melián-González et al. (2021)	Extended UTAUT2 Approach.	Performance expectancy, effort expectancy, social influence, hedonic motivations, habit, perceived innovativeness, attitude towards self-service technologies (SSTs), inconveniences, anthropomorphism, automation.
Ukpabi et al. (2019)	Institutional Theory, Organizational Learning Theory.	Isomorphic pressures, learning capabilities, learning competencies, barriers.
Rafiq et al. (2022)	S-O-R Theoretical Framework.	Perceived usability, interactivity, perceived intelligence, anthropomorphism.
Zhang et al. (2023)	Extended UTAUT2 Approach, Theory of Perceived Risk.	Performance expectancy, social influence, habit, anthropomorphism, personalization, time risk, and privacy risk.
Salloum et al. (2023)	TAM Approach, Flow Theory.	User satisfaction, perceived ease of use, perceived usefulness, flow experience.
Hanji et al. (2023)	Extended UTAUT2 Approach.	Performance expectancy, effort expectancy, habit, personal innovativeness, and awareness of chatbot service.
Hanji et al. (2023)	Diffusion of Innovation Model.	Relative advantage, trialability, compatibility, complexity, observability, trust (moderating factor).
Pereira et al. (2022)	TAM, Information System Success (ISS) Model.	Perceived utility, perceived pleasure, information quality, service quality, ease of use, need for interaction with service employee (moderating factor), satisfaction, brand attachment.
Dhiman and Jamwal (2022)	Task-Technology Fit Theory, Expectation-Confirmation Model (ECM).	Task-technology fit, confirmation, perceived usefulness, satisfaction.

TABLE 5.2 *(Continued)*

Authors	Theory	Factors
Li et al. (2021)	Extended Post-Acceptance Model of IS Continuance, Chatbot Quality Dimensions.	Understandability, reliability, responsiveness, assurance, interactivity, confirmation, satisfaction.
Lei et al. (2021)	Computer-Mediated Communication (CMC), Interpersonal Communication Theories.	Media richness, social presence, task attraction, social attraction, trust.

"smart tourism," "adoption theory," and "algorithm bias," were inserted by the authors to procure access to the relevant publications that were used in the progression of the chapter. Not all chapters collected for this study are selected; the selection criteria such as scope, journal metrics and altmetrics, impact factors, Q-indexing, reliability, relevance, and dependability of the publishers were employed to ensure the quality and accuracy of the source. Several news portals like "The Business Standard," "The Daily Star," and "The Financial Express" were also analyzed to get the current trends of using chatbots in the Bangladesh tourism sector. Moreover, a handful of valid, reliable, and reputed websites were explored for this purpose.

5.4 SMART TOURISM AND CHATBOT FEATURES

Chatbot is a disruptive technology which is changing the landscape of the tourism industry around the word. AI chatbots are well regarded by tourists because of the impersonal nature of the services any time they require it. On the other hand, the major advantage for the service providers to use chatbot service is the reduction of repetitive tasks of humans to provide error-free services to the customers, thus reducing the operational costs of providing that service.

In order to cater to the demand of global customers, organizations are leveraging the untiring and exhilarating nature of chatbot technology to serve customers round the clock. Information and technology-based tourism should assimilate AI chatbots in its system in such a way which will make its activities more efficient and the tourists' journey and whole experience more effortless. To increase the functionality and effectiveness of AI chatbots, two primary features should be incorporated in the system which include:

1. **Versatility in Language Generation and Comprehension:** One of the major features of AI chatbots is its ability to understand and converse in diverse languages. As tourism is a global phenomenon, the multilingual capability of chatbots is the core reason for incorporating it in the current technological stockpile of the tourism sector of a nation (Sidor, 2023).
2. **Omnichannel Tourists' Experience:** AI chatbots can be utilized to provide seamless, customized, and consistent services to smooth the tourist experience with the destination. A tourist destination has several mediums of online communication, like websites, social media, and blogs. The compatibility of AI chatbots with

these channels is the prerequisite for providing holistic travel experiences to the tourists (Sidor, 2023) (Figure 5.4) (Table 5.3).

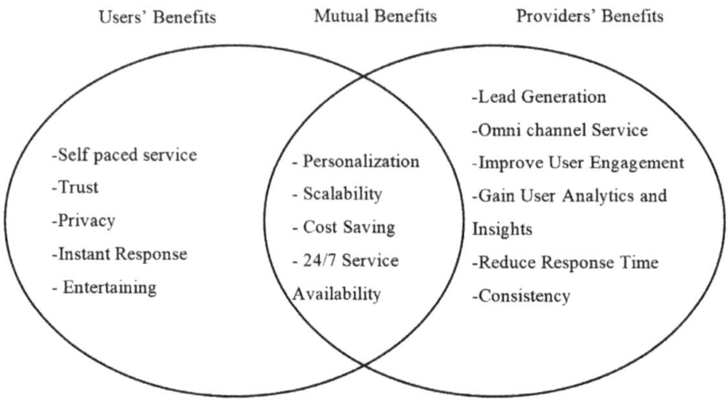

FIGURE 5.4 Advantages of using chatbot from customers and service providers' point of view in tourism sector.

Source: Authors' constructed.

TABLE 5.3 Motivations for Chatbot Use

Motivators	Description
Productivity	Several studies found the ease of use and usefulness features of chatbots brings effectiveness as well as efficiency in their job (Cicco et al., 2021; Pillai and Sivathanu, 2020)
Entertainment	The initial motive of generating chatbot was entertaining people through mimicking human conversation and pleasing users (Shawar and Atwell, 2007).
Social/ Rational	Anthropomorphism defines chatbots as a human being which has social value. Although chatbots are artificial entity, sometimes people perceive a companionship and engagement with the chatbots when they converse with it. Convenience and efficiency aspects of chatbots provide users with a rationale perspective of using it.
Novelty/ Curiosity	People are born to be curious. In order to explore new inventions, people tend to be very curious. Often users' use chatbot to explore its aptness.

Source: Adopted: Brandtzaeg and Følstad (2017).

5.5 APPLICATIONS OF CHATBOTS IN SMART TOURISM

The informative, empathic, and interactive human-like conversation nature of chatbots gives tourists the assurance of being a reliable tour guide for them at an unknown destination (Orden-Mejía, 2022). The unprecedented Covid-19 pandemic also facilitated the need for a non-human tourist guide in providing tourism and hospitality services in the context of social distancing (Hasan et al., 2021). Amalia and Suprayogi (2019) also pointed out that millennials use chatbot features in social media platforms like Messenger to disseminate information in an interactive form, such as:

1. **Destination:** Currently, tourists use several websites, apps, and travel information and service-related portals to get their desired services. An occasional tourist or those lacking knowledge of traveling often get confused about how to use those services wisely to make an efficient decision regarding travel. In order to search for a compatible location for the tourist, get an in-depth idea regarding the best tourist places of that location, and find travel-related services of that location, tourists use AI chatbots to have all the services of a tourist destination (Alotaibi et al., 2020).
2. **Hospitality:** The hospitality sector has come a long way, incorporating digital services at various user points. Intelligent artificial agents are adding a new dimension as they are providing multifaceted services to the user (Alotaibi et al., 2020). Instead of searching through many websites or installing various apps for traveling recommendations, booking hotels or transportation, or navigating cityscapes or sights, these agents are providing full service at one stop (Alotaibi et al., 2020; Essien and Chukwukelu, 2022). Moreover, the users were satisfied as they could interact in natural language (Alotaibi et al., 2020). With the advancement of deep learning, AI bots suggest they can forecast demand for tourist locations and understand travel trends, as well as provide more personalized service to the tourists (Essien and Chukwukelu, 2022). Chatbots, when injected with emotion, bear semblance to humanized interaction, leading to better user engagement (Pamungkas, 2019). As chatbots are widely used in customer service, the tone of these conversational agents also influences user satisfaction. Thus, emotionally aware chatbots (EAC) or incorporating emotion information in the textual communication are found to have positive interaction as well as reduce stress among the users (Pamungkas, 2019).

Chatbots can be incorporated along with human voice-call services in the hotels, restaurants, and catering (HoReCa) sector due to the increase in online food home-delivery services (de Esteban Curiel et al., 2023). Carlisle et al. (2020) also pointed out that ordering online food delivery, online reservations, and online booking facilitated the usage of chatbots in HoReCa channels. Sławińska (2020) also argued that the incorporation of chatbots (automated systems) for providing customer services lessens human biases and errors in the HoReCa sector, thus providing superior customer services.

5.6 THEORY INTO PRACTICALITY: CASE OF GOZAYAAN

GoZayaan, one of the Online Travel Aggregators (OTA) of Bangladesh, started its journey in 2017. According to Mr. Ridwan Hafiz, founder and CEO of GoZayaan, his previous involvement in the e-commerce sector of Bangladesh and traveling to numerous countries around the world led him to start the venture with business-to-consumer (B2C) air ticketing services for domestic and international passengers. He dreamed of a travel company that could be the Klook, Expedia, Agoda, or Booking.com of Bangladesh. Gradually, they expanded the horizon of their operations into different travel activities like visa processing, hotel booking, and tour packages. Seeing the consistent growth and prospect of the global market, GoZayaan acquired the first and largest Pakistani online travel platform, FindMyAdventure, back in 2022 (Hasan, 2022). From the initial days, the major challenge for GoZayaan was shifting the conventional travel agency mindset of travelers into adopting digital platforms like travel websites or apps. They overcame this obstacle by executing diverse sales promotions like discounts, cashbacks, and promo codes. They customized packages for the travelers. Thirdly, by associating with IPDC Finance Ltd., GoZayaan introduced travel loans with EMI facilities (Kamal and Jui, 2020). Finally, by streamlining the payment gateway of the platform, GoZayaan acquired the users' trust and engagement with their platform. With their extensive research and vast knowledge of the travel industry of Bangladesh, GoZayaan realized that people crave traveling short or long distances. But with their limited knowledge and time, they cannot make a tour plan that maximizes their satisfaction from the tour. Considering the gap, GoZayaan brings the first-ever digital travel assistant "Zayaan" to assist the travel enthusiasts of Bangladesh. Zayaan is an AI-powered conversational agent that answers all travel-related queries. Zayaan is also developed to recommend personalized travel experiences based on travelers' past travel history.

5.7 BIASES OF USING AI CHATBOTS IN TOURISM

Despite its wide acceptance as conversational agents, chatbots are not without bias. There have been cases of learning bias in chatbots while training with biased datasets (e.g., search results of CEOs showing only male CEOs in the industry). The hidden social bias in the chatbots' response is mostly due to the unintentional bias in their embedded NLP (Lee and Fung, 2019). The presence of latent bias can be seen in the health sector, where patients try to self-diagnose and find healthcare professionals near their locations. For instance, while searching for the right ophthalmologists in the US, these AI chatbots (ChatGPT, Bing Chat, Bard) all showed implicit bias (inaccuracy) and gender bias in their recommendations (e.g., less likely recommendation for female ophthalmologists, recommended professors more than assistant professors, made inaccurate suggestions of medical doctors instead of ophthalmologists) (Oca et al. 2023).

Gender stereotyping in human behavior can be transmitted to chatbots due to users' preconceptions and biases, whereas these conversational agents may mimic such behavior for personalized interaction. Users may judge the chatbot's performance according to the visible gender cues (masculine/feminine avatar) or ethnic cues (diversity) (McDonnell and Baxter, 2019). The CASA (Computers-as-Social Actors) theory intimated how humans tend to apply human social scripts in their interactions with chatbots, so even the subtle disembodied cues may result in biased perception in users' digital content consumption (McDonnell and Baxter, 2019) (Table 5.4).

To establish diversity representation in chatbots within a learning environment, Schlimbach et al. (2023) suggested designing pedagogical conversational agents (PCAs) to be socially inclusive (considering social and verbal cues). For instance, people with handicaps are perceived as less competent than non-disabled individuals. So, when avatars were portrayed as handicapped (showcasing wheelchairs), this representation influenced the human-chatbot interaction, resulting in a negative perception of the chatbot's competence. However, this socially inclusive design of PCAs in digital education aided in formulating a positive perception of minorities by the students. On the same note, in recruitment and team formation, people prefer those who are perceived as similar to them, resulting in affinity bias. Therefore, in human-chatbot interaction, social agents (avatars) who resemble the particular user have a better chance of being selected as a team member, reflecting unconscious affinity bias (Trainer et al., 2020).

TABLE 5.4 Classification of Chatbot Biases in Tourism

Broad Classification of Biases	Authors	Latent Biases/Themes
Social bias	Lee et al. (2019)	Age, race, ethnic, language bias
Gender stereotype bias	McDonnell and Baxter (2019); Oca et al. (2023)	Occupational bias, gender cues (masculine/feminine avatar); stereotypical product preference biases
Implicit bias	Oca et al. (2023)	Inaccurate recommendation; response and selection bias
Representation bias	Schlimbach et al. (2023)	Diversity representation, socially inclusive avatars,
Affinity bias	Trainer et al. (2020)	Association bias; creator bias; confirmation bias

5.8 CONCLUSION

AI chatbots are now governing our social engagement. As human-to-human interaction is lengthy and has a temporal limitation, human-to-computer interaction and its advancement are escalating interest among possible travelers. With the prospect of emotionally aware chatbots in natural language, textual interaction will be personalized and will be leaping from assistants to companions. This chapter adds to the chatbot literature and verbalizes its practices, applications, and biases in the context of the hospitality and tourism (H&T) industry in Bangladesh. The drive toward smart tourism (consisting of smart cities and smart destinations) renders the exploration of AI chatbots as travel agents valuable. The tourism sectors have been celebrating the services provided by chatbots for personalizing and structuring conversations at customer contact points, substituting human employees with intelligent bots at the frontline. As smart tourism is the process of bridging the physical and digital domains of travel experiences through digital technology, the application of chatbots, that is, travel bots, assists in personalizing the service interaction through communicating, informing, as well as entertaining travelers in their natural language, establishing a two-way dialogue between humans and computers. This computers-as-social-agents role of chatbots makes it meaningful for travel/destination/city/booking websites to establish convenience, consistency, innovativeness, and low-pressure self-service, mutually benefiting both the service providers' and users' goals.

KEYWORDS

- **adoption**
- **artificial intelligence**
- **chatbot**
- **digital technology**
- **hospitality**
- **natural language processing (NLP)**
- **smart tourism**

REFERENCES

Adamopoulou, E., & Moussiades, L. (2020a). An overview of chatbot technology. In *IFIP International Conference on Artificial Intelligence Applications and Innovations* (pp. 373–383). Cham: Springer International Publishing.

Adamopoulou, E., & Moussiades, L. (2020b). Chatbots: History, technology, and applications. *Machine Learning with Applications*, *2*, 100006.

Agarwal, R., & Wadhwa, M. (2020). Review of state-of-the-art design techniques for chatbots. *SN Computer Science*, *1*(5), 246.

Ahas, R., Mooses, V., Kamenjuk, P., & Tamm, R. (2019). Retrofitting Soviet-era apartment buildings with 'smart city' features: The H2020 SmartEnCity project in Tartu, Estonia. In *Housing Estates in the Baltic Countries* (pp. 357–375). Springer.

Alotaibi, R., Ali, A., Alharthi, H., & Almehamdi, R. (2020). AI chatbot for tourist recommendations: A case study in the city of Jeddah, Saudi Arabia. *International Journal of Interactive Mobile Technologies*, *14*(19), 18–30.

Belanche, D., Casaló, L. V., Schepers, J., & Flavián, C. (2021). Examining the effects of robots' physical appearance, warmth, and competence in frontline services: The Humanness-Value-Loyalty model. *Psychology & Marketing*, *38*(12), 2357–2376.

Bellini, P., Cenni, D., Nesi, P., & Paoli, I. (2017). Wi-Fi based city users' behavior analysis for smart city. *Journal of Visual Languages & Computing*, *42*, 31–45.

Bhuiyan, M. S. I., Razzak, A., Ferdous, M. S., Chowdhury, M. J. M., Hoque, M. A., & Tarkoma, S. (2020). BONIK: A blockchain empowered chatbot for financial transactions. In *2020 IEEE 19th International Conference on Trust, Security and Privacy in Computing and Communications (TrustCom)* (pp. 1079–1088). IEEE.

BIDA. (2021). Tourism sector. Available at: https://bida.gov.bd/publications (accessed on 8th July 2025).

Bienstock, C. C., DeMoranville, C. W., & Smith, R. K. (2003). Organizational citizenship behavior and service quality. *Journal of Services Marketing*, *17*(4), 357–378.

Brandtzaeg, P. B., & Følstad, A. (2017). Why people use chatbots. In *Internet Science: 4th International Conference, INSCI 2017, Thessaloniki, Greece* (pp. 377–392). Springer International Publishing.

Brandtzaeg, P. B., & Følstad, A. (2018). Chatbots: Changing user needs and motivations. *Interactions*, *25*(5), 38–43.

Carlisle, S., Ivanov, S., & Dijkmans, C. (2023). The digital skills divide: Evidence from the European tourism industry. *Journal of Tourism Futures*, *9*(2), 240–266.

Chowdhury, A. (2023). What is Smart Bangladesh really? *a2i*. Available at: https://a2i.gov.bd/what-is-smart-bangladesh-really/ (accessed on 8th July 2025).

De Cicco, R., Iacobucci, S., Aquino, A., Romana Alparone, F., & Palumbo, R. (2021). Understanding users' acceptance of chatbots: An extended TAM approach. In *International Workshop on Chatbot Research and Design* (pp. 3–22). Cham: Springer International Publishing.

de Esteban Curiel, J., Antonovica, A., & Rodríguez Herráez, B. (2023). Drivers for economic recovery: Analysis of family expenditure on catering services. *British Food Journal*, *125*(13), 81–98.

Dhiman, N., & Jamwal, M. (2023). Tourists' post-adoption continuance intentions of chatbots: Integrating task–technology fit model and expectation–confirmation theory. *Foresight*, *25*(2), 209–224.

Essien, A., & Chukwukelu, G. (2022). Deep learning in hospitality and tourism: A research framework agenda for future research. *International Journal of Contemporary Hospitality Management*, *34*(12), 4480–4515.

Fortea, L. (2020). New technologies, digitalization and trends in HoReCa. *HoReCa4VET for Sustainable Development*, 56–67.

Gretzel, U., Sigala, M., Xiang, Z., & Koo, C. (2015). Smart tourism: Foundations and developments. *Electronic Markets*, *25*(3), 179–188.

Hanji, S. V., Hungund, S., Blagov, E., Desai, S., & Hanji, S. S. (2023). Examining the factors influencing diffusion and adoption of AI chatbots in tourism and travel industry. In *International Working Conference on Transfer and Diffusion of IT* (pp. 150–160). Cham: Springer Nature Switzerland.

Hanji, S. V., Navalgund, N., Ingalagi, S., Desai, S., & Hanji, S. S. (2023). Adoption of AI chatbots in travel and tourism services. In *International Congress on Information and Communication Technology* (pp. 713–727). Singapore: Springer Nature Singapore.

Harrison, C., Eckman, B., Hamilton, R., Hartswick, P., Kalagnanam, J., Paraszczak, J., & Williams, P. (2010). Foundations for smarter cities. *IBM Journal of Research and Development*, *54*(4), 1–16.

Hasan, M. (2022). GoZayaan acquires Pakistani travel-tech platform. *The Daily Star*. Available at: https://www.thedailystar.net/business/organisation-news/news/gozayaan-acquires-pakistani-travel-tech-platform-2963116 (accessed on 8th July 2025).
Hasan, R., Koles, B., Zaman, M., & Paul, J. (2021). The potential of chatbots in travel and tourism services in the context of social distancing. *International Journal of Technology Intelligence and Planning*, *13*(1), 63–83.
Honeywill, R. (2016). What is an intelligent city? Available at: https://rosshoneywill.com/articles/what-is-an-intelligent-city/#:~:text=An%20Intelligent%20City%20is%20characterised,social%20intelligence%20and%20digital%20innovation (accessed on 8th July 2025).
Hussain, S., Ameri Sianaki, O., & Ababneh, N. (2019). A survey on conversational agents/chatbots classification and design techniques. In *Web, Artificial Intelligence and Network Applications: Proceedings of the Workshops of the 33rd International Conference on Advanced Information Networking and Applications (WAINA-2019)* (pp. 946–956). Springer International Publishing.
Kamal, R. S., & Jui, U. M. (2020). Go Zayaan: Feeding the wanderlust of Bangladeshi travellers. *The Business Standard*. Available at: https://www.tbsnews.net/companies/go-zayaan-feeding-wanderlust-bangladeshi-travellers-51751 (accessed on 8th July 2025).
Kautish, P., & Sharma, R. (2021). Study on relationships among terminal and instrumental values, environmental consciousness and behavioral intentions for green products. *Journal of Indian Business Research*, *13*(1), 1–29.
Khan, M. S., Woo, M., Nam, K., & Chathoth, P. K. (2017). Smart city and smart tourism: A case of Dubai. *Sustainability*, *9*(12), 2279.
Lai, S. T., Leu, F. Y., & Lin, J. W. (2019). A banking chatbot security control procedure for protecting user data security and privacy. In *Advances on Broadband and Wireless Computing, Communication and Applications: Proceedings of the 13th International Conference on Broadband and Wireless Computing, Communication and Applications (BWCCA-2018)* (pp. 561–571). Springer International Publishing.
Lappeman, J., Marlie, S., Johnson, T., & Poggenpoel, S. (2023). Trust and digital privacy: Willingness to disclose personal information to banking chatbot services. *Journal of Financial Services Marketing*, *28*(2), 337–357.
Lee, N., Madotto, A., & Fung, P. (2019). Exploring social bias in chatbots using stereotype knowledge. In *WNLP@ ACL* (pp. 177–180).
Lei, S. I., Shen, H., & Ye, S. (2021). A comparison between chatbot and human service: Customer perception and reuse intention. *International Journal of Contemporary Hospitality Management*, *33*(11), 3977–3995.
Li, L., Lee, K. Y., Emokpae, E., & Yang, S. B. (2021). What makes you continuously use chatbot services? Evidence from Chinese online travel agencies. *Electronic Markets*, 1–25.
Li, W., Batty, M., & Goodchild, M. F. (2020). Real-time GIS for smart cities. *International Journal of Geographical Information Science*, *34*(2), 311–324.
Lidén, A., & Nilros, K. (2020). Perceived benefits and limitations of chatbots in higher education. *Bachelor's dissertation*, University of Linnaeus. Available at: https://www.divaportal.org/smash/get/diva2:1442044/FULLTEXT01.pdf (accessed on 8th July 2025).
McDonnell, M., & Baxter, D. (2019). Chatbots and gender stereotyping. *Interacting with Computers*, *31*(2), 116–121.

Melián-González, S., Gutiérrez-Taño, D., & Bulchand-Gidumal, J. (2021). Predicting the intentions to use chatbots for travel and tourism. *Current Issues in Tourism*, *24*(2), 192–210.

Oca, M. C., Meller, L., Wilson, K., Parikh, A. O., McCoy, A., Chang, J., Sudharshan, R., Gupta, S., & Zhang-Nunes, S. (2023). Bias and inaccuracy in AI chatbot ophthalmologist recommendations. *Cureus*, *15*(9), e45911.

Okuda, T., & Shoda, S. (2018). AI-based chatbot service for financial industry. *Fujitsu Scientific and Technical Journal*, *54*(2), 4–8.

Orden-Mejia, M., & Huertas, A. (2022). Analysis of the attributes of smart tourism technologies in destination chatbots that influence tourist satisfaction. *Current Issues in Tourism*, *25*(17), 2854–2869.

Pereira, T., Limberger, P. F., Minasi, S. M., & Buhalis, D. (2022). New insights into consumers' intention to continue using chatbots in the tourism context. *Journal of Quality Assurance in Hospitality & Tourism*, 1–27.

Pillai, R., & Sivathanu, B. (2020). Adoption of AI-based chatbots for hospitality and tourism. *International Journal of Contemporary Hospitality Management*, *32*(10), 3199–3226.

Rafiq, F., Dogra, N., Adil, M., & Wu, J. Z. (2022). Examining consumer's intention to adopt AI-chatbots in tourism using partial least squares structural equation modeling method. *Mathematics*, *10*(13), 2190.

Rahman, A. M., Al Mamun, A., & Islam, A. (2017). Programming challenges of chatbot: Current and future prospective. In *2017 IEEE Region 10 Humanitarian Technology Conference (R10-HTC)* (pp. 75–78). IEEE.

Salem, F. (2016). A smart city for public value: Digital transformation through agile governance – The case of "Smart Dubai." Dubai: Governance and Innovation Program, Mohammed Bin Rashid School of Government, World Government Summit. Available at: https://ssrn.com/abstract=2733632 (accessed on 8th July 2025).

Salloum, S. A., Shwedeh, F., Alfaisal, A. M., Alshaafi, A., Aljanada, R. A., Al Sharafi, A., Raghad, & Dabash, A. (2023). Understanding and forecasting chatbot adoption: An SEM-ANN methodology. *Migration Letters*, *20*(S11), 652–668.

Schlimbach, R., Stoppel, A., Lampka, L., & Robra-Bissantz, S. (2023). A picture is worth a thousand words–Exploring bias in inclusive chatbots. In *18th International Conference on Wirtschaftsinformatik*, Paderborn, Germany.

Shawar, B. A., & Atwell, E. (2007). Chatbots: Are they really useful? *Journal for Language Technology and Computational Linguistics*, *22*(1), 29–49.

Sidor, J. (2023). The 9 must-have chatbot features to transform online support. *The 9 Must-Have Chatbot Features to Transform Online Support*. Available at: https://getzowie.com/blog/what-makes-good-chatbot-nine-must-have-features (accessed on 8th July 2025).

Sławińska, K. (Ed.). (2020). *New technologies, digitalization and trends in HoReCa for sustainable development*. Stowarzyszenie na Rzecz Innowacji i Edukacji. Available at: https://shorturl.at/oJLM6 (accessed on 8th July 2025).

Statista. (2023). *Tourism Worldwide*. Statista. Available at: https://www.statista.com/topics/962/global-tourism/#editorsPicks (accessed on 8th July 2025).

Suhaili, S. M., Salim, N., & Jambli, M. N. (2021). Service chatbots: A systematic review. *Expert Systems with Applications*, *184*, 115461.

Terpening, E., & Littleton, A. (2016). The 2016 state of social business: Social's shift from innovator to integrator. *Altimeter, a Prophet Company*.

Trainer, T., Taylor, J. R., & Stanton, C. J. (2020). Choosing the best robot for the job: Affinity bias in human-robot interaction. In *Social Robotics: 12th International Conference, ICSR 2020, Golden, CO, USA, November 14–18, Proceedings 12* (pp. 490–501). Springer International Publishing.

Uddin, R., Arefin, M. A., Sima, S. A., & Arman, A. (2023). Mapping tourists' mind to introduce and nurture data driven decision-making culture in the tourism sector for creating informal employment opportunities. *UNDP Bangladesh.* Available at: https://www.undp.org/bangladesh/blog/mapping-tourists-mind-introduce-and-nurture-data-driven-decision-making-culture-tourism-sector-creating-informal-employment#:~:text=In%20Bangladesh%2C%20tourism%20sector%20contributes,contributed%20by%20the%20tourism%20sector (accessed on 8th July 2025).

Ukpabi, D. C., Aslam, B., & Karjaluoto, H. (2019). Chatbot adoption in tourism services: A conceptual exploration. In *Robots, Artificial Intelligence, and Service Automation in Travel, Tourism and Hospitality* (pp. 105–121). Emerald Publishing Limited.

UNB. (2022). PM unveils vision to build Smart Bangladesh by 2041. *The Business Standard.* Available at: https://www.tbsnews.net/bangladesh/pm-hasina-unveils-vision-build-smart-bangladesh-2041-550018 (accessed on 8th July 2025).

Zhang, B., Zhu, Y., Deng, J., Zheng, W., Liu, Y., Wang, C., & Zeng, R. (2023). "I am here to assist your tourism": Predicting continuance intention to use AI-based chatbots for tourism. Does gender really matter? *International Journal of Human–Computer Interaction, 39*(9), 1887–1903.

Zumstein, D., & Hundertmark, S. (2017). Chatbots–An interactive technology for personalized communication, transactions and services. *IADIS International Journal on WWW/Internet, 15*(1), 96–109.

CHAPTER 6

SMART TOURISM AND ECOTOURISM DESTINATIONS: A CONCEPTUAL FRAMEWORK OF ECO-SMART DESTINATIONS

BAPON CHANDRA KURI*

Department of Tourism and Hospitality Management, Gopalganj Science & Technology University, Gopalganj, Bangladesh

*Corresponding author

ABSTRACT

The purpose of this book chapter is to probe how tourists' ecotourism experiences may be improved via the use of intelligent destination technology. Moreover, this chapter explored numerous concepts of eco-smart destinations, technological sophistication, marketing strategies, benefits, and challenges. The qualitative approach has been utilized to unearth the concept of "Eco-Smart Destination" and its utilization in the eco-tourism area. Both primary (expert opinion) and secondary (Google Scholar, books, journals, newspapers, and articles) sources of data have been gathered to find the contribution of eco-smart destinations for the tourists, community, business stakeholders, and government. The findings showed that the digital transformation of eco-tourism enhances visitors' experiences, and generates financial benefits for the community, business, and government. Nevertheless, excessive accessibility in eco-smart destinations will hamper the overall cultural and natural ecosystem of the destination. Therefore, the practical and theoretical implications of the eco-smart destination life cycle illustrate how to manage destination resources and technological utilization in the ecological area.

Smart Tourism: Managing Technological Revolution in the Tourism Industry. Santus Kumar Deb (Ed.)

DOI: 10.1201/9781779641090-6

6.1 INTRODUCTION

Visiting and enjoying superior natural areas while simultaneously safeguarding tourists from detrimental effects is becoming a viable and profitable concept. Consequently, a term that is appropriate and conjures up positive associations with the term "eco," such as ecology, ecosystem, ecosphere, and eco-sensitive, has been merged with the tourism industry. Moreover, smart technology, e-tourism, virtual reality, augmented reality, and artificial intelligence have received more attention from researchers to explore the contribution of digital tools in the tourism and hospitality industry (Alkhatib et al., 2024). Therefore, this chapter provides a four-pillar framework for eco-smart tourism and conceptually connects the ideas of ecotourism and smart destinations. It specifically proposes that eco-smart tourism projects can be strengthened by supporting three distinct models by providing smart destinations with sensing, opening, sharing, governing, and innovative capacities through smart tourism infrastructure and governance. Building technology infrastructure to deliver data that can be converted into business models and destination management insights on the one hand, and superior tourism experiences on the other, is what is referred to as "smart tourism" in academic circles and industry practice (Ercan, 2023). Similarly, ecotourism can enhance the visitor's experiences by conserving and preserving cultural and natural resources sustainably (Lai et al., 2006). However, this chapter combines these two concepts in a judgmental way to prove how eco-smart tourism can be an alternative source of entertainment. Furthermore, technological advancement may deteriorate the overall sustainability of any destination, therefore a conceptual eco-smart life cycle has been proposed to understand the in-depth view of the eco-smart destination.

There have been many proposals for the regeneration of cities, ranging from Agenda 21, which was presented at the 1992 UN conference in Rio de Janeiro, to the 2015 UN Agenda 2030 recommendations (Colglazier, 2015). The European Commission adopted an integrated approach to the cities of the future as early as 2010 with the "European Green Capital Award," which recognized the achievements of a select few European towns chosen based on metrics that helped to define policies and initiatives for "green cities." (Demaziere, 2020). In addition, the "Green City" (Brilhante et al., 2018) must be a place that prioritizes eco-innovation, a green economy, and improved governance to combat climate change and employ resources efficiently. The advent of new paradigms, both technical and environmental, has a significant influence on how to link with smart citizens, the economy, society, and government (Yan et al., 2021). Globally, the idea of "urban smartness" (Caragliu et al., 2012) has grown in importance and prominence as a key political objective. While two-thirds of US cities have made investments in smart technologies, one-third of UK cities have started smart development processes, and European governments also seem to be increasingly focused on creating smart and sustainable cities (Calzada, 2016). The term "smart city," which originally only referred to technology and energy supply, is becoming more and more ambiguous every day as a means of meeting actual, emerging needs (Winkowska et al., 2019). Most recently, a health emergency has arisen to reorganize and redesign our cities (Vishnivetskaya et al., 2019).

However, these research articles focused on how smart cities can bring privilege to urban people. They ignored the importance of the digital transformation of ecotourism in the field of travel and tourism industry. Therefore, the study of smart governance, economy, society, and environmental sustainability is required to be analyzed conveniently. Furthermore, future cities practice smart governance (Pereira et al., 2018) which aims to improve the social, economic, and environmental well-being of its residents. In light of the most recent advancements, including the green economy (Bina, 2013), the economy of sustainable development (Lavrinenko et al., 2019), and the circular economy (Suárez et al., 2019) prioritize ecological quality and resilience of cities to ensure the sustainability of future cities. Eco-smart cities are technologically advanced urban areas that prioritize environmental preservation, technological innovation, and high standards of living (Cheng et al., 2011). The terms "Ecological City," (Dong et al., 2016) "Information City,"(Stock, 2011), and "Leisure City" (Thibault et al., 2014) are all embraced by this metropolis, signifying the eventual transition from an industrial to a post-industrial society. This chapter uses the concept of "Eco-Smart Destination" and its implication in tourism areas which refers to how a city integrates its structural, cognitive, and public-political elements while placing sustainability, innovation, accessibility, and the environment at the center of its new vision. First and foremost, the "Eco-Smart Destination" is a peaceful, natural, and healthy place to live and travel. Second, an artificial environment that is highly integrated with the natural environment is referred to as an "Eco-Smart Destination." Third, it is situated in remote unscathed natural areas and aims to achieve harmony between the digital and natural environments.

6.2 LITERATURE REVIEW

The main goal of ecotourism is to promote responsible tourism for the community people and environment while visiting natural landmarks (Wondirad et al., 2020). It concentrates on raising awareness of the environment and maintaining ecosystems within tourist attractions (Phelan, 2020). The goal of this type of travel is to avoid or minimize negative effects on the environment, resources, and ecosystems in a given area. The impact on the local population is one of the primary factors taken into account (Donohoe and Needham, 2006; Fennell, 2001; Wallace, 2019). However, few research works have been done on smart tourism and its implications on the destination. Therefore, this book chapter minimizes the gap by illustrating how the digital transformation of ecotourism can be an alternative source of tourism, therefore, we recreate a theory about "Eco-smart destination."

Mondino et al. (2019) stated that traveling to mostly natural locations where one may learn about the environment without endangering it. The review of ecotourism-related literature concentrates on various ecotourism viewpoints. It appears that Western perspectives have dominated earlier research to explore the essence of ecotourism for a country's economic growth. Therefore, scholars have not yet delved into other nations, including the Asian context, since this chapter took several examples from developing countries, like Bangladesh. Studies on ecotourism differ among nations because of several variables, including laws, customs, social structures, and natural areas (Wondirad, 2019). These research works ignore the implication of smart tools and attachment in ecological destinations. Therefore, we proposed an eco-smart economy, eco-smart society, green politics, and eco-smart technology in the paradigm of an ecotourism destination.

The process of dramatically transforming enterprises through the application of technology is known as digital transformation (Bastidas et al., 2020). Additionally, this transformation process is advancing at an incredibly fast rate (Cobo et al., 2011). There is now a plethora of diverse methods to boost productivity and efficacy in the workforce (Chatterjee et al., 2022). Numerous industries, including media, entertainment, financial services, retail, healthcare, and insurance, have witnessed the digital revolution (Cuomo et al.,2021). However, after the covid 19 crisis, most destinations now adopted digital involvement to track and minimize the crisis by introducing smart tourism on the spot (Akhtar et al., 2021; Deb and Nafi, 2020). They did not shade their experiments on how the ecotourism industry internalizes smart technology to capitalize on tourist experiences.

This chapter has demystified the embryonic theme of eco-smart destinations for the upliftment of the tourism industry. The widespread digital transformation has radically changed the travel and tourism sector (Cheng et al., 2018). The tourist business can implement novel and inventive solutions due to the swift advancement of the digital realm (Deb et al., 2022). Peer-to-peer communication is extremely important in the technologically advanced tourism sector and this kind of communication, together with the widespread use of smart devices, has completely changed scheduling, finances, and administration (Del et al., 2015). It has also created new opportunities to introduce cutting-edge sales and marketing technologies in the travel and tourism sector (Encalada et al., 2017). The tourism industry has benefited from the digital revolution by having access to new solutions such as cloud-based booking, information exchange, and experience sharing via digital platforms (Islam et al., 2023). Customers can virtually do anything with their smartphones because of the integration of smart gadgets, including ordering room service, checking in, and opening their doors, therefore, travelers can arrange their whole journey without hassling to anyone (Gómez et al., 2018). The digital transformation yields several advantages, including increased customer loyalty, increased standardization and centralization of operational procedures, and the creation of new, crucial opportunities for businesses through the reconfiguration of business models (Hein et al., 2020).

While there are numerous examples of businesses that have successfully used digital technologies to create new business models, other businesses are having difficulty with the process of digital transformation (Iansiti et al., 2014). As it requires huge investment and technical expertise to operate the entire system. Despite of this, digital technology has brought about a shift in how businesses interact with their clientele. When it came to booking a ticket, people used to physically visit travel agencies, but in modern times, the use of digital technologies (such as wearables, smartphones, virtual reality, and artificial intelligence) has drastically changed the travel industry (Mariani et al., 2023). These technologies have led to the development of global distribution networks and computerized reservation systems, which allow travel agencies, such as tour operators and airlines, to better serve their customers' needs and requirements in less time at reduced cost (Matt et al., 2016). Customers' expectations have altered as a result of the growth and dispersion of digital technology in the tourism sector (Hinings et al., 2018). For example, social networking, limitless mobile internet, transparency, and real-time video sharing give a unique essence of a smart destination (Li et al. 2016). However, the goal of the tourism sector is to capture consumers' attention by offering unique concepts and deals that facilitate sharing, operations, and communication (Almeida et al., 2020). More innovations in the tourism industry are being forced by new economic models based on collaborative consumption, such as the sharing economy (Happet et al., 2018).

In this modern era, the eco-smart tourism concept is now evolving by incorporating tourist, community, government, and business stakeholders in the realm of the technological web and globalization network. For example, websites for booking accommodations, such as Booking.com, function deeply; they utilize customers' locations to present them with current, personalized offers (Pencarelli, 2020). New economic systems have been created as a result of collaborative consumption-based businesses like Uber and Airbnb continuing to expand and offer greater services to customers (Podzharaya et al., 2022). Every day, new and inventive technologies are shaped to enhance the adventure and excitement of the tourism sector like digital payment methods. These cashless transactions and online documentation directly stop our deforestation process to produce paper and any other primary resources (Rana et al., 2022). However, the idea of eco-smart tourism is presumably equally a result of people's increasing general interest in the natural world and their understanding of how crucial it is to preserve its quality (Shoval et al., 2016). For example, Deb et al. (2024) emphasize the digital transformation in the hospitality industry, and knowledge sharing is a vital factor to disseminate information worldwide.

However, there is a need to provide a more precise definition of the term due to the abundance of synonyms and the growing acceptance of ecotourism and smart tourism as the umbrella term for this type of travel. Nevertheless, it is clear that the phrases "ecotourism" "smart tourism" and "eco-smart city" have many different connotations when one looks through the definitions assigned to them (Winkowska et al., 2019). There has been little research on the use of ICTs in ecotourism, even though a lot of previous research has focused on ICTs in tourism (Zubizarreta et al., 2016). However, ICTs can be a useful teaching and learning tool that can help to create sustainable environments at ecotourism destinations

(Gössling 2020; Nafi and Ahmed, 2017). This is an important way to develop sustainable tourism, which addresses the need to conserve local ecosystems. In light of this, it can be said that ICTs may advance the social, economic, and environmental aspects of ecotourism and that local involvement results in advantages for the community. However, the definition of eco-smart destination has not been explained in the previous literature, this study has defined the connotation of "eco-smart destination" with several diagrams.

6.3 RESEARCH METHODOLOGY

Using primary and secondary data analysis as its main method, this study will investigate the concept of eco-smart destinations. This is how the process is broken down:

1. **Primary Source:** The experts of the tourism and hospitality industry give their opinion, on how to develop eco-smart destinations for ecotourism areas.
2. **Academic Literature:** A comprehensive review of academic journals, conference proceedings, and books focusing on ecotourism, smart destinations, and the intersection of these two fields will be conducted. Reputable databases like Science Direct, Scopus, and Google Scholar will be used for efficient searching.
3. **Industry Reports and Publications:** Reports from organizations like the World Tourism Organization (UNWTO), The International Ecotourism Society (TIES), and research bodies focused on smart tourism will be consulted. These reports often provide valuable insights into industry trends and best practices.
4. **Government Websites and Policy Documents:** Government websites and policy documents related to ecotourism development and smart destination initiatives will be examined. This will offer details on specific strategies implemented in different regions and their effectiveness.
5. **News Articles and Case Studies:** News articles from reputable sources can offer current perspectives on the evolving relationship between ecotourism and smart destinations. Additionally, case studies of successful ecotourism destinations utilizing smart technologies will be analyzed to understand best practices.

An analysis of the data is based on recent views of the tourism industry and relies on scholars' statements.

6.4 CURRENT WORLD SCENARIO OF ECOTOURISM AND MARKET SIZE OF ECOTOURISM

The concept of ecotourism became more obvious when the UN declared the attainment of the sustainable development goal of 2030 (Lee et al., 2016). The global ecotourism scene is quite promising and has a lot of room to develop. The estimates suggest the global ecotourism market was valued at around USD 220.94 billion in 2022. The market is expected to experience significant growth, with projections ranging from a CAGR (Compound Annual Growth Rate) of 7.94%–14.5% by 2030. This translates to a potential market size exceeding USD 400 billion by the end of the decade which includes (Samal et al., 2023):

1. **Growing Demand:** Experiences that are both pleasurable and ecologically conscious are in greater demand from travelers. This promotes ecotourism, which emphasizes cultural heritage preservation and conservation.
2. **Sustainability Focus:** As people become more conscious of how tourism affects natural environments and contributes to climate change, they adopt ecotourism-based, sustainable travel strategies. Moreover, environmental, social, economic, and technological perspectives are the key issues for ecotourism growth and demand.
3. **Government Initiatives:** Several governments are putting rules and initiatives in place to encourage responsible tourism behavior in protected areas and nearby communities because they understand the economic benefits of ecotourism.

Therefore, ecotourism now takes a digital format to capitalize on the community's financial status, tourist experiences, digital hotel, and transportation management, and updated recreation facilities.

6.5 SMART DESTINATIONS: A BOOMING TREND

The concept of smart destinations is rapidly gaining traction worldwide, transforming how destinations and tourist hubs manage visitors and resources. Here's a breakdown of the current situation, market size, and tourist numbers. Smart destination practices can significantly enhance the ecotourism experience in several ways. First, the smart destination trend is witnessing significant growth across the globe when ecotourism takes a digital form. Second, destinations are implementing smart technologies to enhance visitor experiences and optimize resource management, and sustainability. However, the global smart tourism market is estimated to be worth around USD 412.2 billion in 2023 (Podzharaya et al., 2022). Experts predict significant growth, with projections ranging from CAGR 12.2%–17.5% by 2030. This translates to a potential market size exceeding USD 1 trillion within the next decade. Global tourism is expected to rebound strongly following the pandemic (Novera, et al., 2023). The UNWTO forecasts international tourist arrivals to reach between 910 million and 1.18 billion in 2023 (Novera, et al., 2023). This growth, coupled with the increasing demand for sustainable travel experiences, fuels the adoption of smart destination practices. Smart destinations are focusing on various areas like:

1. **Real-time Information:** Providing tourists with real-time information on public transport, attractions, and wait times through mobile apps and interactive displays.
2. **Traffic Management:** Implementing smart traffic systems to reduce congestion and improve air quality at the smart destination.
3. **Sustainability:** Utilizing sensors to monitor energy and water consumption and promoting sustainable practices amongst tourists and businesses.
4. **Accessibility:** Developing inclusive technologies to cater to visitors with disabilities.

The smart destination concept is still evolving, and there's no single standardized approach. Therefore, different destinations are implementing solutions tailored to their specific needs and resources. Challenges like data privacy concerns, cybersecurity risks, and the need for infrastructure development require careful consideration when implementing smart destination technologies.

6.6 ECO-SMART TOURISM AND ITS GOALS

Digital transformation is considered essential about the fact that digital technology enables an individual to find the solution to their traditional problems and is also inclined towards having digital solutions rather than traditional clarifications. The transformation implies that digital usages characteristically empower new kinds of development, creativity, and innovation in a particular domain as opposed to just enhancing and supporting conventional methods. However, an eco-smart destination implies the composition of ecotourism and technological involvement among stakeholders of ecological destinations. Equation 6.1 and figure demonstrate the entire process of digital transformation of ecotourism.

$$\text{(Ecotourism + Green Technological Tools + Smart Tourism)} \\ = \text{Eco-Smart Tourism}$$

EQUATION 6.1 Eco-Smart Tourism

Source: Authors Compilation

Ecotourism and smart destinations share a common goal of creating a more sustainable and enriching travel experience. Here's how ecotourism effectively fulfills the demands of eco-smart destinations:

1. **Reduced Environmental Impact:** Ecotourism prioritizes low-impact activities and responsible practices, minimizing waste generation and resource consumption. This aligns perfectly with the sustainability goals of smart destinations, which use technology to optimize resource management and promote eco-friendly behavior.
2. **Conservation Efforts:** Ecotourism revenue directly supports conservation efforts in protected areas. Smart destinations can leverage data collected through sensors and visitor tracking to identify areas that need additional protection and promote responsible ecotourism activities in these regions.
3. **Real-time Information:** Smart destination apps can provide ecotourists with real-time information on sustainable transportation options, eco-friendly accommodation, and responsible tour operators. This empowers ecotourists to make informed choices that align with both their travel interests and a sustainable approach.
4. **Educational Tools:** Smart destinations can utilize modern technologies to offer immersive educational experiences about local ecosystems and conservation efforts. This fosters a deeper

appreciation for nature among ecotourists and complements the educational components often found in ecotourism initiatives.

5. **Responsible Behavior:** Smart destinations can collect data on visitor flow in sensitive areas. This data can be used to develop targeted campaigns encouraging responsible behavior, such as staying on designated trails, minimizing waste generation, and respecting wildlife habitats. Ecotourism principles often emphasize these practices, making data a valuable tool for reinforcement.
6. **Economic Benefits:** Ecotourism directly benefits local communities by promoting locally-owned businesses and homestays. Smart destination platforms can leverage data to connect ecotourists directly with these local businesses, ensuring economic benefits reach those who need them most.
7. **Certification Programs:** Collaboration between smart destinations and ecotourism certification programs can ensure that ecotourism operators adhere to genuine sustainability practices.
8. **Community Involvement:** Smart technology can be used to empower local communities to participate in ecotourism initiatives and decision-making processes within smart destinations.

The concept of "Eco-Smart Destination" aims to expand the meaning of modern high-tech eco-tourism destination in two aspects: ecology and technical attachment. Moreover, it creates various types of networks, each of which is of special significance, such as ecological networks, eco-smart living networks, eco-smart government networks, eco-smart transport networks, and eco-smart economic and knowledge networks.

6.7 THE IMPLICATION OF THE ECO-SMART CONCEPT IN ECOTOURISM DESTINATIONS

Overall, smart destinations represent a promising approach to managing tourism in the future. As the market grows and technologies evolve, we can expect even more innovative solutions that benefit tourists and their destinations. Digital transformation of ecotourism triggers the sustainable management of environmental, social, economic, and political issues with the help of technological advancement. The following key points illustrate how eco-smart destinations can ensure a win-win situation among the stakeholders of the tourism and hospitality industry:

1. **Improved Information and Planning:**
   a. **Digital Platforms:** Ecotourists can access real-time information on local attractions, sustainable accommodation options, eco-friendly transportation choices, and responsible tour operators through mobile apps and interactive websites.
   b. **Personalized Recommendations:** Smart destinations can leverage data to provide personalized recommendations for ecotourism activities based on visitors' interests and preferences. This can help tourists discover hidden gems and lesser-known sustainable experiences.
2. **Enhanced Visitor Engagement and Education:**
   a. **Interactive Learning:** Smart destinations can utilize augmented reality (AR) and virtual reality (VR) technologies to provide immersive educational experiences about local ecosystems, conservation efforts, and indigenous cultures.
   b. **Real-time Monitoring:** Smart sensors can track visitor flow in sensitive areas, allowing for better management and protection of fragile ecosystems. Interactive displays can educate visitors on responsible behavior within these areas.
3. **Promoting Sustainable Practices:**
   a. **Resource Management:** Smart systems can monitor energy and water consumption in eco-lodges and tourist facilities, promoting resource efficiency and reducing environmental impact.
   b. **Waste Management:** Smart waste management solutions can encourage recycling and responsible waste disposal among both tourists and local communities.
   c. **Reduced Carbon Footprint:** Smart transportation options like bike-sharing programs, electric vehicle charging points, and real-time public transportation information can help ecotourists minimize their carbon footprint during their stay.
4. **Empowering Local Communities:**
   a. **Direct Booking Platforms:** Smart platforms can connect ecotourists directly with local guides, homestays, and artisans, promoting fair trade practices and ensuring economic benefits reach local communities.

b. **Cultural Preservation:** Smart technologies can be used to showcase local traditions, art forms, and languages, promoting cultural exchange and fostering respect for indigenous heritage.

Overall, smart destination practices can create a more enriching and responsible ecotourism experience. By providing access to information, promoting sustainable choices, and empowering local communities, these practices contribute to a positive long-term impact on both the environment and the local way of life.

6.8 ECO-SMART TECHNOLOGICAL TOOLS AND MARKETING STRATEGY

The key to a successful eco-smart destination lies in striking a balance between smart economic benefits, smart environmental protection, and cultural preservation. This requires careful planning, community involvement in technological sophistication, and robust certification programs to ensure eco-smart tourism practices are truly sustainable.

In today's modern world, technology plays a pivotal role in promoting sustainable tourism practices. Here's a glimpse into how tech is transforming the way we travel responsibly:

1. **Smart Destination Management Systems:** These systems collect data on tourist flow, energy consumption, and waste generation in destinations. This data can be used to optimize resource allocation, promote energy efficiency in tourist facilities, and implement targeted waste management strategies.
2. **Sharing Economy Platforms:** Online platforms like Airbnb can encourage responsible travel by promoting homestays and locally-owned accommodations, reducing the pressure on large-scale resorts that can have a larger environmental footprint.
3. **Carbon Footprint Calculators:** Online tools can help travelers calculate the carbon footprint of their trip and encourage them to offset emissions through carbon-neutral initiatives.
4. **Direct Booking Platforms:** Technology can facilitate direct connections between tourists and local communities. This allows travelers to support locally-owned businesses and ensures economic benefits reach those who need them most.

5. **Technological Infrastructure:** This forms the backbone of a smart destination. It includes things like:
    a. **Sensor Networks:** These collect real-time data on everything from air quality and traffic congestion to tourist footfall at popular attractions.
    b. **Big Data Analytics:** This collected data is then analyzed to understand tourist behavior and preferences, resource utilization, and potential areas for improvement.
    c. **Mobile Apps:** These provide visitors with personalized recommendations, real-time transportation information, wait times, and hidden gems to explore. They can also facilitate contactless ticketing and payments.
    d. **Mobile Apps and Platforms:** Ecotourists can leverage mobile apps to discover sustainable travel options, book eco-friendly accommodations, and access real-time information on responsible behavior in specific areas. These platforms can also promote responsible tour operators and local businesses.
    e. **360 Visualizations:** This feature gives a unique opportunity for the tourist to explore every micro detail from 360 angles from a remote location. For example, Club Med is now the first global tourism operator to present a collection of 360° videos showcasing its resorts, providing a novel and interactive experience of immersive virtual reality.
6. **Sustainability Focus:** Smart destinations use technology to promote eco-friendly practices. Examples include:
    a. **Smart Grids:** These optimize energy usage in public spaces and tourist areas, reducing the environmental footprint.
    b. **Waste Management Systems:** Smart sensors can monitor waste accumulation and optimize collection routes, minimizing waste and promoting a cleaner environment.
    c. **Traffic Management Systems:** Real-time traffic data helps tourists choose eco-friendly modes of transportation like cycling or public transit, reducing congestion and emissions.
7. **Tourist Knowledge:** Technology streamlines the tourist journey and personalizes exploration. Imagine exploring a coral reef virtually before diving in or using AR to identify local flora and fauna on a nature hike. VR/AR can enhance ecotourism

experiences by providing immersive educational content without disturbing sensitive environments.

a. **Interactive Information Kiosks:** These provide multilingual information on attractions, cultural events, and local businesses.
b. **Augmented Reality (AR) Experiences:** AR apps can overlay historical information on landmarks, enhancing cultural understanding.
c. **Virtual Reality (VR) Experiences:** These allow visitors to pre-experience activities and attractions, helping them plan their itinerary and manage expectations.

8. **Community Engagement:** Smart destinations don't just cater to tourists. They also use technology to:
   a. **Communicate with Residents:** Provide updates on city events, infrastructure improvements, and potential disruptions caused by tourism.
   b. **Promote Local Businesses:** Mobile apps can highlight locally owned restaurants, shops, and cultural experiences.
   c. **Gather Feedback:** Smart destinations use surveys or social media to understand resident concerns and preferences regarding tourism's impact on their community.

By working together, ecotourism and smart destinations can pave the way for a future of travel that benefits both tourists and the destinations (Figure 6.1).

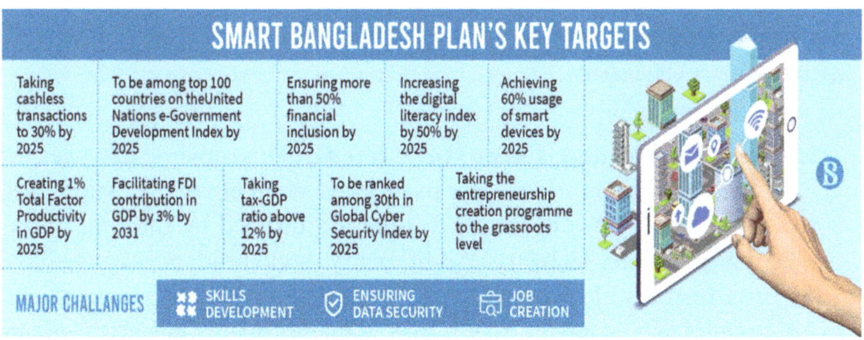

FIGURE 6.1 Smart Bangladesh.

Source: The Business Standard. https://www.tbsnews.net/economy/smart-bangladesh-govt-wants-raise-citizens-income-thru-tech-upskilling-742094

6.9 DIGITAL TRANSFORMATION OF ECO-TOURISM

Smart ecotourism technologies include advanced ICTs and other tactical tools used in the development of ecotourism solutions. Ecotourism could be greatly increased by high-tech information and communication technologies. These technologies can be used to manage destination resources, visualize sites and attractions, offer tourism education, collaborate with local businesses at the destination, provide information to tourists, and plan transportation routes to preserve and protect the natural environment (Figure 6.2).

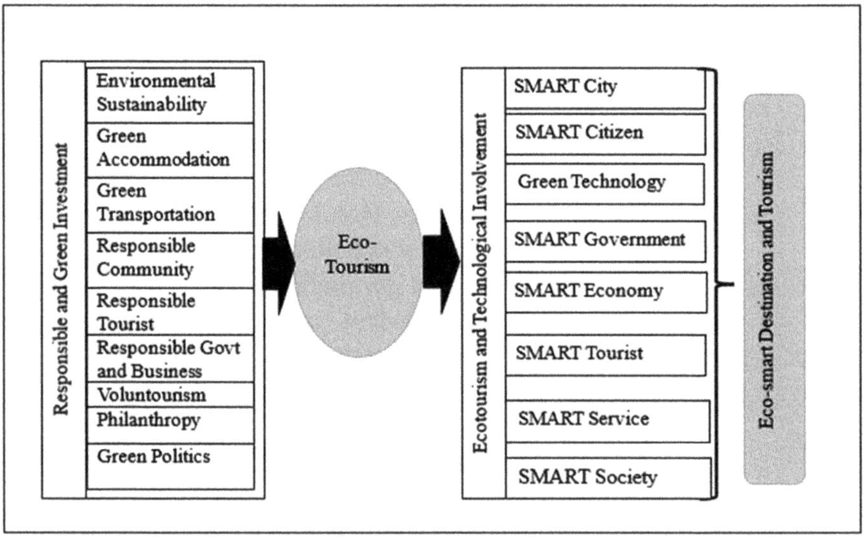

FIGURE 6.2 Digital transformation of eco-tourism.

Source: Authors Compilation.

1. **Responsible and Green Investment:** This covers the uses of renewable energy like solar panels and wind energy for supplying necessary electricity in the destination utilized by the transportation and hotel industry. Additionally, stakeholders must behave in a responsible way which contributes to protecting natural and manmade resources for future generations. Similarly, philanthropic activities in the destination will accelerate free economic influx for the community. Even, the "Green Politics" concept can be accentuated in a feasible way that can promote and protect

tourism resources without any argument where political parties must agree to preserve the resources sustainably. The government must embrace the “Green concept” to ensure profit maximization for the community people and enrich the tourist experience by assigning environmental rules and regulations at tourist spots.

2. **Ecotourism:** This term emphasizes environmental sustainability, green accommodation, green transportation, responsible community, responsible tourism and business. When we talk about ecotourism, we mean ecological tourism, where ecology refers to both the environment and society. Now, ecotourism embraces the e-tourism concept, where visitors are accustomed to emerging technologies and make regular use of them. It is composed of many distinct components which include social media, reservation systems, and applications for interacting and collaborating.
3. **Ecotourism and Technological Involvement:** “Smart” refers to the application of cutting-edge technologies to enhance different facets of society, including robots, drone technology, 3D printing, artificial intelligence, blockchain, big data, and the Internet of Things (IoT). For instance, the four main pillars of the “Smart Bangladesh” ecosystem are “Smart Citizen,” “Smart Government,” “Smart Society,” and “Smart Economy. “With a “digital first” mentality, the “Smart Citizen” initiative aims to empower Bangladeshi individuals by launching campaigns and putting in place extensive digital literacy initiatives. The “Smart Government” ought to establish fully paperless offices and highly customized service platforms in high-priority domains such as public safety, healthcare, education, and agriculture. Bangladesh automated more than 2,000 government functions, which was a strong start toward digitalizing government services, but a significant increase is still required to meet the 2041 deadline. Furthermore, a “smart society” is a civilization that employs innovation and technology to improve living conditions for its citizens. Citizens will embrace ethics, morals, and digital tolerance—a cashless transaction, signifying the use of online commerce by tourists and the intelligent community. Moreover, intelligent instruments can monitor general alterations in the flora and fauna of the ecotourism site; consequently, appropriate preemptive actions can safeguard environmental resources and expedite the establishment of wildlife breeding facilities (Hasan et al., 2024).

4. **Eco-smart Destination:** Once, the eco-destination encompasses the smart technologies in the tourist spots it will provide a unique essence to explore the eco-smart tourism concept virtually. However, this concept will be accepted universally, only when people, society, country, and stakeholders embrace technological sophistication to run the ecotourism destination. Originating from the notion of a smart city, the term "smart destination" has gained traction in response to the evolving demands of travelers. Even, sustainable tourism business development requires eco-friendly techniques and smart destinations. Moreover, eco-friendly practices can help conserve and preserve the natural environment by integrating technology in travel locations.

Finally, the eco-smart destination where both eco-tourism and technological involvement drive the country to transform the spot into an eco-smart socio-economy, eco-smart government, eco-smart destination, and eco-smart technology.

6.10 ECO-SMART DESTINATION BENEFITS AND CHALLENGES

The goal of technical progress is not only to provide a better quality of life through improved services but also to establish a chain that will guarantee effective service delivery. ICTs aid in the development of interactive services that link tourists with local businesses to provide prompt service. The benefits of eco-smart destination:

1. **Environmental Conservation and Smart Management:** Ecotourism prioritizes the conservation of natural areas. Revenue generated from ecotourism activities can be directed toward park management, wildlife protection programs, and habitat restoration.
2. **Smart Economic Benefits for Local Communities:** Ecotourism encourages responsible travel practices that directly benefit local communities. This can involve employing local guides, supporting locally-owned eco-lodges, and purchasing traditional crafts and products. This economic boost can contribute to poverty alleviation and empower local communities to participate in conservation efforts.
3. **Smart Education and Awareness:** Ecotourism fosters environmental awareness among both tourists and residents. Educational programs and immersive experiences can cultivate a deeper appreciation for nature and the importance of sustainable practices.

4. **Cultural Preservation and Virtual Reality:** Ecotourism can promote cultural exchange and respect for indigenous traditions. Tourists gain insights into local cultures and ways of life, while local communities benefit from the preservation of their heritage.
5. **Smart Waste Management and Reduced Environmental Impact:** Ecotourism generally promotes low-impact activities that minimize environmental damage. This can include minimizing waste generation, utilizing sustainable transportation options, and avoiding activities that disrupt wildlife habitats.

ICTs can therefore be utilized to educate people and raise awareness of ecotourism attractions through the use of digital media by digitally mimicking fragile or difficult-to-access sites. They can also take the place of some tourism-related activities that might cause the depletion of natural resources—the challenges of transforming ecotourism into an eco-smart destination are:

1. **Greenwashing:** Unfortunately, some tourism operators engage in "greenwashing," where they present themselves as eco-friendly without adhering to actual sustainable practices. This can mislead tourists and undermine the integrity of ecotourism.
2. **Over-Tourism:** Ecotourism's popularity can lead to over-tourism in certain destinations. This can put a strain on local resources, create infrastructure issues, and negatively impact the very ecosystems ecotourism aims to protect.
3. **Displacement of Local Communities:** In some cases, ecotourism development can lead to the displacement of local communities, particularly when large-scale resorts are built. This raises ethical concerns and underscores the importance of ensuring local communities benefit from tourism projects.
4. **Leakage of Revenue:** Revenue generated from ecotourism may not always reach local communities. Large corporations might own eco-lodges or control tour operations, resulting in a significant portion of the profits leaving the local economy.
5. **Limited Infrastructure:** Remote ecotourism destinations might lack basic infrastructure like proper waste management or sustainable energy sources. Addressing these limitations is crucial for ensuring the long-term sustainability of ecotourism projects.

6. **Accessibility:** Not all destinations have the necessary infrastructure to fully embrace smart technologies. Bridging the digital divide is crucial to ensure equitable access to the benefits of tech-driven sustainable tourism.
7. **Data Privacy:** As technology plays a bigger role, concerns around data privacy need to be addressed. Transparency and robust data security measures are essential to ensure responsible data collection and utilization.
8. **Technological Expertise:** Building capacity and training local communities in using new technologies is important to ensure they are not left behind in the shift towards sustainable tourism practices.

There is a big issue in preparing the general public to use future technologies. For instance, Smart Bangladesh has made significant progress toward a paperless workflow, and maintaining data security is yet another significant barrier.

6.11 ECO-SMART SUSTAINABILITY PILLARS

Overall, ecotourism serves as a perfect complement to smart destinations by prioritizing sustainability, promoting responsible behavior, and utilizing technology to enhance the visitor experience. Additionally, ecotourism effectively fulfills the demands of smart destinations seeking a more responsible and enriching tourism model (Figure 6.3).

The four pillars and their benefits from the demand side and supply side perspectives are illustrated in the subsections.

6.11.1 ECO-SMART DESTINATION

The green and environmentally friendly destination with all types of logistic technological support where community, business people, and tourists are connected with the technological web, software, and tools. Accommodation sectors use renewable energy to generate power and artificial intelligence to serve the tourists including online reservations, developing e-profiles, and virtual reality to enjoy real-time experiences. The transportation industry utilizes solar panels to transport passengers from one location to another location. This sector initiated Ubar and Patho services where tourists install software to frequently communicate with the service provider. Even, smart tourist police provide an emergency number to call for assistance. Moreover, tourists are staying under the umbrella of safety issues by Google surveillance method. Even, local authentic food is of top priority for the tourist to enjoy the meal, therefore AR gives a unique perception of how the food was prepared.

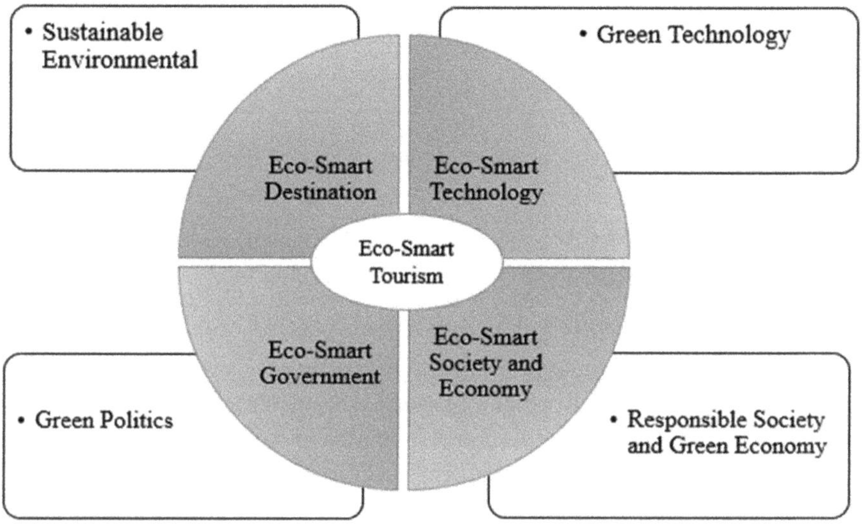

FIGURE 6.3 Eco-smart sustainability pillars.

Source: Authors compilation.

$$\text{Ecotourism} + \text{Smart (community} + \text{tourist} + \text{business} + \text{DMO)} = \text{Eco-smart Destination}$$

EQUATION 6.2 Eco-Smart Destination.

Source: Authors compilation.

6.11.2 ECO-SMART COMMUNITY

The community people are the prime stakeholders for restoring, nurturing, and practicing natural and cultural authenticity for decades. Therefore, their indigenous knowledge contributes to preserving their lifestyle, credo, cultural ceremony, artifacts, and environmental sustainability from generation to generation. When community people come in touch with digital attachments that will help them to update and store their indigenous knowledge for the next generation. However, this technological association not only improves their knowledge but also makes a way to initiate eco-smart tourism projects at the destination.

Ecological Indigenous knowledge + Smart community = Eco-smart Society

EQUATION 6.3 Eco-Smart Society.

Source: Authors compilation.

6.11.3 ECO-SMART TECHNOLOGY

Smart citizens, smart economy, smart government, and smart society will all be established on the foundation of digital connectivity, which will be essential leading factors for building an eco-smart destination. For example, artificial intelligence, 5G service, virtual reality, augmented reality, and robotics are the main agenda for digital transformation. Environmental technology, sometimes referred to as "green" or "clean" technology, is the creation of new sophistication to preserve, observe, or minimize detrimental effects on the environment and resource usage. Therefore, eco-smart technology refers to the use of technological tools, and software that makes a smart country by reducing harmful effects on people's health and environment.

Green Technology + Smart Destination = Eco-smart technology

EQUATION 6.4 Eco-Smart Technology.

Source: Authors compilation.

6.11.4 ECO-SMART ECONOMY

Google payment, e-pay, bikes, rocket bar code scanning method, and ATM service are the key indicators of a smart economy. This paperless transaction ensures both environmental sustainability and technological involvement among stakeholders of the destination. The government has prioritized cashless transactions, increasing per capita income, and decreasing poverty in its efforts to create a smart economy.

Green Investment + Cashless Transaction + Smart Documentation = Eco-smart economy

EQUATION 6.5 Eco-Smart Economy.

Source: Authors compilation.

6.11.5 ECO-SMART GOVERNMENT

Green politics and e-government are mostly enabled by advances in technology and better services. As a result, the emergence of intelligent/smart technologies like AI, ML, Big Data, IoT, Cloud Computing, mobile applications (Mobile Apps), and other newly developed ICTs has been seen as a great chance for governments to improve their operations and procedures in the most efficient manners. The goal of the smart government market is to increase citizen involvement, efficiency, and transparency through the digital transformation of the public sector. However, smart government utilizes digital identification, blockchain, artificial intelligence, and e-governance platforms to improve governance and service delivery. Eco-smart destinations apply technological involvement in the ministry of the government, regulatory, and statutory sectors to provide a one-stop online service by eliminating paper-based documents. Governments may solve complex social concerns and adjust to changing demands and trends through technology-driven policy innovations, regulatory reforms, and data-driven decision-making.

Green politics + Smart (Ministry + Service + Transparency) = Eco-smart Government

EQUATION 6.6 Eco-Smart Government.

Source: Authors compilation.

Moreover, "Eco-smart destination" provides a unique "eco-smart tourist experience" which refers to the idea that personalized, context-aware, and real-time experiences by utilizing green technological tools in their entire trip.

6.12 ECO-SMART DESTINATION LIFE CYCLE

The eco-smart destination life cycle consists of 10 stages which cover multi-stage concepts are (Table 6.1):

TABLE 6.1 Eco-Smart Destination Lifecycle

Stage 1	Identification of eco-destination
Stage 2	Community involvement
Stage 3	Green investment and development
Stage 4	Eco and Responsible Tourism
Stage 5	Digital transformation of eco-tourism
Stage 6	Eco-smart destination
Stage 7	Mass Tourism
Stage 8	Environmental degradation and misuse of technology
Stage 9	Stop the Tourism
Stage 10	Leave the destination for decades

Source: Authors compilation.

1. **Identification of Eco-Destination:** This stage delineates the identification of unscathed natural and cultural resources of any particular destination. Only drifters, explorers, or any institution of any governmental bodies take the responsibility to figure out nature-based tourism authenticity. However, in this stage engagement is not more than 1%, which is the reason tourist arrivals exist below the timeline. The community people are solely dependable on what they have like, collecting foods from natural resources, cultivating products by using indigenous knowledge, and so on. In this stage, these community people are detached from technological attachments and have no idea about tourism.
2. **Community Involvement:** When an eco-destination is identified with specific community people who are nurturing their natural and cultural resources. The private and public stakeholders make

them understand the paramount importance of tourism which will bring a financial influx for the community people.

3. **Green Investment and Development:** To make a green investment in tourism, investors must put environmental sustainability first. It includes making investments in environmentally friendly lodging, supporting ethical wildlife tourism, and supporting low-impact modes of transportation. For example, eco-resorts will be developed with renewable energy sources, sustainable waste management, green technological tools, green transportation, the use of jute bags, and eco-friendly recreation facilities. However, the government will restrict the tourist arrival until it accomplishes its investment and development. Therefore, tourist influx is located below the timeline.
4. **Eco and Responsible Tourism:** Opportunities are created for local communities by sustainable tourism. It emphasizes on contributing to the local community people by providing time, talent, and treasurer. Tourist arrival in this stage is gradually growing, therefore, it has existed above the timeline.
5. **Digital Transformation of Eco-Tourism:** In this phase, tourism stakeholders must be converted into smart forms by the proper utilization of technological tools and advancement. Digital training and literacy are required to educate the tourism stakeholders about technological features. Even, Digital destination marketing will be the key factor in promoting and branding eco-smart destinations.
6. **Eco-Smart Destination:** This destination is embellished with natural, cultural, and technological sophistication to cater to both tourists and the community. However, the unique feature of eco-smart destinations where tourist will enjoy the scenic beauty and capture crucial moments by adding additional facets to their videos and photos. This opportunity makes the destination more allured to the tourist. After several years, the eco-smart destination moves towards mass tourism.
7. **Mass Tourism:** When the eco-smart destination is thriving towards mass tourism then it automatically creates a huge tourist influx in the destination, therefore, destination carrying capacity and technological service will be dismantled.
8. **Environmental degradation and misuse of technology:** The long-term effects of digital transformation on the economy, politics, and society that could result in electronic imperialism in popular

tourism destinations including online cybercrime and e-casinos. Even, environmental resources are exhausted, and synthetic cultural products are taken place in the replacement of traditional authenticity due to mass tourism. Excessive dependence on technology leads the destination towards artificial society and environment.

9. **Stop the Tourism:** In this stage, the eco-smart destination will strive to get back its original form. Therefore, accessibility for the tourists will be restricted for the years. Specific tourism organizations will get access to evaluate the destination's current status, ensuring zero accessibility in eco-smart destinations, creating alternative sources of income for the community people for decades, and renovating the ecological balance by converting eco-smart destinations into ecological facilities.
10. **Leave the Destination for Decades:** This phase will ignite the renovation of ecological resources where community people move towards the first phase by utilizing minimal technological tools and uses of indigenous knowledge to recreate the ecological balance. This is how the destination will get its natural authenticity.

Finally, it is highly noticeable that when ecotourism transforms into an eco-smart destination then tourist arrival is propagating. Therefore, more digital marketing strategy is required to make it acceptable for everyone in the world. Figure 6.4 shows that it will take 5–10 years to reach the ecotourism stage from the identification of an eco-tourism destination. It will require 2–5 years to jump into the mass tourism phase and take 20 years to get back to its original stage.

6.13 MANAGERIAL AND PRACTICAL IMPLICATIONS FOR DMO, ECOTOURISM OPERATORS, AND TECHNOLOGY DEVELOPERS

The convergence of ecotourism and smart destinations presents exciting opportunities to create a more responsible and enriching tourism sector. Here are some key managerial and practical implications that have been recommended to unlock this synergy:

1. **For Destination Management Organizations (DMOs):**
   a. **Develop a Smart Ecotourism Strategy:** Conduct a comprehensive assessment of the destination's natural and cultural resources. Furthermore, integrate ecotourism principles with

smart destination technologies to create a strategic plan that promotes responsible tourism practices.

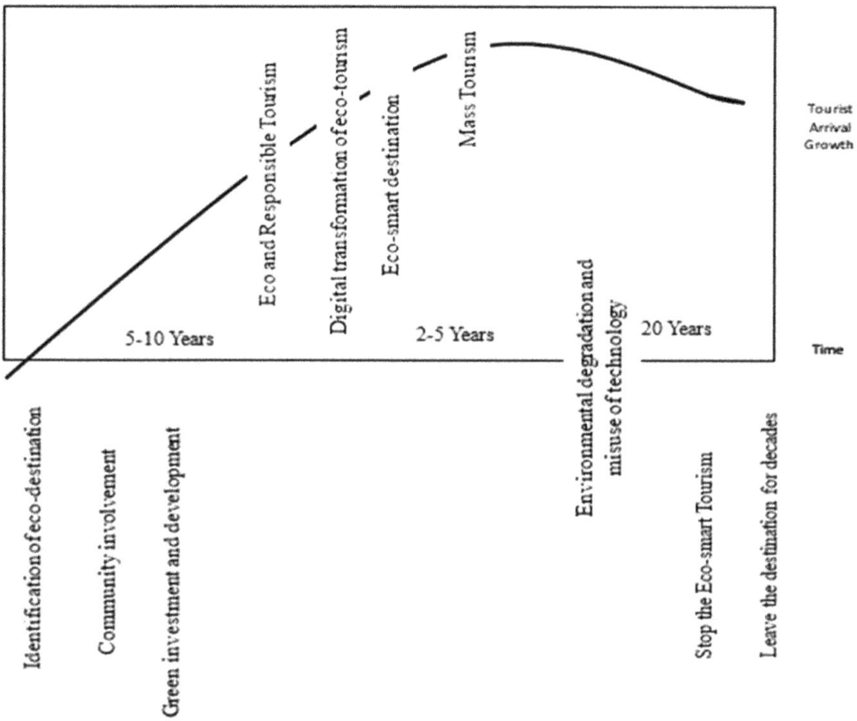

FIGURE 6.4 Eco-smart destination life cycle (tourist arrival and time matrix).
Source: Authors compilation.

b. **Invest in Sustainable Infrastructure:** Develop eco-friendly transportation options like bike-sharing programs and public electric vehicle charging stations. Additionally, promote the use of renewable energy sources in tourist accommodations.
c. **Partner with Ecotourism Operators:** Collaborate with certified ecotourism operators to ensure responsible tour practices and empower local communities. Moreover, utilize smart platforms to connect tourists directly with these businesses.
d. **Embrace Data-Driven Decision Making:** Leverage data collected through smart technologies to monitor tourist flow in

sensitive areas and identify opportunities to improve resource management and promote responsible behavior.

e. **Promote Education and Awareness:** Implement educational programs about local ecosystems, conservation efforts, and responsible travel practices.

2. **For Ecotourism Operators:**
   a. **Embrace Smart Technologies:** Utilize mobile apps to provide ecotourists with real-time information on sustainable activities, responsible behavior in natural areas, and local community initiatives.
   b. **Invest in Green Technology:** Upgrade facilities to promote energy and water efficiency by utilizing sustainable waste management solutions and educating tourists on waste reduction practices.
   c. **Engage with Smart Destination Platforms:** Register with smart destination platforms to connect with potential ecotourists and promote responsible ecotourism packages.
   d. **Support Data Collection Efforts:** Collaborate with smart destination initiatives to provide anonymized data on tourist behavior in eco-sensitive areas. This data can be used to refine ecotourism practices and promote responsible behavior.
   e. **Promote Cultural Exchange:** Integrate local culture and traditions into ecotourism experiences through capitalizing smart technologies to showcase local arts and crafts and empower local communities to participate in tourism activities.
3. **For Technology Developers:**
   a. **Focus on Sustainability:** Design smart destination technologies with sustainability in mind by developing solutions that promote energy efficiency, resource management, and waste reduction in tourism operations.
   b. **Ensure Accessibility:** Make smart destination technologies accessible to all visitors, including those with disabilities.
   c. **Prioritize Data Privacy:** Uphold robust data security measures and ensure transparency in data collection and usage.
   d. **Develop User-Friendly Interfaces:** Design user-friendly mobile apps and interactive platforms that cater to diverse tech experience levels among ecotourists.

e. **Partner with Ecotourism Stakeholders:** Partner with ecotourism operators and DMOs to develop technologies that address the specific needs and challenges of ecotourism destinations.

By implementing these recommendations, all stakeholders can contribute to a future where ecotourism and smart destinations work hand-in-hand to create a more sustainable and enriching travel experience for all. This collaborative approach can ensure the long-term health of our planet's natural wonders while fostering cultural exchange and economic prosperity for local communities.

6.14 CONCLUSION

In conclusion, the convergence of ecotourism and smart destinations presents a compelling vision for the future of travel. Ecotourism, with its focus on environmental responsibility and cultural preservation, aligns perfectly with the sustainability goals of smart destinations that leverage technology for resource optimization and visitor management. This synergy offers a multitude of benefits, likesmart technologies can elevate ecotourism experiences by providing real-time information, interactive educational tools, and promoting responsible behavior. Moreover, ecotourism principles, coupled with data-driven insights from smart destinations, can minimize environmental impact, optimize resource management, and promote responsible visitor behavior. Similarly, ecotourism and smart destinations working together can ensure that economic benefits from tourism reach local communities through direct booking platforms and support for locally-owned businesses. Furthermore, bridging the digital divide and fostering responsible data practices are essential for successful implementation. We can create a future where travel fosters cultural exchange, environmental responsibility, and a deeper appreciation for the wonders of our planet by embracing innovation, collaborating effectively, and prioritizing sustainability. The future holds immense potential for travelers seeking enriching experiences, destinations striving for sustainability, and local communities looking to benefit from eco-smart tourism practices.

KEYWORDS

- **artificial intelligence (AI)**
- **augmented reality**
- **eco-smart destination**
- **eco-smart society**
- **ecosystem**
- **smart citizen**
- **smart tourism**
- **virtual reality**

REFERENCES

Akhtar, N., Khan, N., Mahroof Khan, M., Ashraf, S., Hashmi, M. S., Khan, M. M., & Hishan, S. S. (2021). Post-COVID-19 tourism: Will digital tourism replace mass tourism? *Sustainability*, *13*(10), 5352.

Alkhatib, A. W., & Valeri, M. (2024). Can intellectual capital promote the competitive advantage? Service innovation and big data analytics capabilities in a moderated mediation model. *European Journal of Innovation Management*, *27*(1), 263–289.

Almeida-Santana, A., David-Negre, T., & Moreno-Gil, S. (2020). New digital tourism ecosystem: Understanding the relationship between information sources and sharing economy platforms. *International Journal of Tourism Cities*, *6*(2), 335–345.

Bastidas-Manzano, A., Sánchez-Fernández, J., & Casado-Aranda, L. (2020). The past, present, and future of smart tourism destinations: A bibliometric analysis. *Journal of Hospitality and Tourism Research*, *45*, 529–552.

Bina, O. (2013). The green economy and sustainable development: An uneasy balance? *Environment and Planning C: Government and Policy*, *31*(6), 1023–1047.

Brilhante, O., & Klaas, J. (2018). Green city concept and a method to measure green city performance over time applied to fifty cities globally: Influence of GDP, population size and energy efficiency. *Sustainability*, *10*(6), 2031.

Calzada, I. (2016). Smartness for prosperity: Brazil and UK. *Future Cities Catapult*.

Caragliu, A., & Del Bo, C. (2012). Smartness and European urban performance: Assessing the local impacts of smart urban attributes. *Innovation: The European Journal of Social Science Research*, *25*(2), 97–113.

Chatterjee, S., Chaudhuri, R., Vrontis, D., & Jabeen, F. (2022). Digital transformation of organization using AI-CRM: From micro foundational perspective with leadership support. *Journal of Business Research*, *153*, 46–58.

Cheng, M., & Foley, C. (2018). The sharing economy and digital discrimination: The case of Airbnb. *International Journal of Hospitality Management*, *70*, 95–98.

Cheng, S., & Xie, S. (2011). Making eco-smart cities in the future. In *2011 International Conference on Consumer Electronics, Communications and Networks* (pp. 1380–1383). IEEE.

Cobo, M. J., López-Herrera, A. G., Herrera-Viedma, E., & Herrera, F. (2011). Science mapping software tools: Review, analysis, and cooperative study among tools. *Journal of the American Society for Information Science and Technology*, *62*(7), 1382–1402.

Colglazier, W. (2015). Sustainable development agenda: 2030. *Science*, *349*(6252), 1048–1050.

Cuomo, M. T., Tortora, D., Foroudi, P., Giordano, A., Festa, G., & Metallo, G. (2021). Digital transformation and tourist experience co-design: Big social data for planning cultural tourism. *Technological Forecasting and Social Change*, *162*, 120345.

Deb, S. K., & Nafi, S. M. (2020). Impact of COVID-19 pandemic on tourism: Recovery proposal for future tourism. *GeoJournal of Tourism and Geosites*, *33*.

Deb, S. K., Kuri, B. C., & Nafi, S. M. (2024). Application of knowledge management in tourism and hospitality industry: A sustainable approach. In *Knowledge Management and Knowledge Sharing: Business Strategies and An Emerging Theoretical Field* (pp. 99–116). Cham: Springer Nature Switzerland.

Deb, S. K., Nafi, S., & Valeri, M. (2022). Promoting tourism business through digital marketing in the new normal era: A sustainable approach. *European Journal of Innovation Management*. https://doi.org/10.1108/EJIM-04-2022-0218 (accessed on 8th July 2025).

Del Chiappa, G., & Baggio, R. (2015). Knowledge transfer in smart tourism destinations: Analyzing the effects of a network structure. *Journal of Destination Marketing and Management*, *4*(3), 145–150.

Demaziere, C. (2020). Green city branding or achieving sustainable urban development? Reflections of two winning cities of the European Green Capital Award: Stockholm and Hamburg. *TPR: Town Planning Review*, *91*(4).

Dong, H., Fujita, T., Geng, Y., Dong, L., Ohnishi, S., Sun, L., & Fujii, M. (2016). A review on eco-city evaluation methods and highlights for integration. *Ecological Indicators*, *60*, 1184–1191.

Donohoe, H. M., & Needham, R. D. (2008). Internet-based ecotourism marketing: Evaluating Canadian sensitivity to ecotourism tenets. *Journal of Ecotourism*, *7*(1), 15–43.

Encalada, L., Boavida-Portugal, I., Ferreira, C., & Rocha, J. (2017). Identifying tourist places of interest based on digital imprints: Towards a sustainable smart city. *Sustainability*, *62*, 120–131.

Ercan, F. (2023). Smart tourism destination: A bibliometric review. *European Journal of Tourism Research*, *34*, 3409–3409.

Fennell, D. A. (2001). A content analysis of ecotourism definitions. *Current Issues in Tourism*, *4*(5), 403–421.

Gómez, L. M., Fernández, L. R., & Navío-Marco, J. (2018). Application of communication technologies (ICT) within the tourism industry in the European Union. *Tourism: An International Interdisciplinary Journal*, *66*(2), 239–245.

Gössling, S. (2020). Technology, ICT and tourism: From big data to the big picture. *Journal of Sustainable Tourism*, *29*(5), 849–858.

Happ, É., & Ivancsó-Horváth, Z. (2018). Digital tourism is the challenge of future–A new approach to tourism. *Knowledge Horizons. Economics*, *10*(2), 9–16.

Hasan, S., Halim, M. A., Md. Nafi, S., & Amin, M. R. (2024). Perception and attitude towards wildlife-based tourism: Bangladesh perspective. *Global Sustainability Research*, *3*(1), 61–76. https://doi.org/10.56556/gssr.v3i1.733 (accessed on 8th July 2025).

Hein, A., Schreieck, M., Riasanow, T., Setzke, D. S., Wiesche, M., Böhm, M., & Krcmar, H. (2020). Digital platform ecosystems. *Electronic Markets*, *30*, 87–98.

Hinings, B., Gegenhuber, T., & Greenwood, R. (2018). Digital innovation and transformation: An institutional perspective. *Information and Organization*, *28*(1), 52–61.

Iansiti, M., & Lakhani, K. R. (2014). Digital ubiquity: How connections, sensors, and data are revolutionizing business. *Harvard Business Review*, *92*(11), 90–99.

Islam, Q. T., Ahmed, J. U., & Sayed, A. (2023). Digitization and integration of Sustainable Development Goals (SDGs) in emerging economies. In *Fostering Sustainable Businesses in Emerging Economies: The Impact of Technology* (pp. 23–38).

Lai, P. H., & Nepal, S. K. (2006). Local perspectives of ecotourism development in Tawushan Nature Reserve, Taiwan. *Tourism Management*, *27*(6), 1117–1129.

Lavrinenko, O., Ignatjeva, S., Ohotina, A., Rybalkin, O., & Lazdans, D. (2019). The role of green economy in sustainable development (case study: The EU states). *Entrepreneurship and Sustainability Issues*, *6*(3), 1113.

Lee, B. X., Kjaerulf, F., Turner, S., Cohen, L., Donnelly, P. D., Muggah, R., & Gilligan, J. (2016). Transforming our world: Implementing the 2030 Agenda through Sustainable Development Goal indicators. *Journal of Public Health Policy*, *37*, 13–31.

Li, F., Nucciarelli, A., Roden, S., & Graham, G. (2016). How smart cities transform operations models: A new research agenda for operations management in the digital economy. *Production Planning and Control*, *27*(6), 514–528.

Mariani, M. M., Borghi, M., & Laker, B. (2023). Do submission devices influence online review ratings differently across different types of platforms? A big data analysis. *Technological Forecasting and Social Change*, *189*, 122296.

Matt, C., Hess, T., Benlian, A., & Wiesbock, F. (2016). Options for formulating a digital transformation strategy. *MIS Quarterly Executive*, *15*(2), 123–139.

Mondino, E., & Beery, T. (2019). Ecotourism as a learning tool for sustainable development: The case of Monviso Transboundary Biosphere Reserve, Italy. *Journal of Ecotourism*, *18*(2), 107–121.

Nafi, S. M., & Ahmed, T. (2017). Sustainable tourism in Saint Martin Island: An observation on young tourist perception and awareness level. *Journal of Humanities and Social Science*, *22*(10), 73–80.

Novera, C. N., Ahmed, Z., Kushol, R., Wanke, P., & Azad, M. A. K. (2022). Internet of Things (IoT) in smart tourism: A literature review. *Spanish Journal of Marketing-ESIC*, *26*(3), 325–344.

Pencarelli, T. (2020). The digital revolution in the travel and tourism industry. *Information Technology and Tourism*, *22*, 455–476.

Pereira, G. V., Parycek, P., Falco, E., & Kleinhans, R. (2018). Smart governance in the context of smart cities: A literature review. *Information Polity*, *23*(2), 143–162.

Phelan, A., Ruhanen, L., & Mair, J. (2020). Ecosystem services approach for community-based ecotourism: Towards an equitable and sustainable blue economy. *Journal of Sustainable Tourism*, *28*(10), 1665–1685.

Podzharaya, N., & Sochenkova, A. (2022). Growth of modern technologies usage in tourism due to COVID-19: Examples and forecasts. In *2022 11th Mediterranean Conference on Embedded Computing* (pp. 1–6). IEEE.

Rana, R. L., Adamashvili, N., & Tricase, C. (2022). The impact of blockchain technology adoption on tourism industry: A systematic literature review. *Sustainability*, *14*, 7383.

Samal, R., & Dash, M. (2023). Ecotourism, biodiversity conservation and livelihoods: Understanding the convergence and divergence. *International Journal of Geoheritage and Parks*, *11*(1), 1–20.

Shaikh, A. (2023). Smart Bangladesh: Govt wants to raise citizens' income through tech upskilling. *The Business Standard.* Available at: https://www.tbsnews.net/economy/smart-bangladesh-govt-wants-raise-citizens-income-thru-tech-upskilling-742094 (accessed on 8th July 2025).

Shanker, D. (2008). ICT and tourism: Challenges and opportunities. In *Proceedings of The Conference on Tourism in India – Challenges Ahead*, 15–17 May. 2008, IIMK.

Shoval, N., & Ahas, R. (2016). The use of tracking technologies in tourism research: The first decade. *Tourism Geographies*, *25*(2), 264–286.

Stock, W. G. (2011). Informational cities: Analysis and construction of cities in the knowledge society. *Journal of the American Society for Information Science and Technology*, *62*(5), 963–986.

Suárez-Eiroa, B., Fernández, E., Méndez-Martínez, G., & Soto-Oñate, D. (2019). Operational principles of circular economy for sustainable development: Linking theory and practice. *Journal of Cleaner Production, 952–961.*

Thibault, A., & Lavigne, M. A. (2014). The "leisure city": What are we talking about? *World Leisure Journal*, *56*(1), 2–5.

Vishnivetskaya, A., & Alexandrova, E. (2019). "Smart city" concept. Implementation practice. In *IOP Conference Series: Materials Science and Engineering* (Vol. *497*(1), pp. 012019). IOP Publishing.

Wallace, R. (2019). Ecotourism in Asia: How strong branding creates opportunity for local economies and the environment. In *Positioning and Branding Tourism Destinations for Global Competitiveness* (pp. 192–211). IGI Global.

Winkowska, J., Szpilko, D., & Pejić, S. (2019). Smart city concept in the light of the literature review. *Engineering Management in Production and Services*, *11*(2), 70–86.

Wondirad, A. (2019). Does ecotourism contribute to sustainable destination development, or is it just a marketing hoax? Analyzing twenty-five years contested journey of ecotourism through a meta-analysis of tourism journal publications. *Asia Pacific Journal of Tourism Research*, *24*(11), 1047–1065.

Wondirad, A., Tolkach, D., & King, B. (2020). Stakeholder collaboration as a major factor for sustainable ecotourism development in developing countries. *Tourism Management*, *78*, 104024.

Yan, J., Yang, J., Zhu, F., & Teng, Z. (2021). Green city and government ecological environment management based on Zigbee technology. *Environmental Technology & Innovation*, *23*, 101711.

Zubizarreta, I., Seravalli, A., & Arrizabalaga, S. (2016). Smart city concept: What it is and what it should be. *Journal of Urban Planning and Development*, *142*(1), 04015005.

CHAPTER 7

PROMOTING SMART TOURISM USING DIGITAL MARKETING AND QR CODES

SANTUS KUMAR DEB*

Department of Tourism and Hospitality Management, University of Dhaka, Dhaka, Bangladesh

*Corresponding author

ABSTRACT

The purpose of the study is to find out the impact of digital marketing and QR codes on the development of tourism. Data from secondary sources, like review chapters, were used with a systematic approach to gather and analyze information from relevant academic journals and sources. To accomplish the qualitative part of the chapter, 10 industry experts, such as tour operators, travel guides, hotel owners, tourism destination managers, and travel agency owners, were interviewed through focus group discussions (FGD). Ninety-minute time slots and semi-structured questionnaires were used for this purpose. This study found that digital marketing makes it possible to gather and analyze customer behavior data, which aids in the decision-making process for travel agencies, and QR codes are utilized for contactless ticketing, reservations, and check-ins at all types of establishments, including museums, events, and accommodations. The tourism industry has seen significant changes due to the fact that it is increasingly marketing itself through online media, especially with regard to smart tourism.

Smart Tourism: Managing Technological Revolution in the Tourism Industry. Santus Kumar Deb (Ed.)

DOI: 10.1201/9781779641090-7

7.1 INTRODUCTION

In today's digital era, the tourism industry has witnessed a significant transformation in the way it promotes its products and services. The emergence of technology has revolutionized marketing strategies, shifting from traditional to digital methods. Digital marketing indicates promoting products and services besides attracting customers in the tourism market by the application of digital media (Yasmin et al., 2015). Yasmin et al. (2015) describe digital marketing as the use of an electronic platform to promote items and services. Digital marketing is a very new scientific discipline that has grown rapidly and is often seen as the future of marketing (Magano et al., 2020).

The application of digital marketing techniques has become crucial for tourism businesses to increase their reach and attract potential tourists (Deb et al., 2022). By incorporating digital techniques, tourism businesses can utilize various media and tools to promote their products, establish their brand, attract new consumers, and ultimately boost their profits. Digital marketing techniques have opened up new opportunities for tourism businesses, enabling them to expand their reach and stay competitive in the modern tourism industry. Schal et al. (2005) stated that digital marketing is a technology that enables a company to engage with its customers on a regular basis. This technique not only facilitated the promotion of tourism goods but also saved time for potential tourists in their decision-making process (Floričić et al., 2023). Besides, there are many tools and techniques used in digital marketing. This chapter focuses on the uses of digital marketing tools or strategies in the tourism industry to meet the goal of smart tourism.

Buhalis and Park (2021) defined digital marketing as the use of any digital platform by an organization to advertise, market, or promote goods and services to customers. Digital marketing makes use of many websites, mobile devices, social media, search engines, and related platforms. The enormous field of digital marketing has a wealth of opportunities, especially for skilled persons who know how to take use of it. Through sincere emotional encounters that fortify brand loyalty, it acts as a bridge between brands, their audiences, and customers, enabling them to understand their brands better (Buhalis and Park, 2021). The digital economy has experienced a rapid expansion of innovations in recent years. Modern technologies and methods, including cloud computing, open-source systems, platform technology, high-speed and efficient search engines, smartphones, and other devices based on wireless and wired network infrastructure, have all significantly changed the perception of many industries (Bassano et al., 2019).

Smart tourism has become a prominent field of study in the tourism industry, fueled by advancements in technology and shifts in consumer behavior. As per the research conducted by Buhalis and Amaranggana (2015), smart tourism can be described as the application of information and communication technology (ICT) to create inventive tools and solutions for the tourism industry (Figure 7.1).

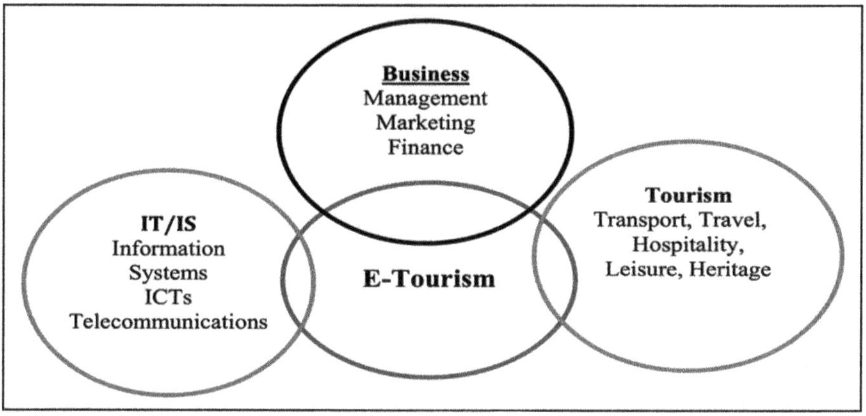

FIGURE 7.1 e-Tourism concept and domains.

Source: Adapted from Buhalis and Amaranggana (2013).

The definition emphasizes the crucial impact of technology on shaping the tourism experience. In the early 21st century, the widespread availability of Internet access and the introduction of mobile devices brought about a significant transformation in the way individuals strategize and arrange their travel arrangements. In their study, Dwyer et al. (2016) highlight the significant impact of smartphones and mobile apps on travelers. They emphasize how these technological advancements have empowered individuals by granting them immediate access to information and services, enabling them to make informed decisions in real-time, and ultimately enhancing their overall travel experience.

7.2 BACKGROUND OF DIGITAL MARKETING

In the 1990s, the term "digital marketing" was first used to describe the shift in marketing towards the digital era. The introduction of the internet and the Web 1.0 platform marked the beginning of the digital age, with users unable to share information (Khwaja et al., 2020). The first clickable banner appeared online in 1993, and Hot Wired bought banner adverts to use in their advertising. New technologies began to appear in the digital marketplace in 1994, leading to the debut of Yahoo in 1996. Google was founded in 1998, followed by Microsoft's MSN search engine and Yahoo's web search. The Internet bubble burst in 2006, leaving smaller search engines behind or completely destroyed, making room for industry titans like Google and Yahoo. The first significant upswing in digital marketing occurred in 2006 when search engine traffic increased to over 6.4 billion in a single month (Sethi, 2019).

Web 2.0 shifted people from passive users to active participants, allowing communication between businesses and other users. Internet terms like "super information highway" started to be used, leading to an exponential expansion in information flow and channels used by digital marketers. By 2004, Internet advertising and marketing generated around USD 2.9 billion in revenue in the US alone. Social networking sites quickly appeared, with MySpace and Facebook being the first to appear (Fuentes et al., 2017).

Another significant turning point in digital marketing history was the cookie. Marketers started searching for alternative strategies to profit from the nascent technology, such as monitoring user behaviors to customize marketing materials and promotions. Over time, cookies have evolved to provide marketers with more options for gathering actual user data. Social media posting is the most popular online activity in the United States, with the average American spending 37 minutes per day on it (Zhang et al., 2019). According to Bassano et al. (2019), the top three social networking platforms for marketers are LinkedIn, Twitter, and Facebook.

7.3 EVOLUTION OF E-TOURISM TOWARDS SMART TOURISM

In the tourism industry, technological advancements serve as the foundation for new methods and innovative business models. Computerized Reservation Systems (CRS) developed in the 1970s, Global Distribution

Systems (GDS) emerging at the end of the 1980s, the Internet in the 1990s, and mobile technologies in the 2000s are among the most significant technological advances. In recent years, developments in information and communication technologies have generated drastic and unanticipated shifts in society (Ma and Li, 2021).

The World Tourism Organization defines smart tourism as clean, green, ethical, and high-quality service at all levels. The concept appears to be explained in the second approach in 2015, which includes information and communication technology as well as the concept of intelligent destination (*Source:* UNWTO, 2019). According to the UN World Tourist Organization Conference on Smart Destinations (2019), the notion of smart destinations represents the future of tourist development. Smart destinations have been defined as intelligent, sustainable, and competitive tourism destinations that use data-linked resources, such as geo-referenced data, massive data, and the Internet of Things, with all stakeholders involved.

The European Union's Smart Tourism Directive, which aims to "promote smart tourism in the EU, network and strengthen tourism destinations, and facilitate the exchange of best practices," gave rise to the concept of smart tourism. Let us look at some examples of smart tourism (Table 7.1).

TABLE 7.1 Summary of Smart Tourism: Foundations and Developments

Focal Point	E-tourism	Smart Tourism
Specialty	Digital	Bridging digital and physical
central part of technology	Websites	Sensors and Smartphones
Voyage phase	Pre- and post- travel	During trip
Lifeblood	Information	Big data
Concept	Interactivity	Technology-mediated co-creation
Exchange	B2B, B2C, C2C	Public-private-consumer collaboration

Source: Gretzel et al. (2015).

7.4 LIAISON BETWEEN DIGITAL MARKETING FOR TOURIST DECISION MAKING

To begin, an "online presence" needed to be established and propagated in order for digital tourism to exist in Hungary. In 2012, the former Hungarian Tourism Co. summed up the share of "online population" in each exporting market. In 1997, 1% of the Hungarian population had access to the Internet (Molz, 2012). Digital marketing has transformed traditional marketing strategies in the tourist business (Buhalis and Law, 2008). This move to digital platforms has been fueled by shifting traveler patterns, with more visitors depending on online resources and social media for trip preparation (Buhalis and Law, 2008). Tourism organizations may boost their online presence and engage with potential passengers using techniques including SEO, content marketing, and social media advertising. Kumar et al. (2016) also mentioned that Facebook, Instagram, Twitter, and YouTube have all become important communication outlets for the tourist business. Real-time interaction on social media allows tourist organizations to directly contact passengers, give rapid support, respond to concerns, and resolve difficulties.

In addition, social media platforms have had a profound impact on consumer behavior within the tourism sector. In their study, Kim and Kim (2020) highlight the significant impact of social media on travel aspirations, destination choices, and travel planning. They emphasize the role of user-generated content and peer recommendations in this process. Over the past few years, there has been a growing interest in the idea of smart destinations. This approach emphasizes the use of technology to improve tourism management and promote sustainable practices in various destinations. As Xiang et al. (2015) suggest, smart destinations utilize ICT to optimize tourism operations, elevate visitor experiences, and mitigate adverse environmental effects (Figure 7.2).

Furthermore, intelligent tourism initiatives hold the potential to support the development of sustainable tourism through the promotion of responsible travel practices and the cultivation of community engagement. In their study, Floričić et al. (2023) propose that the implementation of smart tourism strategies can effectively address the challenge of balancing economic growth, environmental conservation, and social well-being in tourism destinations. This approach aims to ensure the long-term sustainability and viability of these destinations. In the future, the development of smart tourism will be influenced by the progress made in emerging technologies such as artificial intelligence, augmented reality, and blockchain (Gretzel et al., 2015). Nevertheless, there are various challenges that need to be tackled in order to fully unlock the potential of smart tourism. An important challenge that arises in the modern tourism industry is the imperative to safeguard data privacy and security amidst the growing digitization of the field. As Gretzel et al. (2020) argue, the gathering and utilization of personal data give rise to concerns regarding privacy infringement and data breaches. This underscores the significance of implementing strong cybersecurity measures and regulatory frameworks.

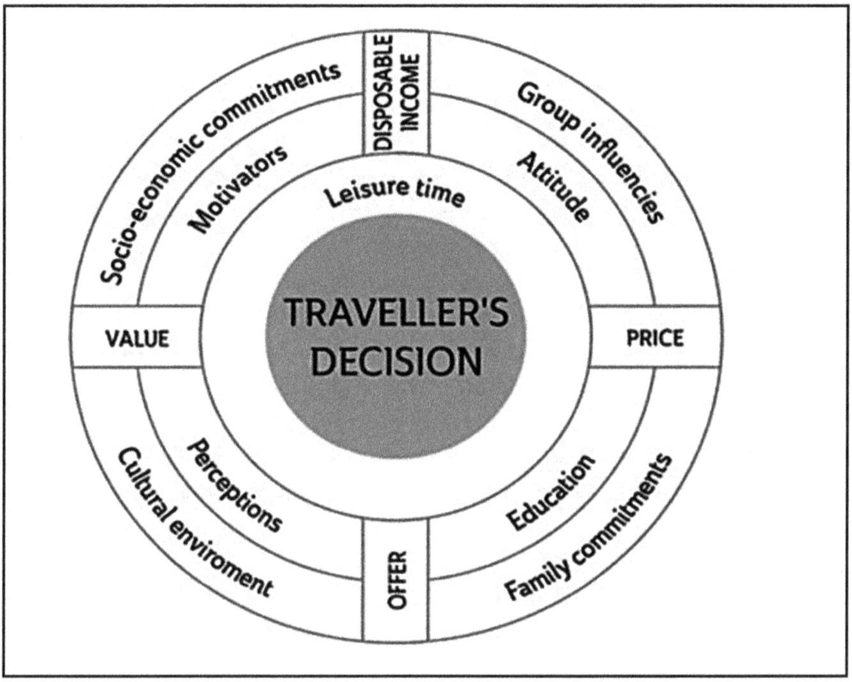

FIGURE 7.2 Influencing factors of tourist's decision.

Source: Adapted from Zsarnóczky (2017).

In addition, there is an increasing awareness of the digital divide and disparities in technology access, which could worsen inequalities in tourism development and contribute to social exclusion. The importance of inclusive approaches to smart tourism, which prioritize digital literacy and accessibility, is emphasized by UNWTO (2020). This is especially crucial in marginalized communities. Ultimately, the coming together of technology, consumer habits, and concerns for sustainability marks the development of smart tourism. Through the utilization of ICT, smart destinations can greatly improve the tourism experience, support sustainable development, and encourage inclusive growth (Gössling, 2021). Addressing challenges related to data privacy, digital inclusion, and ethical considerations is of utmost importance in order to fully realize the benefits of smart tourism.

7.5 LITERATURE REVIEW AND THEORETICAL BACKGROUND OF THE STUDY

It is now widely acknowledged that digital marketing is an essential component of the modern tourist and hospitality industries. Digital marketing encompasses a wide range of online methods that are designed to promote destinations, lodgings, and experiences through digital media. Digital marketing is defined by academics such as Buhalis and Amaranggana (2015) as the utilization of internet-based technology to facilitate marketing activities such as promotion, advertising, and distribution. This definition applies to the tourism industry. This shift towards digital marketing methods is a reflection of the changing purchasing patterns of consumers in a world that is becoming increasingly linked. When it comes to promoting goods, services, and brands through digital media, digital marketing comprises a wide variety of online techniques and tactics that are targeted at achieving this goal. In the context of tourism and hospitality, digital marketing plays an essential part in recruiting clients, keeping them engaged, and retaining them. The term "digital marketing" in the tourism industry refers to the use of internet-based technology to facilitate marketing operations such as advertising, promotion, and distribution, as stated by Buhalis and Law (2008).

7.5.1 DIGITAL MARKETING FOCAL POINT FOR SMART TOURISM

The intersection of smart tourism and digital marketing signifies a significant change in how destinations are managed and marketed. This chapter explores different aspects of this intersection, providing insights into new areas of study and their impact on those involved in the tourism industry such as:

1. **The Influence of Digital Marketing Strategies on the Adoption of Smart Tourism:** Channels for digital marketing, such as social

media, search engine optimization (SEO), and email marketing, are crucial in promoting intelligent tourism destinations and experiences. Buhalis and Amaranggana (2013) highlight the significance of digital marketing in promoting destinations and engaging visitors. As an illustration, the "Scotland is Now" campaign by Visit Scotland effectively employed captivating visual content and storytelling on various social media platforms to highlight the wide range of attractions and experiences Scotland has to offer. This approach successfully captured the attention of a worldwide audience, potentially enticing them to visit.

2. **Personalization and Customization in Smart Tourism Marketing:** Through the use of personalization and customization techniques, destination marketers have the ability to craft marketing messages and offerings that are specifically tailored to the unique preferences and interests of each individual traveler (Zhao et al., 2015). In their study, Kim and Kim (2020) emphasize the importance of personalization in improving the relevance and effectiveness of digital marketing messages. As an example, Airbnb employs data analytics to customize accommodation suggestions by considering traveler demographics, previous booking records, and stated preferences. This approach enhances the overall booking process and increases customer satisfaction (Narangajavana Kaosiri et al., 2019).
3. **Utilization of Augmented Reality (AR) and Virtual Reality (VR) in the realm of Tourism Marketing:** AR and VR technologies provide a captivating and interactive experience for travelers, which can greatly influence their decisions when it comes to choosing a destination and planning their travels. Wang et al. (2016) highlight the significant impact that AR and VR can have on destination marketing, showcasing their potential for transformation. A noteworthy case is the "VR Mauritius" campaign, in which the Mauritius Tourism Promotion Authority created virtual reality experiences to highlight the island's stunning landscapes, rich cultural heritage, and thrilling adventure opportunities. This innovative approach enabled potential travelers to virtually explore the destination and make more informed decisions when booking their trips.

4. **Intelligent Destination Branding and Positioning:** Efficient destination branding and positioning play a crucial role in attracting tourists and cultivating destination loyalty. In their study, Salamzadeh et al. (2022) assert that destination branding plays a crucial role in determining the competitiveness of a destination. As an example, Dubai's "Be My Guest" campaign presents the city as a warm and inviting destination, highlighting its wide range of cultural experiences, luxurious amenities, and top-notch hospitality services. This is achieved through well-coordinated digital marketing efforts that encompass social media, partnerships with influencers, and engaging content experiences.
5. **Evaluating the Efficacy of Digital Marketing Campaigns in Smart Tourism:** Measuring the effectiveness of digital marketing campaigns allows destination marketers to assess the return on investment, enhance campaign performance, and strategically allocate resources. In their study, Xiang et al. (2021) highlight the significance of creating comprehensive measurement frameworks to evaluate the influence of digital marketing strategies on tourist behavior and destination performance. Google Analytics offers valuable insights into website traffic, user engagement, and conversion rates. This tool enables destination marketers to effectively track the impact of their digital marketing strategies and pinpoint areas that can be enhanced (Buhalis, 2019).
6. **Ethical and Privacy Considerations in Smart Tourism Marketing:** With the reliance on data-driven marketing practices in smart tourism, it is crucial to prioritize ethical and privacy considerations. In his work, Sigala (2020) highlights the importance of ethical guidelines and regulatory frameworks in protecting consumer privacy within the realm of digital marketing communications. As an illustration, the General Data Protection Regulation (GDPR) in the European Union enforces transparent data practices and grants consumers more control over their personal data.
7. **Exploring the Impact of Digital Marketing in Smart Tourism from a Cross-cultural Lens:** Consumer responses to digital marketing efforts in smart tourism are influenced by cross-cultural differences. Chiao et al. (2018) emphasize the

significance of comprehending cultural nuances and preferences in order to enhance destination-marketing strategies. As an example, Western travelers tend to appreciate interactive and visually engaging content, whereas Asian travelers place a higher emphasis on safety, cleanliness, and shopping experiences.

The convergence of smart tourism and digital marketing presents a multitude of possibilities for innovation and expansion within the tourism sector. Through the exploration of these innovative study areas and a deep understanding of their implications, individuals in academia and industry can acquire valuable insights into the intricate relationship between technology, marketing, and destination management. This knowledge will contribute to the development of sustainable and impactful tourism experiences in the digital age.

Moreover, digital marketing is a multifaceted strategy used in the travel and tourism industry to reach consumers. It does this by connecting with people through social media, SEO, emails, and content. This allows businesses to reach a wider audience, build their brand, and make more bookings. Additionally, via the use of tools like data analytics and digital ads, they can precisely engage with consumers based on their preferences and online behavior. Interestingly, a poll found that 35% of visitors look to social media for inspiration for their trips, with younger people—moreover, 50% of those under 40—particularly depending on these platforms for trip ideas are (Dezney Technology Solution, 2023):

- Draw in more attention and raise consciousness.
- Create curiosity about the location, which will increase reservations.
- Clearly convey special attributes to attract potential tourists and support the expansion of the travel sector.

Specific target marketing strategies focus on determining the tastes, habits, and demographics of your customers; apply data analytics (Buhalis and Amaranggana, 2015). Segment the market based on their online behaviors and interests, target particular client categories more specifically. To develop the marketing tactics, digital interaction is considered. Measure the effectiveness of real-time digital marketing strategies for a better outcome (Table 7.2).

TABLE 7.2 Social Media for Tourism

Social Media Tools	Description	Example
Media sharing sites	Enable users to upload, arrange and exchange multimedia information with selected individuals and/or communities.	YouTube, Instagram, Flickr, Snapchat
Blogs/Microblogs	Enable writers to post on the site, hoping that someone will read them.	Twitter, Tumblr, Weibo
Virtual/Online Communities or Discussion forums	Provide interactive resources on a website; individuals may share personal information, preferences, and opinions.	Lonely Planet, Yahoo Answers, Reddit, Quora
Social Networking sites	Enable people to create social connections and interests between friends and acquaintances and connect to brands online.	Facebook, LinkedIn, Google+
Consumer Review Networks	Enable people to identify, view and exchange information on brands, goods, services, travel destinations, etc., by means of which positive feedback becomes social evidence of product/service quality.	TripAdvisor, Yelp, Zomato
Virtual Worlds	Offer computer-simulated environments in which people in a virtual world can interact.	Second Life, Onverse

Source: Adapted from Deb and Mallik (2023).

Smart marketing and promotion involve using data-driven strategies and innovative techniques to effectively reach and engage target audiences. It focuses on leveraging technology, analytics, and consumer insights to optimize campaigns, personalize messaging, and maximize ROI. When using the term "marketing objectives," we are referring to a certain set of objectives that are created to provide your company a direction to advance and produce results within a certain amount of time. Whitmore et al. (2015) describes SMART goals as: Specific goals must be written in a clear manner and refer to a specific component of the organization. Measurable goals are quantifiable and thus trackable. Attainable – SMART goals must be genuinely achievable given your present knowledge, skills, and time restrictions. Relevant – SMART goals must relate to the business in order to be successful, but the aim is meaningless if it is centered on a result that the company cannot handle. Time-bound – A deadline or period of time by which a goal must be accomplished is required. The objective becomes a changing target and is unlikely to be accomplished without that specific cut-off date.

Increasing online visibility, generating leads, and driving conversions are often the three primary focuses of digital marketing strategies targeted towards the tourism and hospitality industries. An example of this would be Erdmann et al. (2022), who emphasize the significance of SEO and social media marketing (SMM) in terms of boosting website traffic and enhancing brand awareness. According to Hays et al. (2013), potential travelers are engaged through the use of content marketing tactics such as blogging and storytelling. These strategies are implemented in order to deliver important information and to engage with potential travelers.

Functions of digital marketing in the tourism and hospitality industries include the following:

1. **Promotion and Advertising:** According to Buhalis and Law (2008), digital channels such as social media, search engines, and display advertising platforms are utilized in order to market travel destinations, accommodations, and experiences.
2. **Engagement of Customers and Relationship Management:** Digital marketing makes it easier to communicate directly with customers using social media interactions, email newsletters, and personalized messages (Sigala, 2020).
3. **Research and Analysis of the Market:** According to Pencarelli (2020), digital tools and analytics platforms make it possible for tourism enterprises to collect data on customer behavior, preferences, and market trends, which in turn helps them make strategic decisions.
4. **Distribution and Booking:** Digital platforms provide distribution channels for hotel reservations, aircraft bookings, and tour package bookings, which streamlines the booking process for clients who are travelling (Dwyer et al., 2016).
5. **Customer Engagement and Relationship Management:** Digital marketing facilitates direct communication with customers through social media interactions, email newsletters, and personalized messaging (Yoo et al., 2016).
6. **Market Research and Analysis:** Digital tools and analytics platforms enable tourism businesses to gather data on consumer behavior, preferences, and market trends, informing strategic decision-making (Toporowski and Lademann, 2014).

For tourism to flourish, an online presence is essential. In order to interact with prospective travelers, tourist attractions, hotels, and tour operators require expert websites and a dynamic social media presence. It is one of the most important needs for businesses in the digital age. It is an individual or company's Internet persona. It includes being present in emails, social networks, and marketplaces in addition to the Internet. A component of the Internet presence is the own brand (arimetrics). When it comes to digital marketing, there is a wide variety of methods and tactics to choose from, each of which has its own set of goals and particular approaches. In this modern world, people want to consume goods and services while saving time. So, business organizations want to promote their products and services globally within a short time and save costs. Especially travel agencies use digital marketing for their promotion. Due to this type of marketing, the customer and the business are benefited. This marketing could be personalized, social media, content-based, mobile and website-based, online-based, and E-word of mouth, and so forth.

7.5.2 STRATEGIES OF DIGITAL MARKETING UTILIZED IN THE TOURISM AND HOSPITALITY INDUSTRY:

1. **Email Marketing:** Email is still a very important tactic to use to reach customers. It is the best way to make deals and send more content, and it's important to keep email lists growing. According to (Deb and Mallik, 2023), email marketing is defined as the practice of communicating with historical, current, and prospective customers using email newsletters, promotions, and personalized messaging. A planned sequence of emails intended to nurture leads and existing clients in an effort to boost interaction and boost revenue is known as an email marketing campaign. Every email has a unique call to action that encourages readers to do one of the following: register, schedule a call, read on, or add an item to their basket. Email marketing campaigns can be utilized to tell previous guests about sales, occasions, and new attractions in an effort to get them back (Thomas et al., 2022).
2. **Search Engine Optimization (SEO):** In order to make a website's pages more easily accessible, relevant, and popular for user search

queries—and consequently rank higher in search engine results—it is necessary to optimize its technical setup, content relevancy, and link popularity. This process is known as SEO, or search engine optimization (Erdmann et al., 2022). Optimizing web content for search engines helps potential tourists find relevant information about a destination. The tourism stakeholders should focus on SEO activities that have a significant impact, such as targeting long-tail keywords, optimizing for mobile audiences, strategically incorporating keywords in titles, and utilizing analytical tools to assess user behavior (Wang and Park, 2023). They need to ensure that the content is optimized for human readers, as Google favors fresh content. Alfiana et al. (2023) stated that SEO was defined as the process of optimizing the content and structure of a website in order to improve its visibility and rankings in search engine results pages (SERPs).

3. **Search Engine Marketing (SEM):** According to Bhandari and Sin (2023), SEM is defined as the process of paying for adverts to appear in search engine results. This is often accomplished through pay-per-click (PPC) advertising strategies (Wang et al., 2017).
4. **Social Media Marketing (SMM):** Buhalis and Amaranggana (2015) stated that SMM is defined as the utilization of social media platforms like Facebook, Instagram, and Twitter in order to communicate with users, develop brand awareness, and generate traffic to a website. "Social media marketing" entails utilizing social media sites to advertise goods, services, or content. To accomplish marketing objectives, it usually entails tasks like producing and disseminating interesting content, communicating with the public, executing targeted advertisements, and assessing performance indicators (Hays et al., 2013). Businesses can use a variety of tools and services provided by social media platforms such as Facebook, Instagram, Twitter, LinkedIn, and others to connect and interact with their target audience (Alfiana et al., 2023). Effective social media marketing may raise revenues, improve website traffic, create leads, and raise brand awareness. The essential actions in a marketing plan for social media:

a. **Establish Specific Objectives:** decide what you hope to accomplish with your social media marketing, such as raising sales, generating leads, or growing brand awareness;
b. **Pick the Correct Platforms:** decide which social media channels best serve your target market and advance your company's objectives. Popular options include LinkedIn, Twitter, Instagram, Facebook, and more;
c. **Establish your Target Audience:** create a character for your ideal client and modify your messaging and content to appeal to them;
d. **Produce high-quality content:** lay out a schedule of posts, pictures, videos, and narratives, whereas material needs to be interesting, pertinent, and helpful to your readers.

5. **Blogs and Content Marketing:** Content marketing is a marketing tactic that involves producing and disseminating pertinent content, including emails, videos, social media postings, and articles, in order to build audiences, create authority, raise brand awareness, and increase sales. Top-notch media, including images, videos, and blog articles, can highlight a location's charm and individuality. In order to reach a larger audience, this content might be delivered through a variety of online channels (Deb and Mallik, 2023). It is focused on creating and sharing relevant and consistent content to attract and engage a target customer. Content-based marketing could be videos, blogs, infographics, and so on. In Bangladesh, there are many bloggers making different types of videos and blogs on the tourism industry. Kiersten Rich, known as "The Blonde Abroad," is writing a series of blog posts that provide valuable tips and information related to a prominent solo female travel blogger and influencer. Content marketing and blogging are defined as the process of producing and disseminating material that is both educational and engaging through the use of blogs, articles, videos, and podcasts in order to attract and maintain the interest of an audience (Hays et al., 2013).

7.6 METHODS AND MATERIALS OF THE STUDY

This review chapter used a systematic approach to gather and analyze information from relevant academic journals. This process began with a thorough search of PubMed, Google Scholar, ScienceDirect, and Web of Science. The qualitative research design is ideal for investigating complex phenomena and getting in-depth insights into the experiences and viewpoints of individuals and organizations (Schönherr, 2023). For example, Schönherr (2023) experimented through a qualitative study on tourist SMEs, STOs, and DMOs to better understand organizational learning in the context of sustainable tourism and digital transformation and uses QR codes in tourism destinations for branding and content marketing. The study used a qualitative research approach to integrate current information and link it to the themes of sustainable tourism and digital transformation (Schönherr, 2023).

Furthermore, qualitative research techniques such as in-depth interviews, observation, and recording have been used in studies that focus on the role of players in promoting rural tourist marketing in the digital era (Fajri et al., 2022). To accomplish the qualitative part of the chapter, 10 (ten) industry experts, like tour operators, travel guides, hotel owners, tourism destination managers, and travel agency owners, were interviewed through focus group discussions (FGD). Whereas 90 minutes and a semi-structured questionnaire were used for this purpose.

The criteria required peer-reviewed publications that provided substantive insights into the applications, benefits, challenges, or future trends of digital marketing tools in smart tourism initiatives. Studies that did not meet these criteria and non-English publications were excluded to ensure unbiased and better results. Data extraction and analysis was key to the review. Methodically extracting and synthesizing relevant information and findings from selected publications created a comprehensive overview. After that, the selected chapters were critically assessed. This assessment considered author credibility, research methodology rigor, and findings relevance to the topic. Critical analysis of synthesized data revealed research strengths and weaknesses, highlighting areas for future study.

7.7 APPLICATIONS OF DIGITAL MARKETING STRATEGIES IN TOURISM AND HOSPITALITY INDUSTRY

The number and size of online travel agents, also known as OTAs, are expanding at an exceptionally fast pace. Digital marketing offers numerous significant benefits compared to traditional physical marketing, primarily due to its distinctive attributes that enable a wider reach and personalized targeting. Digital marketing enables marketers to effectively reach a vast global audience without any obstacles, while also facilitating personalized and targeted communication at both the group and individual level (Pitana and Pitanatri, 2016). Tourism and digital marketing are intricately linked in the contemporary era. Digital marketing is crucial for promoting tourism destinations, attracting tourists, and enhancing the overall travel experience.

Interactive displays are a form of smart attractions that offer visitors engaging and dynamic means of acquiring knowledge about the historical, cultural, and distinctive aspects of attractions or landmarks. These displays make use of interactive touch screens, gesture recognition, and AR technology to enhance the learning experience.

Multimedia presentations encompass a range of audiovisual elements, such as videos, animations, and virtual tours, which serve to enhance visitors' comprehension of the attraction's historical context and cultural significance.

The utilization of technology in interactive storytelling enables smart attractions to effectively engage visitors by immersing them in compelling narratives that are intricately connected to the historical, mythical, and anecdotal aspects associated with the specific site.

Self-guided tours offer visitors the opportunity to explore their surroundings independently, utilizing mobile applications and wearable devices. These technological tools provide users with up-to-date information, personalized suggestions, and interactive features to enhance their experience.

The implementation of technology is utilized to enhance the accessibility of attractions for individuals with disabilities. This is achieved through the provision of features such as sign language interpretation, audio descriptions, and tactile exhibits.

Community engagement in the context of smart landmarks encompasses various strategies aimed at involving the local community. These strategies include the utilization of social media campaigns, the promotion of user-generated content, and the organization of interactive events. By implementing these approaches, smart landmarks seek to cultivate a sense of ownership and connection among community members (Table 7.3).

TABLE 7.3 Summary of Digital Marketing Strategies Used in Tourism and Hospitality Industry

Methods	Description	Source
Email Marketing	Email marketing is a crucial strategy for reaching customers, promoting deals, and sending content. It involves communication through newsletters, promotions, and personalized messaging. Email marketing campaigns nurture leads and existing clients, with unique call-to-actions. They can also inform previous customers about sales, events, and new attractions to encourage repeat business.	Deb and Mallik (2023)
Search Engine Optimization (SEO)	SEO is the process of optimizing a website's technical setup, content relevance, and link popularity to improve its visibility and ranking in search engine results. Tourism stakeholders should focus on SEO activities like targeting long-tail keywords, optimizing for mobile audiences, and using analytical tools to assess user behaviour.	Erdmann et al. (2022)
Search Engine Marketing (SEM)	Search engine marketing (SEM) is the payment for advertisements to appear in search engine results.	Bhandari and Sin (2023)
Social Media Marketing (SMM):	**Social Media Marketing in Tourism** About 70% of customers when they engage in a brand expect to social media conversational service from the tourism service provider every time. While on social media users prefer on both personal and business accounts and also expect brands to behave fast and personalize response or feedback when they are chatting with concern customers. Social media helps to extensive reach through facebook, Intagram, Twitter, Whatsapp, and Tiktok with the billions of users in the tourism business. In tourism industry uses AI for the better targeting and results of tourism offering insights, automation, and campaign. To sum up, digital marketing is essential to growing the travel and tourism sector. Businesses can distinguish themselves from rivals and reach a larger audience with the correct marketing plan.	Buhalis and Amaranggana (2015); Xifra and Huertas (2008)
Blogs and Content Marketing	Content marketing is a strategy that involves creating and sharing relevant content, such as emails, videos, and social media posts, to build audiences, build authority, raise brand awareness, and increase sales. It can be delivered through various online channels	Deb and Mallik (2023)

TABLE 7.3 *(Continued)*

Methods	Description	Source
Affiliate Marketing	Affiliate marketing entails collaborating with third-party websites and influencers to promote products or services in exchange for a commission on sales generated.	Gretzel et al. (2020)
Market Analytics	Market analytics uses data analytics technologies to monitor website traffic, user behaviour, and marketing campaign effectiveness, aiding decision-making for travel agencies by analyzing customer behaviour data.	Rosário and Dias (2023)
Display Marketing	Display marketing involves promoting advertisements on websites, apps, and social media platforms to target specific audiences through banner ads, video commercials, or interactive ads.	Wang and Zheng (2016)
Pay Per Click (PPC)	Pay-per-click (PPC) advertising involves advertisers paying a fee each time their ad is clicked, allowing them to display ads in search engine results or on websites.	Buhalis and Amaranggana (2015)

7.8 MAIN CHALLENGES AND OPPORTUNITIES OF APPLYING DIGITAL MARKETING FOR SMART TOURISM

The concept of "Smart Tourism" has its roots in technology. Thus, defining Smart Technology is the first step in trying to characterize it. The phrase "smart technology" refers to a collection of particular technologies and technology-driven phenomena that enable data and communication in previously unattainable ways. The core of smart technology is sensor technology, but other essential components of a smart technology infrastructure include widely available Wi-Fi, near-field communication (NFC), smartphones and mobile connectivity, radio-frequency identification (RFID), sophisticated data warehouses, and data mining algorithms.

The following are the main challenges for applying digital marketing for smart tourism:

1. **Data Privacy and Security:** A significant quantity of data on travelers, including their location, tastes, and spending patterns, must be gathered and used in order to practice "Smart tourism." While offering individualized experiences and services depends on this data, privacy and security issues are also brought up by it (Németh and Schmidt, 2007). Due to the heavy reliance of the tourist industry on personal data, user privacy must be upheld. It becomes more complicated to navigate through different data protection standards (Pitt, 2009).
2. **Cross Device Consistency:** One of the most prevalent challenges is ensuring a consistent experience across various devices. Whether using a desktop computer, tablet, or smartphone, the user should always get the same high-quality experience (Lim and Winkenbach, 2019).
3. **Changing Tourist Behavior:** Travelers are increasingly arranging and scheduling their excursions with the use of modern technologies. To fulfill the demands of these new customers, tourism-related enterprises must modify their digital marketing tactics (Deb and Mallik, 2023).
4. **Adapting to Rapid Technological Changes:** Rapid changes are occurring in the digital world. It's difficult to stay on top of cutting-edge technology like voice search, AI, and AR/VR (Bhandari and Sin, 2023).

The following are the main opportunities for applying digital marketing for smart tourism:

1. **Mobile Optimization:** Since most people access information via mobile devices, mobile optimization is not an option—it is a need. This entails using location-based services and maintaining a website that is optimized for mobile devices (Ge et al., 2018).
2. **Enhanced Convenience:** Travelers' vacation planning and booking processes may be facilitated by digital marketing (Deb and Mallik, 2023). For instance, tourist firms to respond to inquiries and offer support around-the-clock may use Chatbots.

7.9 USAGE QUICK RESPONSE (QR) CODE IN SMART TOURISM

The use of QR codes, which stand for "Quick Response," has become widespread in today's society. These codes provide a means of accessing information through smartphones that is both convenient and effective. QR codes are becoming increasingly popular in the tourism industry as a means of enhancing the experiences of visitors, enhancing the accessibility of information, and augmenting marketing strategies (Lou et al., 2017). QR codes have emerged as a versatile tool for connecting the digital and physical worlds. They provide a seamless way to access information using smartphones, which has led to their widespread adoption (Canadi et al., 2010). In the tourism industry, QR codes are being utilized to provide travelers with convenient access to information about their destinations, to enhance the experiences that visitors have at attractions and landmarks, and to facilitate contactless transactions and interactions (Fino et al., 2013).

7.9.1 EXPANDED QUICK RESPONSE (QR) CODE IN TOURISM SECTOR

Given the dynamic nature of tourism and the ability to modify QR codes as needed, placing them in easily accessible spots for scanning offers individuals instant access to AR content and other virtual details that can be conveniently stored on their mobile devices. Additionally, the widespread use of smartphones for translation purposes facilitates the accessibility of information for foreign visitors unfamiliar with the local language (Figure 7.3).

FIGURE 7.3 Quick response (QR) code for visitors.

Source: Supercode https://supercode.com/use-case/qr-codes-for-tourism.

7.9.2 MENTIONABLE QUICK RESPONSE (QR) CODES USED IN THE TRAVEL AND TOURISM INDUSTRY

URL QR codes are ideal for sharing website content or directing visitors to your website for information on events, tours, itineraries, transportation lines, and more. Social media QR codes are perfect for raising the number of followers on your social media accounts, allowing travelers to follow pertinent information about their stay and activities such as:

1. **Feedback QR Codes:** These are ideal for receiving feedback in tourist destinations such as hotels, public transit, and events.
2. **Email QR Codes:** These are ideal for queries about customer service, complaints, ideas, and lost and recovered items.
3. **vCard QR Codes:** These are the quickest and most convenient way for travelers to contact customer service, lost and found, or spa providers straight from a hotel, as well as business cards.

4. **Image QR Codes:** These are ideal for exhibiting past event photographs and encouraging attendees to return the next time.
5. **WhatsApp QR Codes:** These are the easiest way for foreign guests to scan a QR code and get in touch with you as long as they are connected to the internet, and they will not suffer long-distance phone expenses like SMS.

7.9.3 APPLICATIONS OF QUICK RESPONSE (QR) CODES IN SMART TOURISM

QR codes are being utilized in a variety of aspects of the tourism industry in order to improve the accessibility of information, enhance the experiences that visitors have, and streamline the processes that are involved in various operations (Canadi et al., 2010). The following are some of the use cases of QR codes in the tourism industry that are discussed in this section:

1. **Information Regarding the Destination:** QR codes are positioned at tourist destinations, landmarks, and points of interest in order to provide visitors with access to information that is pertinent to their visit, as well as historical evidence and multimedia content (Solima and Izzo, 2018).
2. **Wayfinding and Navigation:** QR codes are incorporated into signage, maps, and directional markers in order to provide assistance to travelers in the process of navigating urban environments, trails, and destinations (Canadi et al., 2010).
3. **Interactive Experiences:** QR codes make it possible to have interactive experiences, such as scavenger hunts, AR tours, and gamified activities, which increase the level of engagement and immersion that visitors have (Fino et al., 2013).
4. **Ticketing and Reservations:** QR codes are utilized for contactless ticketing, reservations, and check-ins at all types of establishments, including museums, events, and accommodations.
5. **e. Marketing and Promotion:** QR codes are a versatile marketing tool that can be used to promote tourism destinations, activities, and services through traditional print materials, advertisements, and social media platforms (Fahmy and Al-Azab, 2016) (Figure 7.4).

FIGURE 7.4 Contactless hotel booking with quick response (QR) code.

7.9.4 CASE STUDIES: EXEMPLIFYING QUICK RESPONSE (QR) CODE IMPLEMENTATION IN TOURISM

Case studies of QR code implementations that have been successful in tourism destinations, attractions, and marketing campaigns are presented in this section. Examples include:

1. **The Louvre Museum:** QR codes are utilized at the Louvre Museum to provide visitors with comprehensive information regarding the artworks, exhibits, and historical artifacts on display, thereby enhancing the overall experience of visiting the museum (Solima and Izzo, 2018).
2. **City Walking Tours:** QR codes are incorporated into self-guided walking tours, which offer visitors the opportunity to access audio guides, maps, and points of interest along the route.
3. **Destination Marketing Campaigns:** Tourism boards and marketing organizations use QR codes in promotional materials, brochures, and advertisements to encourage engagement and drive visitor interest during marketing campaigns (Figure 7.5).

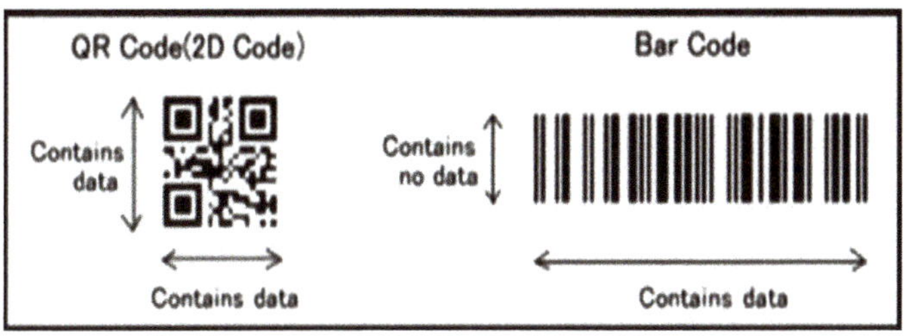

FIGURE 7.5 A quick response (QR) code allows storing more information than a barcode.

7.9.5 QUICK RESPONSE (QR) FOR THE PROMOTION OF TOURISM AND BRANDING

QR codes have emerged as a versatile tool for tourism promotion and branding, offering destination marketers innovative ways to engage with travelers, enhance destination visibility, and create immersive experiences. At the heart of QR code utilization lies its capacity to bridge the physical and digital realms seamlessly, enabling tourists to access a wealth of information, multimedia content, and interactive experiences using their smartphones. Through strategic placement at tourist attractions, landmarks, transportation hubs, and promotional materials, QR codes serve as gateways to destination narratives, historical insights, and practical travel information, enriching the visitor experience and fostering deeper connections with the destination. For instance, QR codes integrated into signage at historical sites can provide visitors with contextual information, audio guides, and virtual tours, enhancing their understanding and appreciation of the cultural significance of the site.

Additionally, QR codes embedded within destination marketing materials such as brochures, posters, and advertisements facilitate direct engagement with prospective travelers, offering them instant access to promotional videos, photo galleries, and special offers. This interactive approach to destination marketing not only captivates the audience but also empowers them to explore the destination on their terms, fostering a sense of discovery and ownership of their travel experience (Figure 7.8).

FIGURE 7.6 Disbursement of destination information through quick response (QR) code.
Source: Dar, 2022.

Moreover, QR codes play a pivotal role in destination branding efforts, enabling tourism organizations to convey key brand messages, values, and unique selling propositions to target audiences effectively. By incorporating QR codes into branding materials such as logos, merchandise, and promotional merchandise, destinations can create memorable touch points that reinforce brand identity and evoke emotional connections with travelers. For example, a destination's logo featuring a QR code can serve as a visual symbol of its commitment to innovation, technology, and seamless visitor experiences, positioning it as a forward-thinking and digitally savvy destination (Solima and Izzo, 2018).

Additionally, QR codes embedded within branded merchandise such as T-shirts, souvenirs, and accessories can serve as conversation starters, encouraging tourists to scan and explore the destination's offerings further (Pike, 2011). This experiential branding approach not only strengthens brand recall but also encourages user-generated content and word-of-mouth promotion, as tourists share their experiences and discoveries with their social networks, amplifying the destination's reach and influence.

Furthermore, QR codes offer destination marketers valuable insights into visitor behavior, preferences, and engagement patterns, enabling data-driven decision-making and targeted marketing strategies. By tracking QR code scans, click-through rates, and user interactions, tourism organizations can gain valuable insights into which marketing materials, attractions, and experiences resonate most with their target audience, allowing them to refine their promotional efforts and allocate resources more effectively (Houliez, 2010). For instance, QR codes embedded within destination marketing campaigns can be customized to track specific metrics such as geographic location, demographic profile, and conversion rates, providing valuable feedback on the campaign's performance and return on investment.

Additionally, QR code analytics can inform segmentation strategies, enabling destinations to tailor their marketing messages and offerings to different audience segments based on their preferences, interests, and behavior patterns (Fahmy and Al-Azab, 2016). This personalized approach to destination marketing not only enhances the relevance and effectiveness of promotional efforts but also strengthens the destination's overall competitive position in the market.

7.9.6 *THE BENEFITS AND THE CHALLENGES OF QUICK RESPONSE (QR) CODES IN TOURISM*

QR codes provide tourism stakeholders a number of benefits, including enhanced visitor experiences, improved accessibility to information, and marketing solutions that are more cost-effective (Lou et al., 2017). However, the implementation of these systems presents a number of challenges, including those pertaining to consumer adoption, data security, and technical compatibility (Solima and Izzo, 2018). As we look to the future, it is anticipated that QR codes will continue to develop and play a significant role in the process of shaping the future of tourism management and marketing. Emerging trends and potential future directions for the use of QR codes in tourism are discussed in this section, including the following:

1. **Integration with Emerging Technologies:** QR codes have the potential to be integrated with emerging technologies such as AR, artificial intelligence (AI), and location-based services in order to provide travelers with experiences that are more immersive and personalized (Fino et al., 2013).
2. **Enhanced Accessibility and Inclusivity:** QR codes can be utilised to improve accessibility and inclusivity for travelers who have special needs, language barriers, or disabilities (Fu and Timothy, 2021). This helps to ensure that tourism experiences are accessible to all individuals.
3. **Data Analytics and Personalization:** QR codes make it possible for tourism stakeholders to collect valuable data on the behaviour, preferences, and demographics of visitors (Stylos et al., 2021). This data can then be used to facilitate targeted marketing campaigns and personalized experiences.

In wrapping up, QR codes offer significant potential for enhancing visitor experiences, improving accessibility to information, and augmenting marketing strategies in the tourism industry. When tourism destinations, attractions, and businesses make effective use of QR code technology, they can provide travelers with experiences that are both memorable and engaging, while simultaneously maximising their operational efficiency and marketing effectiveness (Lou et al., 2017).

7.10 CONCLUSION

The tourism industry has seen significant changes due to the fact that it is increasingly marketing itself through online media, especially as regards Smart Tourism. It has transformed the way in which tourism products and services are marketed, opening up a wider worldwide audience through personalized and individualized communication with travelers. The sector has evolved to provide convenience and personal experiences for passengers through extensive use of technology, the Internet, and mobile devices. There are problems, such as data protection and security challenges, cross-device coherence, and changing behavior of tourists due to rapid changes in technology, that must be addressed (Fahmy and Al-Azab, 2016). Social media involvement can provide value, increasing the destination's image in the minds of tourists (Crespo et al., 2015). DMOs, on the other hand, should use social media consistently in order to project a positive image to all stakeholders.

Finally, it can be mentioned that printing QR Codes on maps of bus and train lines will allow passengers to easily track their location and identify popular destinations. QR codes that direct people to transit apps are also useful. QR Codes also announce future festivals and community gatherings in high-traffic areas; placing a display with festival and event dates together with a QR Code helps transmit intriguing information to visitors who were unaware of the event or who want to learn more. Make it simple for customers to purchase tickets through m-commerce sites and provide scanning promotions, such as 2*1 tickets or a free drink with their purchase when they come (Fu and Timothy, 2021).

In addition, generate QR Codes for theme parks; instead of standing in huge lines for food and rides, scan a QR Code and place guests in a queue, allowing them to explore the park and roam around until it's their turn to ride and eat. This also eliminates the need to wait in line for hours only to discover that five individuals have cut in front of you (Solima and Izzo, 2018).

However, opportunities to improve mobile capabilities, increase convenience, enhance the personal experience, and use Augmented Reality and Virtual Reality in order to create immersive travel experiences accompany these challenges. In order to succeed in Smart Tourism, providing travelers with better options and enhancing the overall travel experience, an integration of Digital Marketing is crucial as the tourism industry (Lou et al., 2017).

KEYWORDS

- **digital marketing**
- **economic development**
- **e-tourism**
- **quick response (QR) code**
- **search engine optimization (SEO)**
- **smart tourism**
- **social media**
- **tourist expectation**

REFERENCES

Alfiana, F., Khofifah, N., Ramadhan, T., Septiani, N., Wahyuningsih, W., Azizah, N. N., & Ramadhona, N. (2023). Apply the search engine optimization (SEO) method to determine website ranking on search engines. *International Journal of Cyber and IT Service Management*, *3*(1), 65–73.

Bassano, C., Barile, S., Piciocchi, P., Spohrer, J. C., Iandolo, F., & Fisk, R. (2019). Storytelling about places: Tourism marketing in the digital age. *Cities*, *87*, 10–20.

Bhandari, R., & Sin, M. V. A. (2023). Optimizing digital marketing in hospitality industries. *Startupreneur Bisnis Digital (SABDA Journal)*, *2*(1).

Buhalis, D. (2019). Technology in tourism—from information communication technologies to etourism and smart tourism towards ambient intelligence tourism: A perspective article. *Tourism Review*, *75*(1), 267–272.

Buhalis, D., & Amaranggana, A. (2014). Smart tourism destinations. *Information and Communication Technologies in Tourism 2014: Proceedings of the International Conference in Dublin, Ireland, 2013*, 21–24.

Buhalis, D., & Amaranggana, A. (2015). Smart tourism destinations enhancing tourism experience through personalization of services. *Information and Communication Technologies in Tourism 2015: Proceedings of the International Conference in Lugano, Switzerland, February 3–6, 2015*.

Buhalis, D., & Law, R. (2008). Progress in information technology and tourism management: 20 years on and 10 years after the internet—The state of tourism research. *Tourism Management*, *29*(4), 609–623. https://doi.org/10.1016/j.tourman.2008.01.005 (accessed on 8th July 2025).

Buhalis, D., & Park, S. (2021). Brand management and cocreation lessons from tourism and hospitality: Introduction to the special issue. *Journal of Product & Brand Management, 30*(1), 1–11.

Canadi, M., Höpken, W., & Fuchs, M. (2010). Application of QR codes in online travel distribution. In *Information and Communication Technologies in Tourism* (pp. 137–148). Springer.

Chiao, H. M., Chen, Y. L., & Huang, W. H. (2018). Examining the usability of an online virtual tour-guiding platform for cultural tourism education. *Journal of Hospitality, Leisure, Sport & Tourism Education, 23*, 29–38.

Crespo, Á. H., Gutiérrez, H. S. M., & Mogollón, J. H. (2015). Perceived influence on behavior of user-generated content on social network sites: An empirical application in the hotel sector. *Revita Española de Investigation de Marketing ESIC, 19*, 12–23.

Dar, V. (2022).The cool code: QR codes boost digitisation & innovation in unique ways. *Financial Express*. Available at: https://www.financialexpress.com/life/lifestyle-the-cool-code-qr-codes-boost-digitisation-innovation-in-unique-ways-2596123/

Deb, S. K. (2021). Proposing the eTourism adoption model (eTAM) in the context of Bangladesh. *Technology Application in the Tourism and Hospitality Industry of Bangladesh*, 3–24.

Deb, S. K., & Mallik, N. (2023). Effects of social media in tourism marketing: Outlook on user generated content. *Journal of Digital Marketing and Communication, 3*(2), 49–65.

Deb, S. K., Mohanty, P. P., & Valeri, M. (2022). Promoting family business in handicrafts through local tradition and culture: An innovative approach. *Journal of Family Business Management*. https://www.emerald.com/insight/search?q=P.P.%20Mohanty (accessed on 8th July 2025).

Dezney Technology Solution (India). (2023). How digital marketing benefits travel & tourism industry. Available at: https://www.linkedin.com/pulse/how-digital-marketing-benefits-travel-tourism-xo44c (accessed on 8th July 2025).

Dwyer, L., Gretzel, U., Huang, Z., Kim, C., Ku, E., Li, G., Shoval, N., Tsang, N., Wang, D., & Xiang, Z. (2016). Smart tourism in China. *Journal of China Tourism Research, 12*(1), 1–84.

Erdmann, A., Arilla, R., & Ponzoa, J. M. (2022). Search engine optimization: The long-term strategy of keyword choice. *Journal of Business Research, 144*, 650–662.

Fahmy, T. M., & Al-Azab, M. R. (2016). Evaluation of the tourist acceptance of quick response (QR) code: Using technology acceptance model. *Minia Journal of Tourism and Hospitality Research (MJTHR), 1*(2), 49–74.

Fajri, D., Damanik, J., Priyambodo, T., & Sutikno, B. (2022). Actor's role in developing creative rural tourism marketing in the digital era: A case study in Ponggok, Central Java. *Proceedings*, 278–292. https://doi.org/10.2991/978-2-494069-73-2_21 (accessed on 8th July 2025).

Fino, E. R., Martín-Gutiérrez, J., Fernández, M. D. M., & Davara, E. A. (2013). Interactive tourist guide: Connecting Web 2.0, augmented reality and QR codes. *Procedia Computer Science, 25*, 338–344.

Floričić, T., Šker, I., & Benassi, H. M. (2023). Enhancing digital promotion of cultural attractions: Assessing websites, online marketing tools and smart technologies. *Business Systems Research: International Journal of the Society for Advancing Innovation and Research in Economy, 14*(1), 72–92.

Fu, Y., & Timothy, D. J. (2021). Social media constraints and destination images: The potential of barrier-free internet access for foreign tourists in an internet-restricted destination. *Tourism Management Perspectives*, *37*, 100771.

Fuentes, C., Bäckström, K., & Svingstedt, A. (2017). Smartphones and the reconfiguration of retailscapes: Stores, shopping, and digitalization. *Journal of Retailing & Consumer Services*, *39*, 270–278.

Ge, T., Liu, Z., & Ma, L. (2018). *Mobile Commerce: Concepts, Methodologies, Tools, and Applications* (Vol. III). IGI Global.

Gössling, S. (2021). Tourism, technology and ICT: A critical review of affordances and concessions. *Journal of Sustainable Tourism*, *29*(5), 733–750.

Gretzel, U., Fuchs, M., Baggio, R., Hoepken, W., Law, R., Neidhardt, J., Pesonen, J., Zanker, M., & Xiang, Z. (2020). e-Tourism beyond COVID-19: A call for transformative research. *Information Technology & Tourism*, *22*(2), 187–203.

Gretzel, U., Sigala, M., Xiang, Z., & Koo, C. (2015). Smart tourism: Foundations and developments. *Electronic Markets*, *25*, 179–188.

Hays, S., Page, S. J., & Buhalis, D. (2013). Social media as a destination marketing tool: Its use by national tourism organizations. *Current Issues in Tourism*, *16*(3), 211–239.

Houliez, C. (2010). Branding places or branding spatial practices? Retail spaces in the age of mixed realities. *Place Branding & Public Diplomacy*, *6*(2), 87–96.

Kapenieks, A., & Zuga, B. (2012). Internet, television and mobile technologies for innovative eLearning. In *Society. Integration. Education. Proceedings of the International Scientific Conference* (pp. 301–308).

Khwaja, M. G., Mahmood, S., & Zaman, U. (2020). Examining the effects of eWOM, trust inclination, and information adoption on purchase intentions in an accelerated digital marketing context. *Information*, *11*(10), 478.

Kim, M., & Kim, J. (2020). Destination authenticity as a trigger of tourists' online engagement on social media. *Journal of Travel Research*, *59*(7), 1238–1252.

Kumar, A., Bezawada, R., Rishika, R., Janakiraman, R., & Kannan, P. (2016). From social to sale: The effects of firm-generated content in social media on customer behavior. *Journal of Marketing*, *80*(1), 7–25. https://doi.org/10.1509/jm.14.0249 (accessed on 8th July 2025).

Lim, S. F. W. T., & Winkenbach, M. (2019). Configuring the last-mile in business-to-consumer e-retailing. *California Management Review*, *61*(2), 132–154. https://doi.org/10.1177/0008125618805094 (accessed on 8th July 2025).

Lou, L., Tian, Z., & Koh, J. (2017). Tourist satisfaction enhancement using mobile QR code payment: An empirical investigation. *Sustainability*, *9*(7), 1186.

Ma, H., & Li, J. (2021). An innovative method for digital media education based on mobile internet technology. *International Journal of Emerging Technologies in Learning (iJET)*, *16*(13), 68–81.

Magano, J., & Cunha, M. N. (2020). Digital marketing impact on tourism in Portugal: A quantitative study. *African Journal of Hospitality, Tourism and Leisure*, *9*(1), 1–19.

Molz, J. G. (2012). *Travel connections: Tourism, technology and togetherness in a mobile world*. Routledge.

Nafi, S., & Ahmed, T. (2018). Travel and leisure activities, motivation and behavior of young Bangladeshi tourists. *European Scientific Journal*, *14*(5), 341–358.

Narangajavana Kaosiri, Y., Callarisa Fiol, L. J., Moliner Tena, M. Á., Rodríguez Artola, R. M., & Sánchez García, J. (2019). User-generated content sources in social media: A new approach to explore tourist satisfaction. *Journal of Travel Research*, *58*(2), 253–265.

Németh, J., & Schmidt, S. (2007). Toward a methodology for measuring the security of publicly accessible spaces. *Journal of the American Planning Association*, *73*(3), 283–297. https://doi.org/10.1080/01944360708977978 (accessed on 8th July 2025).

Pencarelli, T. (2020). The digital revolution in the travel and tourism industry. *Information Technology & Tourism*, *22*(3), 455–476.

Pike, A. (2011). *Brands and Branding Geographies*. Edward Elgar.

Pitt, M. (2009). Security in the retail, hotel and leisure industries. *Journal of Retail & Leisure Property*, *8*(1), 1–1. https://doi.org/10.1057/rlp.2008.30 (accessed on 8th July 2025).

Rosário, A. T., & Dias, J. C. (2023). How has data-driven marketing evolved: Challenges and opportunities with emerging technologies. *International Journal of Information Management Data Insights*, *3*(2), 100203.

Salamzadeh, A., Tajpour, M., Hosseini, E., & Salamzadeh, Y. (2022). Geotourism and destination brand selection: Does social media matter? In *Economics and Management of Geotourism 2022* (pp. 105–124). Springer.

Scharl, A., Dickinger, A., & Murphy, J. (2005). Diffusion and success factors of mobile marketing. *Electronic Commerce Research and Applications*, *4*(2), 159–173.

Schönherr, S. (2023). Organizational learning and sustainable tourism: The enabling role of digital transformation. *Journal of Knowledge Management*, *27*(11), 82–100. https://doi.org/10.1108/jkm-06-2022-0434 (accessed on 8th July 2025).

Sethi, A. (2019). *Chinese Consumers: Exploring the World's Largest Demographic*. Springer Singapore. Available at: http://ludwig.lub.lu.se/login?url=http://search.ebscohost.com/login.aspx?direct=true&db=cat07147a&AN=lub.5310673&site=eds-live&scope=sitehttp://ludwig.lub.lu.se/login?url=https://doi.org/10.1007/978-981-10-8992-3 (accessed on 8th July 2025).

Sigala, M. (2020). Tourism and COVID-19: Impacts and implications for advancing and resetting industry and research. *Journal of Business Research*, *117*, 312–321.

Solima, L., & Izzo, F. (2018). QR codes in cultural heritage tourism: New communications technologies and future prospects in Naples and Warsaw. *Journal of Heritage Tourism*, *13*(2), 115–127.

Stylos, N., Zwiegelaar, J., & Buhalis, D. (2021). Big data empowered agility for dynamic, volatile, and time-sensitive service industries: The case of tourism sector. *International Journal of Contemporary Hospitality Management*, *33*(3), 1015–1036.

Thomas, J. S., Chen, C., & Iacobucci, D. (2022). Email marketing as a tool for strategic persuasion. *Journal of Interactive Marketing*, *57*(3), 377–392.

Toporowski, W., & Lademann, R. (2014). The importance of assortment, pricing, and retail site location for competition in food retailing–Results from marketing research. *Marketing: ZFP–Journal of Research and Management*, *36*(2), 131. http://ludwig.lub.lu.se/login?url=https://search.ebscohost.com/login.aspx?direct=true&db=edsjsr&AN=edsjsr.26426721&site=eds-live&scope=site (accessed on 8th July 2025).

UNWTO. (2019). *International Tourism Highlights*. Available at: https://www.e-unwto.org/doi/pdf/10.18111/9789284421152 (accessed on 8th July 2025).

Wang, H., & Zheng, L. (2016). Sentiment classification of Chinese online reviews: A comparison of factors influencing performances. *Enterprise Information Systems*, *10*(2), 228–244. https://doi.org/10.1080/17517575.2014.947635 (accessed on 8th July 2025).

Wang, T. Y., & Park, J. (2023). Destination information search in social media and travel intention of Generation Z university students. *Journal of China Tourism Research*, *19*(3), 570–588.

Wang, X., Li, X. R., Zhen, F., & Zhang, J. (2016). How smart is your tourist attraction? Measuring tourist preferences of smart tourism attractions via a FCEM-AHP and IPA approach. *Tourism Management*, *54*, 309–320.

Whitmore, A., Agarwal, A., & Da Xu, L. (2015). The Internet of Things—A survey of topics and trends. *Information Systems Frontiers*, *17*, 261–274.

Xiang, Z., Stienmetz, J., & Fesenmaier, D. R. (2021). Smart tourism design: Launching the *Annals of Tourism Research* curated collection on designing tourism places. *Annals of Tourism Research*, *86*, 103154.

Xiang, Z., Tussyadiah, I., & Buhalis, D. (2015). Smart destinations: Foundations, analytics, and applications. *Journal of Destination Marketing and Management*, *4*(3), 143–144.

Yasmin, A., Tasneem, S., & Fatem, K. (2015). Effectiveness of digital marketing in the challenging age: An empirical study. *International Journal of Management Science and Business Administration*, *(5)*, 69–80.

Yoo, K. H., Sigala, M., & Gretzel, U. (2016). Exploring TripAdvisor. In *Open Tourism* (pp. 239–255). Springer.

Zhang, G., Ma, L., Zhang, X., Ding, X., & Yang, Y. (2019). Understanding social media users unfollow intentions: Take WeChat subscriptions as an example. *Online Information Review*, *43*(7), 1081–1097.

Zhao, X., Wang, L., Guo, X., & Law, R. (2015). The influence of online reviews to online hotel booking intentions. *International Journal of Contemporary Hospitality Management*, *27*(6), 1343–1364.

Zsarnóczky, M. (2017). The impact of silver tourism on rural areas. *Roczniki (Annals)*, *2016*(3).

CHAPTER 8

THE TRANSFORMATIVE IMPACT OF ARTIFICIAL INTELLIGENCE IN HOSPITALITY AND TOURISM

MOHAMMAD BADRUDDOZA TALUKDER,[1*] SANJEEV KUMAR,[2] and IVA RANI DAS[3]

[1]*College of Tourism and Hospitality Management, International University of Business Agriculture and Technology, Dhaka, Bangladesh*

[2]*School of Hotel Management and Tourism, Lovely Professional University, Phagwara, Punjab, India*

[3]*Daffodil Institute of IT, Dhaka, Bangladesh*

*Corresponding author

ABSTRACT

The study explores the impact of Artificial Intelligence (AI) on the tourism and hospitality sectors, focusing on its impact on tourist experience, operational efficiency, and industry competitiveness. It evaluates the advantages and disadvantages of AI integration, focusing on critical areas like data security, workplace ethics, and employee skill development. The methodology involves an extensive literature review of AI in the tourism and hospitality domains. Through particular analysis and synthesis of findings from diverse sources, the study assesses the multifaceted impact of AI. Special attention is given to identifying concerns and challenges associated with AI implementation, emphasizing ethical considerations and workforce development. The findings of this study illuminate the transformative potential of AI-driven personalized experiences.

Smart Tourism: Managing Technological Revolution in the Tourism Industry. Santus Kumar Deb (Ed.)

DOI: 10.1201/9781779641090-8

Notably, integrating AI contributes to developing dynamic strategies, heightened operational efficiency, and increased visitor satisfaction. Key questions emerge regarding the ethical dimensions of AI in the hospitality industry, its long-term effects on eco-friendly practices, and the collaborative possibilities between humans and AI. The implications of this research extend to the imperative of constant inquiry into AI's role within the dynamic hospitality sector.

8.1 INTRODUCTION

The hospitality and tourism industries are at the forefront of a revolution that will change the world. We live in a time of giant leaps in technology and unprecedented connectivity. Artificial Intelligence (AI) is central to this revolution (Brem et al., 2021). AI is a force that has spread to every part of our daily lives. As technology keeps improving, the effects of AI are becoming more evident in hospitality and tourism (Li et al., 2021). This chapter aims to reveal the complex web of AI's impact on hospitality and tourism by exploring its historical roots, breaking down its many uses, and thinking about what the future holds for AI. As the tourism industry looks for new ways to meet the changing needs of travelers and the hospitality industry tries to improve the guest experience and operations more efficiently, AI comes into its own as a powerful ally, offering solutions and opportunities that were once thought impossible.

The journey starts with a look back at how technology has changed these fields over time, laying the groundwork for AI to appear. The world has changed as digital systems and intelligent technologies have emerged. AI has become one of the most critical factors in changing how hospitality and tourism work (Jabeen et al., 2022). As we look deeper into the complex web of AI applications in hospitality, from intelligent chatbots that provide seamless guest services to machine learning that creates personalized experiences, it becomes clearer how these applications have changed how the industry works. At the same time, the story moves to the tourism industry, where AI is orchestrating a symphony of innovations that are changing everything from how people plan trips to how tourist destinations are perceived.

Even though there is hope for progress, problems start to show up. Data security and privacy concerns overshadow the bright future of integrating AI (Jaime et al., 2023). People are looking closely at the delicate balance between automation and human interaction because using AI technologies in a way that does not violate ethics is needed. The author takes the reader through case studies, success stories, and ethical problems with AI. It looks into the future and makes guesses about the trends determining the next part of this technological story. The global view opens up the conversation by recognizing that the use of AI varies by region and that cultural differences will always affect its progress (Bareis and Katzenbach, 2022).

This examines how AI is changing hospitality and tourism and tries to break down the present and picture the future. It is a call to action, asking everyone involved to consider using AI to change things, seeing its potential while protecting the most essential parts of hospitality and tourism. We are navigating the complicated dance between AI and the natural human touch. This sets the stage for an exciting story that will shape the future of industries that host our journeys.

8.2 SIGNIFICANCE OF THE STUDY

The novelty and significance of this study are underscored by its in-depth exploration of the transformative influence of AI on the tourism and hospitality sectors. While existing literature has touched upon various aspects of AI in these industries, this study extends the discourse by offering a more nuanced understanding of how AI shapes visitor experiences, operational efficiency, and industry competitiveness. As highlighted by Li et al. (2018), the specific focus on these dimensions provides a unique contribution to the existing body of knowledge.

The study's novelty is further evident in its thorough evaluation of the benefits and drawbacks of AI integration, emphasizing critical areas such as data security, ethical considerations, and employee skill development. Building on the works of Buhalis and Leung (2018) and Sigala (2017), the research delves into these dimensions with a level of granularity that goes beyond mere acknowledgment of AI's influence, providing practical insights into the complexities of its implementation in the hospitality sector.

A distinctive aspect of the study lies in its identification of dynamic strategies and enhanced operational efficiency resulting from AI-driven personalized experiences. This aligns with Neuhofer et al. (2019) findings, highlighting the study's contribution in offering concrete examples of positive outcomes, thereby advancing our understanding of the practical implications of AI in the hospitality context.

Moreover, the study raises thought-provoking questions about the ethical implications of AI in the hospitality sector and its potential long-term effects on eco-friendly practices. These questions, in line with the discussions by Xiang et al. (2017), not only indicate the study's forward-looking approach but also open up novel avenues for further research, enriching the discourse on the ethical considerations associated with AI implementation.

The global perspective presented in the study, illuminating regional differences in adopting AI in the hospitality industry, adds another layer of novelty. By considering diverse viewpoints, the research aligns with Gretzel et al. (2015) call for a more context-aware approach to understanding the global dynamics of AI adoption in tourism and hospitality.

Furthermore, the study's strong emphasis on ethical considerations, including potential biases in algorithms and the societal effects of automation, contributes to the ongoing discourse on responsible AI integration. This aligns with the recommendations of Schwab (2017) for a thoughtful and ethical approach to deploying AI technologies. The study's multifaceted approach, combining an in-depth analysis of AI's impact, exploration of ethical considerations, and a global perspective, contributes significantly to the existing literature. By building upon and extending the findings of prominent scholars, this research provides valuable insights for academia and industry practitioners, offering guidance in navigating the evolving landscape of AI in tourism and hospitality.

8.3 OBJECTIVES OF THE STUDY

The objectives of a chapter on the impact of Artificial Intelligence in hospitality and tourism encompass the specific goals and aims that the chapter seeks to achieve. Here are the potential objectives, such as:

- To Examine the Historical Evolution of Technology in Hospitality and Tourism;
- To Investigate AI Applications in Tourism and Hospitality Services;
- To Conclude with Recommendations for Industry Stakeholders;

8.4 REVIEW OF THE RELEVANT LITERATURE

8.4.1 LITERATURE AND THEORETICAL BACKGROUND

This study is anchored in a rich theoretical framework, incorporating fundamental theories to enhance the understanding of the profound influence of AI on the tourism and hospitality sectors. The Technology Acceptance Model (TAM), proposed by Davis (1989), forms a foundational lens, elucidating the factors influencing the adoption of AI technologies in the hospitality sector. TAM's focus on perceived ease of use and usefulness guides the exploration of how industry professionals and visitors perceive and integrate AI-driven personalized experiences. Complementing this, Rogers' (1962) Innovation Diffusion Theory provides a comprehensive perspective on the stages of AI adoption across different hospitality industry segments. Categorizing adopters into innovators, early adopters, and others, this theory assists in analyzing the diverse patterns and rates of AI integration. Further enriching the theoretical underpinning, Homans' (1958) Social Exchange Theory is employed to explore establishing trust and reciprocity between stakeholders in the context of AI implementation. This theoretical triad collectively justifies and guides the study's exploration of AI's transformative effects, ensuring a nuanced and comprehensive examination of its implications for the tourism and hospitality sectors. The impact of AI on the hotel and tourist sectors is particularly notable. AI has become a disruptive force across all industries. To provide a comprehensive understanding of the current landscape, challenges, and opportunities associated with incorporating artificial intelligence in the hospitality and tourism sectors, the study synthesizes research that has already been conducted (Li et al., 2021).

8.4.2 THE DEVELOPMENT OF TECHNOLOGY IN THE HOSPITALITY AND TOURISM INDUSTRY THROUGHOUT HISTORY

Regarding the hotel and tourism industries, the incorporation of artificial intelligence may be traced back to the more general development of technology in these sectors. Technical improvements have been a significant factor in determining guests' experiences, the efficiency of operations, and the industry's competitiveness (Bharwani and Mathews, 2021). The advancement of artificial intelligence as an essential actor in redefining the dynamics of hospitality and tourism was made possible by the move from old systems to intelligent technologies, which established the framework for its growth (Ruel and Njoku, 2021).

8.5 ARTIFICIAL INTELLIGENCE (AI)'S APPLICATIONS IN THE HOSPITALITY INDUSTRY

The numerous uses of artificial intelligence in the hotel industry, aligned with intelligent chatbots driven by natural language processing, have grown widespread to provide guests with smooth, personalized services. Machine learning algorithms are utilized for dynamic pricing and revenue management (Kastius and Schlosser, 2021). These algorithms optimize hotel prices based on existing demand trends. The incorporation of the Internet of Things (IoT) and intelligent room technologies results in an increase in operational efficiency, which in turn provides guests with a stay that is both more connected and more convenient (Mercan et al., 2021).

8.5.1 THE IMPACT OF ARTIFICIAL INTELLIGENCE (AI) ON TOURISM SERVICES

In the tourism industry, artificial intelligence has revolutionized how passengers plan and enjoy their trips. There are sophisticated algorithms that examine large quantities of data to provide personalized travel suggestions tailored to everyone's tastes. Virtual and augmented reality technologies help create immersive tourism experiences (Fan et al., 2022). These technologies enable prospective tourists to explore places before deciding their trip plans. Intelligent transportation systems, which are powered by artificial intelligence, optimize routes and timetables, resulting in a travel experience that is both more efficient and more entertaining (Bazzan and Klügl, 2022).

8.5.2 CONSIDERATIONS OF ETHICAL IMPLICATIONS AND OBSTACLES

AI in the hotel and tourist industry is not without difficulties, just like any other technical progress (Nam et al., 2021). Several scholars have investigated data security and privacy concerns, highlighting the importance of implementing stringent controls to secure sensitive information about guests. Regarding guest interactions, in particular, ethical questions around artificial intelligence bring to light the delicate balance between automation and the human touch (Du and Xie, 2021). Several studies highlight the significance of using AI in a transparent manner that emphasizes the trust and happiness of guests.

8.5.3 TENDENCIES AND INNOVATIONS OF THE FUTURE

Engaging in speculative conversations about the future of artificial intelligence in the hotel and tourism industries is anticipating the future (Grundner and Neuhofer, 2021). More developments in natural language processing are expected, as well as a rise in the utilization of robotics for different operational duties and an ongoing evolution of tailored experiences driven by artificial intelligence (Pan and Zhang, 2021). Furthermore, the literature emphasizes the need for industry stakeholders to maintain flexibility in responding to constant technological advances.

8.5.4 PERSPECTIVES FROM AROUND THE WORLD ON THE IMPLEMENTATION OF ARTIFICIAL INTELLIGENCE (AI)

The literature examines the worldwide landscape of AI adoption in the hotel and tourist industries, considering regional variances driven by cultural influences, economic concerns, and legal frameworks for the hospitality industry. Dwivedi et al. (2022) explain that exploring different cultures gives a detailed knowledge of how artificial intelligence technologies are implemented and insights into how other civilizations interpret and adopt this technology.

8.5.5 THE INTERACTION BETWEEN ARTIFICIAL INTELLIGENCE (AI) AND HUMANS

The interplay between artificial intelligence and human connection in the hospitality and tourist industries is a significant topic woven into the fabric of literature. A detailed study on how automation and technology influence the conventional human-centered features of guest services has been carried out by academics in several disciplines. There are concerns surrounding the maintenance of the warmth and human touch characteristic of hospitality, even though chatbots and virtual assistants powered by artificial intelligence expedite operations and increase efficiency (Shvaikovskaia, 2022). Several studies highlight the significance of achieving a delicate equilibrium in which AI enhances human talents rather than completely replacing them.

8.5.6 REPERCUSSIONS FOR EDUCATION AND THE DEVELOPMENT OF SKILLS

When preparing the workforce for a future in hospitality dominated by artificial intelligence, the literature acknowledges the importance of education and the development of skills critical to providing training programs that provide workers with the skills necessary to utilize and manage artificial intelligence technology. Furthermore, studies highlight the necessity for educational institutions to modify their course to incorporate topics related to artificial intelligence. This will ensure that the next generation of industry professionals is adequately prepared to navigate the complexities of a technologically enhanced landscape (Gill et al., 2022).

8.5.7 IMPACT OF ARTIFICIAL INTELLIGENCE (AI) ON SOCIETY AND THE ECONOMY

Furthermore, the literature highlights the more significant social and economic influence that artificial intelligence will have on the hotel and tourism industries, in addition to the operational and guest-centric implications. Considering the possibility that automation could result in the loss of jobs, it is essential to have conversations about methods that can be used to retrain and educate the workforce (Tschang and Almirall, 2021). The economic repercussions of AI-driven efficiency compare the prospective cost reductions to the initial expenditures necessary for implementing the technology.

8.5.8 THE ACCELERATION OF ARTIFICIAL INTELLIGENCE (AI) ADOPTION AND THE GLOBAL PANDEMIC

The worldwide pandemic's effect in speeding up the deployment of artificial intelligence in the hotel and tourist industries is notable. During the pandemic, there has been a greater emphasis on adopting artificial intelligence technology due to the necessity of contactless solutions, increased safety measures, and operational efficiency (Gaur et al., 2021). Studies investigate how artificial intelligence has acted as a catalyst for digital transformation and innovation, facilitating the industry's adaptability to issues that have never been seen before.

8.5.9 A DYNAMIC INTERSECTION OF ARTIFICIAL INTELLIGENCE (AI) AND HUMAN TOUCH

Within the realm of hospitality and tourism, the literature presents a dynamic picture of the interaction of artificial intelligence and the human touch in the context of future forecasting. The literature agrees that the human element will continue to be irreplaceable despite the progress of artificial intelligence technologies. Human hospitality may be distinguished by its capacity for genuine and empathic relationships and encounters, demonstrating cultural awareness and intuitive reactions (Lv et al., 2022). Therefore, the challenge comes in utilizing artificial intelligence to improve and strengthen these attributes, cultivating a mutually beneficial relationship between technology innovation and the core of genuine hospitality.

8.5.10 ETHICAL CONSIDERATIONS AND RESPONSIBLE ARTIFICIAL INTELLIGENCE (AI)

The literature highlights the necessity of responsible AI practices in the hospitality and tourist industries, which expands on the previously mentioned issue of ethical considerations. The application of artificial intelligence in managing visitor data, personal preferences, and behavioral patterns calls for a rigorous approach to protecting privacy and security. To build ethical frameworks and principles that will assist industry practitioners in ensuring that artificial intelligence applications put user consent, transparency, and fairness at the forefront of their priorities (Kaur et al., 2022). When it comes to preventing potential biases and prejudice, discussions revolve around the responsible collecting and usage of data. This is done to advocate for an ethically sound integration of artificial intelligence technology.

8.5.11 IMPACT ON THE ENVIRONMENT AND LONG-TERM SUSTAINABILITY

The convergence of AI, sustainability, and environmental effects in the hospitality and tourist industries is a topic that has been receiving increasing attention in contemporary works of literature. One possibility for the sector to align itself with global sustainability goals is using intelligent technology to reduce waste and increase operational energy efficiency (Javaid et al., 2022). The dialog that is taking place here reflects a growing awareness of the environmental footprint that artificial intelligence applications leave behind and the potential for these technologies to contribute positively to the larger sustainability agenda.

8.5.12 USING ARTIFICIAL INTELLIGENCE (AI), INCLUSIVE TOURISM

When academics investigate how artificial intelligence may help make tourism more accessible and inclusive, accessibility and inclusivity are at the forefront of their attention. Applications that are powered by artificial intelligence have the potential to break down barriers and make travel experiences more accessible to a broader demographic (Stone et al., 2022). Some examples of these applications include language translation services, accessibility features in accommodations, and personalized travel recommendations for individuals with diverse needs. The chapter emphasizes artificial intelligence's revolutionary influence on developing inclusive tourism practices, which helps ensure that technological advantages are available to everyone.

8.5.13 COMMUNITY ENGAGEMENT AND THE EMPOWERMENT OF LOCAL COMMUNITIES

Additionally, the literature investigates the potential of artificial intelligence to encourage community participation and increase local players' agencies within the hospitality and tourist sectors. Artificial intelligence applications that support community-driven tourism initiatives assist local businesses and encourage cultural authenticity, which are being investigated as potential avenues for developing more sustainable and mutually beneficial tourism practices between parties (Anaba et al., 2023). The sector can stimulate economic growth at the grassroots level and strengthen the entire socioeconomic fabric of destinations if it uses artificial intelligence in ways that prioritize community engagement.

8.5.14 COOPERATION BETWEEN DIFFERENT INDUSTRIES

When harnessing the potential of artificial intelligence in the hotel and tourist industries, the study underlines the necessity of cross-industry collaboration. It is believed that collaborative efforts between the technology industry, academic institutions, policymakers, and hospitality professionals are essential to fostering innovation, addressing challenges, and ensuring that applications of artificial intelligence align with broader societal goals (Dwivedi et al., 2021). Much research is needed for open conversations and knowledge-sharing platforms that cultivate a collaborative environment. This would allow the industry to manage the intricacies of artificial intelligence integration cooperatively.

8.5.15 HOW POLICIES OF THE GOVERNMENT PLAY A ROLE

Government policies significantly influence the trajectory of artificial intelligence integration in the hotel and tourist industries (Tuo et al., 2021). This chapter investigates the influence of legal frameworks, data protection legislation, and ethical principles on adopting artificial intelligence within the business. To establish an environment in which honest AI activities are encouraged and possible dangers are minimized through rigorous legal measures, chapters investigate the role that governments play in striking a balance between innovation and the protection of consumer rights.

8.5.16 PERSONALIZATION AND THE EMERGENCE OF CHANGING CONSUMER EXPECTATIONS

The literature reveals the changing landscape of customer expectations as it looks further into artificial intelligence's influence on the hotel and tourist industries. The rise of technology driven by artificial intelligence has led passengers to anticipate highly tailored and frictionless experiences (Srivastava et al., 2021). These studies investigate how artificial intelligence algorithms examine enormous databases to comprehend human preferences. This understanding enables the customization of trip itineraries, choices of accommodations, and leisure activities. The capacity of artificial intelligence to anticipate and satisfy customers' requirements in real-time places it in a position to be a significant facilitator in meeting the elevated expectations of the modern traveler.

8.5.17 IMPROVEMENTS IN OPERATIONAL EFFICIENCY AND DECISION-MAKING IN REAL TIME

Artificial intelligence's role in improving real-time decision-making and boosting operational efficiency in the hotel and tourist industry is a significant subject that has emerged from published chapters in recent years. Businesses utilize machine learning algorithms to evaluate complicated information to optimize inventory management, pricing strategies, and resource allocation (Wang and Aviles, 2023). These algorithms provide significant insights that will help enterprises to improve their operations. This real-time adaptation not only helps to enhance internal processes but also contributes to developing an industry that is more flexible and responsive and able to deal with varying needs and unexpected obstacles.

8.5.18 ARTIFICIAL INTELLIGENCE (AI) IN THE MANAGEMENT OF EMERGENCIES AND RESILIENCE

Throughout the literature, artificial intelligence is emphasized as playing an essential part in crisis management and the development of resilience within the hotel and tourist industries. The study shed light on situations in which artificial intelligence-powered solutions have been proven to assist in emergencies such as the worldwide pandemic. Artificial intelligence applications allow for the fast analysis of changing circumstances, the facilitation of contact with stakeholders, and the availability of assistance in making informed decisions during difficult times (Gil et al., 2021). Practitioners in the sector may learn vital lessons from the insights from such study, which can help them better prepare for and navigate future crises.

8.5.19 REGARDING THE TECHNOLOGY OF BLOCKCHAIN AND TRUST IN TRANSACTIONS

Blockchain technology in the hotel and tourist industries is a trending topic currently being investigated in academic literature. It is exploring the potential of blockchain technology to improve the trustworthiness, security, and transparency of financial transactions. Smart contracts, decentralized platforms, and ledgers resistant to tampering all construct a trustworthy and safe environment for conducting financial transactions, exchanging data, and contacting guests (Xu, 2022). According to the study that has been undertaken, blockchain technology is expected to play a crucial part in addressing issues regarding the privacy and security of data.

8.5.20 THE APPLICATION OF ARTIFICIAL INTELLIGENCE (AI) AND NEUROSCIENCE TO IMPROVE EXPERIENCES

Studies that have been conducted recently investigate the potential for brain-computer interfaces and emotion detection technologies to further improve the experiences that guests have. This junction between neuroscience and artificial intelligence is a fascinating one. It is possible for artificial intelligence systems to dynamically change surroundings, services, and products by analyzing emotional reactions and preferences (Huang and Rust, 2021). This allows for the creation of experiences that are immersive and emotionally resonant. This chapter paves the way for a new age of hospitality, in which artificial intelligence not only predicts the practical demands of guests but also caters to the emotional and experiential considerations that contribute to their overall pleasure.

8.5.21 ARTIFICIAL INTELLIGENCE (AI) INTERFACES THAT ARE CONSTANTLY MAKING PROGRESS

In the unfolding narrative of AI in hospitality and tourism, the literature draws attention to the continuous evolution of AI interfaces. Natural language processing, voice recognition, and gesture-based interfaces are explored as avenues to enhance the accessibility and user-friendliness of AI applications. As technology progresses, the industry will likely witness more intuitive and seamless interfaces that blur the lines between human interaction and AI-driven services (Roslan and Ahmad, 2023). These advancements contribute to a more immersive guest experience. They are pivotal in making AI technologies more inclusive for individuals with diverse needs.

8.5.22 ARTIFICIAL INTELLIGENCE (AI) IN CULINARY EXPERIENCES AND FOOD SERVICES

Recent literature sheds light on the intersection of AI and culinary experiences within the hospitality sector. AI applications are being integrated into restaurant operations to optimize inventory management, predict customer preferences, and even innovate in recipe creation (Swink et al., 2022). From automated kitchen processes to AI-driven menu recommendations, these technologies are reshaping the gastronomic landscape. The chapter delves into the implications for culinary creativity, operational efficiency, and the overall enhancement of the dining experience, emphasizing AI's potential to revolutionize the food and beverage sector.

8.5.23 DATA-DRIVEN MARKETING AND GUEST ENGAGEMENT

The literature underscores the pivotal role of AI in data-driven marketing strategies and guest engagement. AI algorithms analyze vast datasets to gain insights into consumer behaviors, preferences, and trends (Bharatiya, 2023). Businesses can tailor their offerings to meet individualized guest expectations through personalized marketing campaigns, targeted promotions, and dynamic pricing strategies. The study explores the nuances of balancing customized marketing and privacy concerns, emphasizing the importance of transparent communication and ethical data practices in maintaining guest trust.

8.5.24 ARTIFICIAL INTELLIGENCE (AI) AND PREDICTIVE ANALYTICS FOR TOURISM PLANNING

The literature focuses on destination management, where AI and predictive analytics are crucial in tourism planning. AI-driven algorithms analyze historical data, social media trends, and external factors to forecast tourist arrivals, peak seasons, and potential challenges (Doborjeh et al., 2022). This proactive approach enables destinations to implement strategic measures for crowd management, resource allocation, and sustainable tourism practices. The integration of AI in tourism planning enhances efficiency. It contributes to the preservation of natural and cultural assets in popular destinations.

8.5.25 CYBERSECURITY MEASURES IN ARTIFICIAL INTELLIGENCE (AI) INTEGRATION

The literature recognizes the growing importance of robust cybersecurity measures in the context of AI integration in hospitality and tourism. The industry faces heightened cybersecurity risks as AI systems become more interconnected and reliant on data exchange (Muheidat and Tawalbeh, 2021). The study delves into the development of secure AI architectures, encryption techniques, and proactive cybersecurity strategies to safeguard against potential threats. The emphasis is on creating resilient systems that can withstand cyber-attacks, ensuring the integrity and confidentiality of sensitive guest and operational data.

8.5.26 ARTIFICIAL INTELLIGENCE (AI) AND SOCIAL RESPONSIBILITY IN TOURISM

An emerging theme in recent literature revolves around social responsibility in tourism, driven by AI applications. Studies explore how AI can be leveraged to promote ethical and sustainable practices within the industry. From carbon footprint calculations and eco-friendly recommendations to supporting local communities, AI contributes to a more socially responsible approach to tourism (Abdou et al., 2022). The literature delves into case studies where AI-driven initiatives align with global sustainability goals, fostering a sense of responsibility towards the environment and the community hosting travelers.

8.5.27 NEURO-INCLUSIVE DESIGN AND ARTIFICIAL INTELLIGENCE (AI) INTERFACES

In a pioneering exploration, the literature delves into neuro-inclusive design in AI interfaces within hospitality. The study investigates how AI applications can be designed to cater to individuals with diverse cognitive needs, ensuring an inclusive and accommodating experience for all guests. Integrating neuro-inclusive principles into AI interfaces reflects a commitment to accessibility and a departure from one-size-fits-all solutions. This represents a significant stride towards creating environments where individuals with varying cognitive abilities can fully engage with and enjoy the benefits of AI-driven services.

8.5.28 EMBRACING AUGMENTED HUMAN INTELLIGENCE

The literature dives into the concept of augmented human intelligence, positioning AI as a tool to enhance human capabilities rather than replace them. Studies explore how AI can empower hospitality and tourism professionals by automating routine tasks, allowing them to focus on creative problem-solving, guest interactions, and strategic decision-making. The synergistic integration of AI and human intelligence emerges as a theme, highlighting the potential for a collaborative future where technology amplifies rather than diminishes the roles of individuals within the industry (Abdou et al., 2022).

8.5.29 QUANTIFYING THE ECONOMIC IMPACT OF ARTIFICIAL INTELLIGENCE (AI) ADOPTION

The study delves into the economic implications of widespread AI adoption in hospitality and tourism. Studies aim to quantify the impact of AI on revenue generation, cost savings, and overall business performance.

By analyzing key performance indicators and financial metrics, scholars provide insights into the return on investment (ROI) for businesses embracing AI technologies. This economic lens contributes valuable data for industry decision-makers, investors, and policymakers seeking to understand the financial implications of AI integration (Gupta et al., 2022).

8.5.30 THE ROLE OF ARTIFICIAL INTELLIGENCE (AI) IN CRISIS RESILIENCE

Building upon the discussion of crisis management, the literature explores how AI contributes to overall resilience in the face of unpredictable events. AI-driven predictive analytics and scenario planning assist businesses in developing robust contingency plans (Chakraborty et al., 2011). From natural disasters to geopolitical shifts, the ability to anticipate, adapt, and recover from crises positions AI as a pivotal component in ensuring hospitality and tourism enterprises' long-term sustainability and viability.

8.5.31 ARTIFICIAL INTELLIGENCE (AI) AND PERSONAL DATA GOVERNANCE

Ethical considerations in AI adoption extend to the governance of personal data. The literature examines frameworks and best practices for responsible data management, emphasizing the importance of transparency, informed consent, and data ownership. As AI relies on vast datasets for analysis, protecting guest privacy and complying with evolving data protection regulations emerge as critical considerations (Walters and Novak, 2021). The study delves into developing ethical guidelines and policies to navigate the delicate balance between leveraging data for enhanced services and safeguarding individual privacy rights.

8.5.32 ARTIFICIAL INTELLIGENCE (AI) IN CULTURAL PRESERVATION AND HERITAGE TOURISM

A noteworthy dimension explored in recent literature is the role of AI in cultural preservation and heritage tourism. The study investigates how AI technologies contribute to preserving cultural artifacts, languages, and historical sites. AI-driven applications such as language translation services, virtual reconstructions of historical sites, and personalized cultural experiences cater to a growing demand for authentic and immersive heritage tourism (Vinnakota et al., 2023). This symbiotic relationship between AI and cultural preservation aligns with the broader societal trend toward fostering appreciation for diverse cultures.

8.5.33 CONSUMER EDUCATION AND ARTIFICIAL INTELLIGENCE (AI) LITERACY

The literature recognizes the need for consumer education and AI literacy to ensure the successful integration of these technologies. As AI becomes more ingrained in hospitality and tourism experiences, studies explore strategies to enhance consumer understanding of AI applications, dispel myths, and foster trust. By promoting AI literacy among guests, the industry can create a more informed and receptive audience that actively participates in and benefits from the evolving landscape of innovative and AI-driven services (Venkatesan and Lecinski, 2021).

8.5.34 ARTIFICIAL INTELLIGENCE (AI)-POWERED PREDICTIVE MAINTENANCE IN INFRASTRUCTURE

The integration of AI extends beyond guest services to infrastructure maintenance. The literature explores AI-powered predictive maintenance as a transformative approach for optimizing the lifespan and efficiency of hospitality assets. By analyzing data from sensors and IoT devices, AI algorithms predict equipment failures, allowing for proactive maintenance and minimizing downtime (Durbhaka, 2021). This ensures the seamless operation of critical systems and contributes to sustainable practices by reducing unnecessary resource consumption and waste.

8.5.35 HUMAN-CENTRIC ARTIFICIAL INTELLIGENCE (AI) DESIGN PRINCIPLES

Recognizing the importance of user experience, recent studies delve into human-centric AI design principles within hospitality and tourism.

These principles emphasize the development of AI interfaces that prioritize user understanding, empathy, and adaptability. The literature explores how implementing these principles can lead to interfaces seamlessly integrating into the guest experience, fostering a sense of comfort and familiarity. Designing AI applications with a human touch becomes imperative in maintaining the industry's commitment to personalized and guest-centric services (Battour et al., 2023).

8.5.36 ARTIFICIAL INTELLIGENCE (AI)'S ROLE IN SUSTAINABLE TOURISM PRACTICES

The literature expands its focus on sustainability, exploring how AI can drive the adoption of environmentally conscious practices in the tourism sector. AI applications contribute to sustainable tourism by optimizing transportation routes to reduce carbon emissions, providing data-driven insights for eco-friendly accommodations, and supporting conservation efforts in popular destinations. Integrating AI becomes a strategic tool for aligning the industry with global sustainability goals, balancing tourism growth and environmental preservation (Zhanbayev et al., 2023).

8.5.37 CROSS-INDUSTRY SYNERGIES WITH ARTIFICIAL INTELLIGENCE (AI)

Through AI collaboration, the study explores potential synergies between the hospitality and tourism industry and other sectors. Case studies and analyses showcase how insights and technologies developed in healthcare, entertainment, or retail can be adapted to enhance guest experiences, streamline operations, and foster innovation within the hospitality sector. Cross-industry collaborations harness the collective intelligence of diverse domains, presenting opportunities for unique and unexpected advancements in hospitality and tourism (Yadav et al., 2021).

8.5.38 ARTIFICIAL INTELLIGENCE (AI)-ENHANCED TRAINING AND SKILL DEVELOPMENT

As the industry adapts to the demands of AI integration, the literature examines the role of AI in training and skill development for hospitality professionals. AI-driven training platforms, simulations, and virtual reality applications offer immersive learning, allowing staff to acquire and refine skills in a controlled environment (Dai and Ke, 2022). The literature underscores the potential for AI to revolutionize workforce development, preparing individuals for the evolving demands of an AI-centric hospitality landscape.

8.5.39 ARTIFICIAL INTELLIGENCE (AI) AND PERSONAL WELLNESS IN HOSPITALITY

The potential of AI to contribute to personal wellness within hospitality experiences. From AI-powered wellness apps that provide personalized fitness and nutrition recommendations to innovative room environments that adjust lighting and ambiance based on guest preferences, the literature explores how AI applications can enhance the holistic well-being of guests. AI and wellness intersect with the industry's shift towards providing comprehensive and health-conscious hospitality services (Das, 2023).

8.5.40 ARTIFICIAL INTELLIGENCE (AI)-DRIVEN PERSONALIZED MARKETING STRATEGIES

The literature expands its exploration into the marketing realm, delving into how AI facilitates highly personalized and targeted strategies. AI-driven analytics scrutinize customer data to unveil nuanced insights into preferences, behaviors, and trends (Vuong and Mai, 2023). This wealth of information enables businesses to tailor marketing campaigns precisely, ensuring that promotional efforts resonate with individual guests. The literature showcases successful case studies where AI-powered marketing boosts engagement and conversions and fosters lasting customer loyalty.

8.5.41 EMOTIONAL INTELLIGENCE IN ARTIFICIAL INTELLIGENCE (AI)-ENHANCED GUEST INTERACTIONS

A compelling avenue of research in recent literature revolves around integrating emotional intelligence into AI systems for guest interactions. Studies delve into the development of AI algorithms capable of recognizing and responding to human emotions. This emotional intelligence enhances guest interactions by allowing AI interfaces to adapt their tone, language, and responses based on the emotional cues of the guests (Huang et al., 2021). The study discusses the implications of emotionally aware AI in creating more empathetic and human-like interactions, contributing to guests' overall satisfaction.

8.5.42 ARTIFICIAL INTELLIGENCE (AI) IN EVENT PLANNING AND MANAGEMENT

The literature extends its focus to event planning within the hospitality sector, exploring how AI can streamline and enhance the organization of conferences, meetings, and special events. AI applications assist venue selection, logistics planning, and attendee engagement (Ergen, 2021). AI-powered chatbots handle event-related inquiries, while machine learning algorithms optimize scheduling and resource allocation. The integration of AI in event management not only improves operational efficiency but also elevates the overall experience for both organizers and attendees.

8.5.43 HUMAN-ARTIFICIAL INTELLIGENCE (AI) COLLABORATION IN GUEST SERVICES

A nuanced exploration in recent literature centers on the collaborative synergy between humans and AI in providing guest services. The study delves into scenarios where AI is a supportive partner to human staff, handling routine tasks. In contrast, human professionals focus on delivering personalized and emotionally nuanced services (Dodds et al., 2022). This collaborative model recognizes the unique strengths of both AI and humans, fostering a guest experience that combines the efficiency of technology with the warmth and intuition inherent in human interactions.

8.5.44 THE EVOLUTION OF ARTIFICIAL INTELLIGENCE (AI) ETHICS AND BIAS MITIGATION

Ethical considerations surrounding AI adoption receive heightened attention in the literature, particularly addressing biases within AI algorithms.

The chapter investigates strategies for mitigating biases related to gender, race, and socioeconomic factors in hospitality and tourism applications. The literature explores the development of ethical guidelines, algorithmic audits, and ongoing monitoring practices to ensure that AI systems operate fairly and unbiasedly, aligning with the industry's commitment to inclusivity and diversity (Boza and Evgeniou, 2021).

8.5.45 ARTIFICIAL INTELLIGENCE (AI)-GENERATED CONTENT IN MARKETING AND STORYTELLING

A novel dimension in recent literature is examining AI-generated content in marketing and storytelling within the hospitality sector. AI applications, such as natural language generation, contribute to creating compelling narratives, blog posts, and promotional materials (Ray, 2023). The literature discusses the implications of AI-driven content creation for maintaining brand voice, ensuring authenticity, and engaging audiences in an era where storytelling remains a powerful tool for shaping the identity of hospitality brands.

8.5.46 ARTIFICIAL INTELLIGENCE (AI) AND REVENUE MANAGEMENT STRATEGIES

The literature deepens its exploration into revenue management strategies, emphasizing the role of AI in optimizing pricing and distribution. Dynamic pricing models powered by machine learning algorithms analyze market trends, competitor pricing, and demand fluctuations to recommend real-time adjustments (Salazar et al., 2023). The literature highlights how AI-driven revenue management not only maximizes profitability but also assists businesses in navigating the complexities of a rapidly changing market landscape.

8.5.47 ARTIFICIAL INTELLIGENCE (AI) IN TOURISM FORECASTING AND DESTINATION MANAGEMENT

The literature explores how AI contributes to tourism forecasting and sustainable destination planning by expanding its scope to destination management. AI algorithms analyze historical data, social media trends, and environmental factors to predict tourist influx, enabling destinations to implement strategies for responsible tourism (Rahmadian et al., 2022). The study underscores the potential for AI to guide destination managers in balancing economic growth with environmental conservation and community well-being.

8.5.48 PUBLIC-PRIVATE PARTNERSHIPS IN ARTIFICIAL INTELLIGENCE (AI) ADOPTION

An emerging theme in recent literature is the exploration of public-private partnerships in driving AI adoption within the hospitality and tourism industry. The chapter investigates collaborations between governmental bodies, industry stakeholders, and technology providers to create ecosystems that support responsible and widespread AI integration. The literature discusses the role of these partnerships in fostering innovation, sharing resources, and addressing regulatory challenges to ensure a harmonized approach to AI adoption.

8.6 PROBLEMS THAT COME UP WHEN TRYING TO USE ARTIFICIAL INTELLIGENCE (AI) IN TOURISM AND HOSPITALITY

The fact that many guests' data is being collected and used makes people worry about data security and privacy. Protecting sensitive information and ensuring rules are followed are the most important things which include:

- The problem is that AI technologies come with significant upfront hardware, software, and training costs. It might be hard for smaller businesses to make this first investment.
- It can be hard to combine AI systems with older infrastructure. Ensuring innovative technologies and workflows work with old ones without problems is challenging.
- Teaching employees how to use and understand AI technologies can take much time.
- Some guests may not want to use AI-powered services because they are not sure will work.
- It is essential to know a lot about the law to follow the many international and regional rules about data protection, consumer rights, and the ethical use of AI.

- Ensuring AI is used ethically, that algorithms are not biased, and that concerns about job loss are addressed are all critical issues that need careful thought.

8.7 OPPORTUNITIES THAT COME UP WITH THE USE OF ARTIFICIAL INTELLIGENCE (AI)

AI allows guests personalized and smooth experiences, from automatic check-ins to personalized suggestions that lead to loyalty and complimentary reviews are:

- AI-powered automation streamlines tasks like housekeeping and inventory management, cutting costs and making things run more smoothly.
- Using AI analytics gives helpful information about what guests want, how the business is doing, and how the market is changing, which helps make intelligent decisions.
- AI makes it easier to create new services, like chatbot concierges, voice-activated room controls, and augmented reality experiences, which helps businesses stand out from their competitors.
- AI-driven dynamic pricing strategies raise or lower prices based on changes in demand, market trends, and competitors' costs to make the most money (Figure 8.1).
- AI-driven predictive maintenance reduces downtime by predicting when equipment will break down. This keeps facilities in attractive shape and guests from waiting too long.
- AI helps with sustainability efforts by maximizing energy use, reducing waste, and using eco-friendly methods, which fits the growing trend of eco-friendly tourism.
- Businesses can adapt to changing customer expectations, modern technologies, and best practices in the industry by keeping up with AI trends.

Using AI allows employees to learn new skills and advance in their careers, encouraging a culture of always learning and adapting. Cross-industry synergies can happen when companies in different industries work together or partner with AI providers. This can lead to new developments and shared innovations.

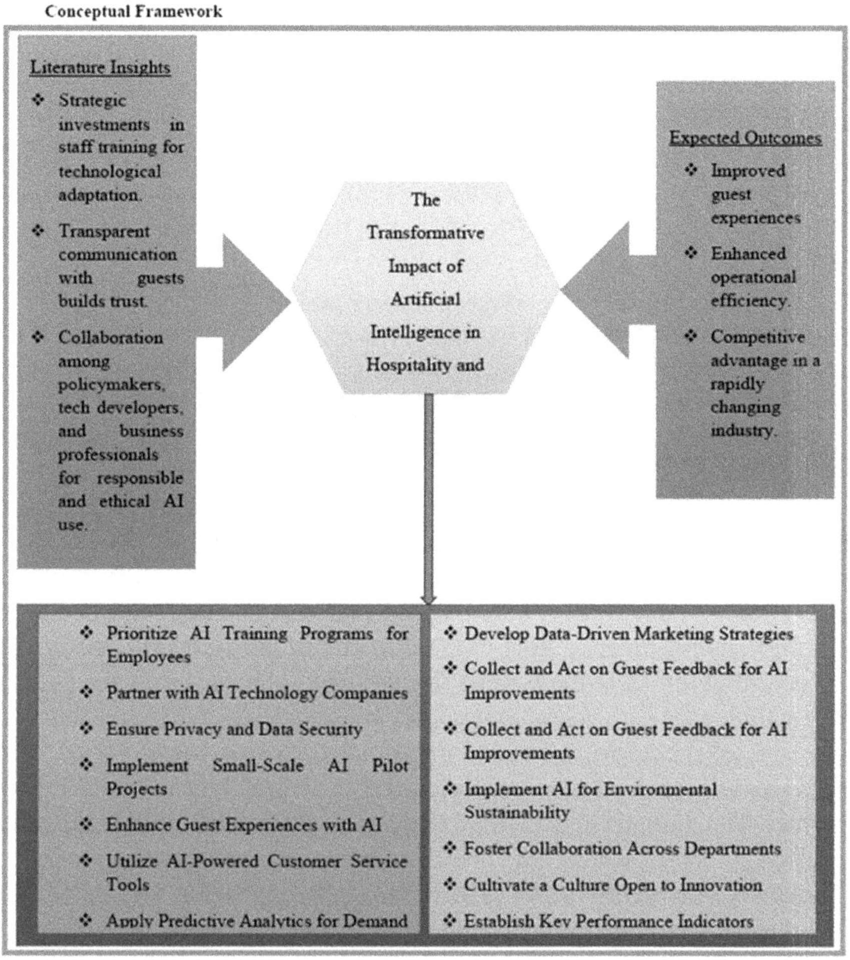

FIGURE 8.1 Conceptual framework.

Source: Authors compilation.

8.8 RESULTS

The integration of cutting-edge technologies, particularly AI, has ushered in a transformative era for the hospitality and tourism industry. Adopting AI-driven solutions has resulted in notable improvements across various facets of the industry. These advancements include heightened customization of services, increased operational efficiency, and the introduction of cutting-edge offerings, collectively redefining the landscape of the hotel sector. AI's impact on visitor experiences is evident in the enhanced personalization and tailored services offered to guests. Furthermore, operational processes have witnessed streamlining, increasing efficiency, and resource optimization. The competitive edge gained through deploying AI technologies positions businesses favorably in the ever-evolving market.

However, the journey towards AI integration is not without its challenges. Concerns about data security have emerged as a significant obstacle, necessitating careful consideration and robust measures to safeguard sensitive information. Workforce upskilling has become a priority to harness the full potential of AI, requiring ongoing learning initiatives to align human skills with technological advancements. Ethical considerations surrounding AI practices have surfaced, emphasizing the need for responsible and transparent implementation.

In navigating these challenges, the study highlights the importance of adherence to regulations, continuous learning initiatives for the workforce, and the establishment of trust with both clients and staff. The successful incorporation of AI in the hospitality industry holds promises for predictive maintenance, revenue optimization, sustainability programs, and overall market competitiveness.

AI is expected to play an increasingly pivotal role as the industry evolves. The findings of this study emphasize the necessity for continuous research to stay abreast of the dynamic changes, address emerging challenges, and identify new opportunities for innovation. The successful future development of the hospitality and tourism industry will hinge on a balanced integration of human touch and technological innovation, ensuring that AI remains a facilitator of enhanced guest experiences rather than a replacement for the human element.

8.9 DISCUSSION

Integrating AI in the hospitality and tourism industry presents a paradigm shift with profound implications. As businesses increasingly adopt AI-driven technologies to enhance operational efficiency and guest experiences, several crucial dimensions warrant thoughtful consideration.

Firstly, ethical considerations surrounding AI practices demand sustained attention. Ensuring transparency in AI algorithms, mitigating biases, and understanding the societal implications of automation are imperative. Ethical AI practices are not only ethically sound but also contribute to building trust among stakeholders, including guests, employees, and regulatory bodies.

The study's acknowledgment of concerns related to data security underscores the need for robust measures to safeguard sensitive information. As AI relies heavily on data, industry players must invest in secure infrastructure and stringent data protection policies to mitigate the risks of data breaches. Striking a balance between innovation and data security will be crucial for the sustained success of AI integration. The discussion also delves into the complex landscape of workforce dynamics. Upskilling the workforce to adapt to AI technologies is a significant challenge. Continuous learning initiatives are necessary to equip employees with the skills to collaborate effectively with AI systems. Additionally, the study suggests that the potential displacement of jobs due to AI adoption should be thoroughly examined. This involves understanding shifts in employment responsibilities, identifying necessary skill sets, and exploring the possibility of creating new roles in response to evolving industry needs.

The psychological aspects of guest interactions with AI-driven services emerge as a critical area for further investigation. The study rightly points out the need to explore how guests perceive and embrace AI in the hospitality sector. This involves understanding the psychological effects of AI integration, the factors influencing trust, and preferences for human versus AI interactions. Such insights are invaluable for tailoring AI applications to align with guest expectations and preferences.

The study's forward-looking perspective suggests several prospective routes for further research. Examining the long-term effects of AI adoption on employment, understanding global standards and laws related to AI in the hotel industry, and exploring the synergy of AI with emerging technologies are highlighted as potential areas for future inquiry. Additionally, the study underscores the importance of dynamic customer journey mapping, cross-industry partnerships, and the sustainability implications of AI adoption.

In conclusion, the discussion emphasizes the need for a holistic and ethical approach to AI integration in the hospitality and tourism sector. The ongoing dialogue surrounding AI's impact on workforce dynamics, guest experiences, and industry practices is vital for navigating the evolving landscape successfully. As the industry continues to embrace AI, continuous research and strategic considerations are crucial to harnessing the benefits while addressing associated challenges and ensuring a harmonious coexistence of human and technological elements.

8.10 RECOMMENDATIONS

If you work in the tourism and hospitality industry and want to use artificial intelligence (AI) to make a big difference, here are some suggestions:

- Make training programs for employees that help them learn about and use AI a top priority. This includes workers at all levels, from those who enable customers to those in charge of making decisions.
- Make deals with companies that use AI technology to access the newest solutions. Working together with tech experts to make AI apps fit the business's needs is essential.
- Follow privacy laws and put in place robust data security measures. Trust from guests is essential, and ensuring that personal information is handled safely is critical for AI to be adopted successfully.
- Start integrating AI with smaller pilot projects to see if it is possible and how it will affect operations. This lets improvements be made over time and lowers the risk of problems with large-scale implementation.
- Use AI to make guests' experiences more unique. Use data analytics to learn what guests like to provide more personalized services, suggestions, and marketing campaigns.
- Add AI-powered chatbots, voice-activated controls, and virtual assistants to improve customer service. Automation can make things run more smoothly, reduce wait times, and make things more efficient.
- Use predictive analytics powered by AI to get accurate predictions about demand. This helps ensure that resources are used effectively, manage inventory, and develop the best pricing strategies.
- Look into using AR and VR to make experiences more immersive and exciting. This can be done with virtual tours, interactive maps, and added travel information.
- Use AI to make marketing strategies that are based on data. AI algorithms can improve targeted advertising, personalized promotions, and dynamic pricing, which makes customers more interested.

- Actively ask guests for feedback on services that use AI. Use this feedback to improve and update AI applications to meet changing guest needs.
- Keep up with new AI trends in the travel and hospitality industry. Stay updated on new ideas and be inspired by them by going to industry conferences and forums and talking to thought leaders.
- Use AI in efforts to be more environmentally friendly. AI can help make things more eco-friendly by managing energy and reducing waste, which fits the growing trend of eco-conscious tourism.
- Get the IT, marketing, operations, and guest services departments to work together more. A unified approach ensures that AI is used throughout the company without any problems, creating a system that works well and is suitable.
- Encourage a culture that is open to new ideas and trying new things. Encourage employees to develop ideas for AI applications and be sure to recognize and reward those who do.
- Set up key performance indicators (KPIs) to see how adopting AI changes things. Review performance metrics regularly to see how well AI applications work and make decisions based on the data.

Based on previous literature, the literature gives valuable suggestions for people involved in the industry. Studies support strategic investments in staff training to keep up with technological changes because AI can change everything. Transparent communication with guests about how AI is used is emphasized as a crucial part of building trust. The literature also stresses how important it is for policymakers, technology developers, and business professionals to collaborate to build an ecosystem that encourages responsible and ethical AI use. By following these suggestions, companies in the tourism and hospitality industry can get the most out of artificial intelligence. This will improve guests' experiences, make operations run more smoothly, and help them stay competitive in an industry that is changing quickly.

8.11 CONCLUSION AND FUTURE RESEARCH DIRECTION

Businesses are progressively integrating cutting-edge technologies to improve visitor experiences, streamline operations, and maintain competitiveness in a constantly changing market, demonstrating the transformative impact of AI on the hospitality and tourism industry. AI-driven solutions have enhanced customization, increased operational efficiency, and cutting-edge services, redefining the hotel industry. Obstacles include worries about data security, the need to upskill the workforce, and the requirement for moral AI practices, highlighting how crucial it is to approach AI adoption with consideration and strategy. To overcome these obstacles, one must be dedicated to following regulations, pursuing continuous learning opportunities, and developing trust with clients and staff. AI offers a wide range of options, from predictive maintenance and revenue optimization to sustainability programs and competitiveness in the market. The industry's future might be shaped by more developments, partnerships, and the creation of novel, groundbreaking applications as companies leverage AI.

8.11.1 PROSPECTIVE ROUTES FOR THE STUDY

Subsequent studies ought to go more deeply into the moral issues surrounding artificial intelligence in the hospitality industry, looking at measures to guarantee transparency, reduce prejudice, and deal with the societal ramifications of automation such as:

- More studies are required to fully comprehend the long-term effects of AI adoption on employment in the hotel industry. Examining shifts in employment responsibilities, necessary skill sets, and the possibility of creating new jobs are all part of this.
- It is critical to look at how guests see and embrace AI-driven services. The psychological effects of AI integration in hospitality encounters, trust-influencing variables, and preferences for human versus AI interactions may be studied throughout the research.
- Studying international standards and laws about artificial intelligence in the hotel industry is essential. This entails being aware of the various regulatory environments and creating structures that guarantee the uniform and appropriate deployment of AI worldwide.
- A critical topic of inquiry is how AI will affect sustainability practices in the hospitality industry over the long run. This involves evaluating how well AI-driven programs work to promote eco-friendly operations and lessen environmental impact.
- In the future, the chapter should examine how AI in hospitality is integrated with emerging technologies like Blockchain and the IoT.

Understanding the synergy of these technologies can result in new uses and improved capabilities.

- Studies that adjust to in-the-moment visitor interactions can investigate dynamic customer journey mapping. This entails realizing how AI can adapt and continually improve visitor experiences in response to changing tastes and actions.
- Looking at AI-related cross-industry partnerships and innovations in the hotel sector is fascinating. It would be interesting to see how artificial intelligence developments from other industries may be applied to improve visitor experiences and operational effectiveness.

AI will continue to influence the hospitality and tourist industries in the future. The study must be done continuously to stay ahead of this change, tackle new issues, and create new opportunities for innovation and superior guest experiences. The next phase of the industry's development will be determined mainly by how human touch and technology innovation work together.

8.12 IMPLICATIONS

8.12.1 THEORETICAL IMPLICATIONS

1. **Advancement of Technology Acceptance Models (TAM):** The study contributes to the theoretical underpinnings of technology acceptance models, particularly TAM, by providing empirical insights into the perceptions and acceptance of AI in the hospitality and tourism sector. The findings can inform the refinement and extension of TAM frameworks to better capture the nuances of AI adoption in a dynamic industry context.
2. **Ethical Frameworks for AI Integration:** The study lays the groundwork for developing ethical frameworks for AI integration in the hospitality industry. It highlights the importance of transparency, bias mitigation, and societal considerations, offering a theoretical basis for future research in the ethical dimensions of AI applications.
3. **Dynamic Customer Journey Mapping:** The study introduces the concept of dynamic customer journey mapping, emphasizing

the adaptability of AI to in-the-moment visitor interactions. This contributes to the theoretical understanding of how AI can continuously enhance and tailor guest experiences in response to changing preferences and behaviors.

8.12.2 PRACTICAL IMPLICATIONS

1. **Guidelines for Artificial Intelligence (AI) Adoption:** The study provides practical guidelines for businesses in the hospitality and tourism industry looking to adopt AI technologies. It emphasizes the significance of data security measures, continuous workforce upskilling, and ethical AI practices, offering a roadmap for responsible AI adoption.
2. **Stakeholder Trust Building:** The emphasis on following regulations, continuous learning initiatives, and establishing trust with clients and staff has direct practical implications for industry practitioners. Fostering trust is a critical component for successful AI implementation, and it can have tangible effects on guest satisfaction and organizational reputation.
3. **Strategic Planning for Sustainability:** The study suggests that AI can influence sustainability practices in the long run. This implies that businesses should strategically integrate AI-driven programs to promote eco-friendly operations and reduce environmental impact, aligning with the industry's recent efforts to address sustainability concerns.
4. **Informed Decision-Making for Industry Leaders:** Industry leaders can leverage the study's insights for informed decision-making. Understanding the challenges, opportunities, and potential pitfalls associated with AI adoption allows decision-makers to navigate the complexities of the evolving landscape, ensuring a strategic and sustainable integration of AI technologies.

In summary, the study offers theoretical advancements and practical guidance, serving as a valuable resource for academics seeking to expand theoretical frameworks and industry professionals aiming to harness the transformative potential of AI in the hospitality and tourism sector. The implications outlined provide a roadmap for responsible, ethical, and strategic AI adoption, facilitating the industry's continued evolution.

KEYWORDS

- **artificial intelligence**
- **data security**
- **hospitality**
- **privacy**
- **technological evolution**
- **tourism**

REFERENCES

Abdou, A. H., Hassan, T. H., Salem, A. E., Albakhit, A. I., Almakhayitah, M. Y., & Salama, W. (2022). The nexus between environmentally sustainable practices, green satisfaction, and customer citizenship behavior in eco-friendly hotels: Social exchange theory perspective. *Sustainability, 14*(19), 12791. https://doi.org/10.3390/su141912791 (accessed on 8th July 2025).

Ahmad, T., Zhang, D., Huang, C., Zhang, H., Dai, N., Song, Y., & Chen, H. (2021). Artificial intelligence in sustainable energy industry: Status quo, challenges and opportunities. *Journal of Cleaner Production, 289*, 125834. https://doi.org/10.1016/j.jclepro.2021.125834(accessed on 8th July 2025).

Almusaed, A., Yitmen, I., & Almssad, A. (2023). Reviewing and integrating AEC practices into Industry 6.0: Strategies for smart and sustainable future-built environments. *Sustainability, 15*(18), 13464. https://doi.org/10.3390/su151813464 (accessed on 8th July 2025).

Anaba, M. I., Masud, M. M., & Ching, G. H. (2023). The role of social identity between community motivation and intention to participate in tourism development in Malaysia. *Environment, Development and Sustainability*, 1–24.

Bareis, J., & Katzenbach, C. (2022). Talking AI into being: The narratives and imaginaries of national AI strategies and their performative politics. *Science, Technology, & Human Values, 47*(5), 855–881. https://doi.org/10.1177/01622439211030007 (accessed on 8th July 2025).

Battour, M., Mady, K., Salaheldeen, M., Elsotouhy, M., Elbendary, I., & Boğan, E. (2023). AI-enabled technologies to assist Muslim tourists in Halal-friendly tourism. *Journal of Islamic Marketing, 14*(5), 1291–1309. https://doi.org/10.1108/JIMA-01-2022-0001 (accessed on 8th July 2025).

Bazzan, A. L. C., & Klügl, F. (2022). *Introduction to Intelligent Systems in Traffic and Transportation.* Springer Nature.

Bharadiya, J. P. (2023). Machine learning and AI in business intelligence: Trends and opportunities. *International Journal of Computer (IJC), 48*(1), 123–134.

Bharwani, S., & Mathews, D. (2021). Techno-business strategies for enhancing guest experience in luxury hotels: A managerial perspective. *Worldwide Hospitality and Tourism Themes*, *13*(2), 168–185. https://doi.org/10.1108/WHATT-09-2020-0121 (accessed on 8th July 2025).

Boza, P., & Evgeniou, T. (2021). Implementing AI principles: Frameworks, processes, and tools. https://dx.doi.org/10.2139/ssrn.3783124 (accessed on 8th July 2025).

Brem, A., Giones, F., & Werle, M. (2021). The AI digital revolution in innovation: A conceptual framework of artificial intelligence technologies for the management of innovation. *IEEE Transactions on Engineering Management*. https://doi.org/10.1109/TEM.2021.3109983 (accessed on 8th July 2025).

Chakraborty, A., Kaza, N., Knaap, G. J., & Deal, B. (2011). Robust plans and contingent plans: Scenario planning for an uncertain world. *Journal of the American Planning Association*, *77*(3), 251–266.

Dai, C. P., & Ke, F. (2022). Educational applications of artificial intelligence in simulation-based learning: A systematic mapping review. *Computers and Education: Artificial Intelligence*, 100087. https://doi.org/10.1016/j.caeai.2022.100087 (accessed on 8th July 2025).

Das, M. P. (2023). Technology and guest experience: Innovations reshaping hotel management. *International Journal for Multidimensional Research Perspectives*, *1*(3), 76–95.

Dingler, T., Kwasnicka, D., Wei, J., Gong, E., & Oldenburg, B. (2021). The use and promise of conversational agents in digital health. *Yearbook of Medical Informatics*, *30*(01), 191–199. https://doi.org/10.1055/s-0041-1726510 (accessed on 8th July 2025).

Doborjeh, Z., Hemmington, N., Doborjeh, M., & Kasabov, N. (2022). Artificial intelligence: A systematic review of methods and applications in hospitality and tourism. *International Journal of Contemporary Hospitality Management*, *34*(3), 1154–1176. https://doi.org/10.1108/IJCHM-06-2021-0767 (accessed on 8th July 2025).

Dodds, S., Russell–Bennett, R., Chen, T., Oertzen, A. S., Salvador-Carulla, L., & Hung, Y. C. (2022). Blended human-technology service realities in healthcare. *Journal of Service Theory and Practice*, *32*(1), 75–99. https://doi.org/10.1108/JSTP-12-2020-0285 (accessed on 8th July 2025).

Du, S., & Xie, C. (2021). Paradoxes of artificial intelligence in consumer markets: Ethical challenges and opportunities. *Journal of Business Research*, *129*, 961–974. https://doi.org/10.1016/j.jbusres.2020.08.024 (accessed on 8th July 2025).

Durbhaka, G. K. (2021). Convergence of artificial intelligence and Internet of Things in predictive maintenance systems–A review. *Turkish Journal of Computer and Mathematics Education (TURCOMAT)*, *12*(11), 205–214. https://doi.org/10.17762/turcomat.v12i11.5862 (accessed on 8th July 2025).

Dwivedi, Y. K., Hughes, L., Baabdullah, A. M., Ribeiro-Navarrete, S., Giannakis, M., Al-Debei, M. M., Dennehy, D., Metri, B., Buhalis, D., & Cheung, C. M. K. (2022). Metaverse beyond the hype: Multidisciplinary perspectives on emerging challenges, opportunities, and agenda for research, practice and policy. *International Journal of Information Management*, *66*, 102542. https://doi.org/10.1016/j.ijinfomgt.2022.102542 (accessed on 8th July 2025).

Dwivedi, Y. K., Hughes, L., Ismagilova, E., Aarts, G., Coombs, C., Crick, T., Duan, Y., Dwivedi, R., Edwards, J., & Eirug, A. (2021). Artificial intelligence (AI): Multidisciplinary perspectives on emerging challenges, opportunities, and agenda for research, practice and policy. *International Journal of Information Management*, *57*, 101994. https://doi.org/10.1016/j.ijinfomgt.2019.08.002 (accessed on 8th July 2025).

Ergen, F. D. (2021). Artificial intelligence applications for event management and marketing. In *Impact of ICTs on Event Management and Marketing* (pp. 199–215). IGI Global. https://doi.org/10.4018/978-1-7998-4954-4.ch012 (accessed on 8th July 2025).

Fan, X., Jiang, X., & Deng, N. (2022). Immersive technology: A meta-analysis of augmented/virtual reality applications and their impact on tourism experience. *Tourism Management*, *91*, 104534. https://doi.org/10.1016/j.tourman.2022.104534 (accessed on 8th July 2025).

Femenia-Serra, F., & Ivars-Baidal, J. A. (2021). Do smart tourism destinations really work? The case of Benidorm. *Asia Pacific Journal of Tourism Research*, *26*(4), 365–384. https://doi.org/10.1080/10941665.2018.1561478 (accessed on 8th July 2025).

Gaur, L., Afaq, A., Singh, G., & Dwivedi, Y. K. (2021). Role of artificial intelligence and robotics to foster the touchless travel during a pandemic: A review and research agenda. *International Journal of Contemporary Hospitality Management*, *33*(11), 4079–4098. https://doi.org/10.1108/IJCHM-11-2020-1246 (accessed on 8th July 2025).

Gil, Y., Garijo, D., Khider, D., Knoblock, C. A., Ratnakar, V., Osorio, M., Vargas, H., Pham, M., Pujara, J., & Shbita, B. (2021). Artificial intelligence for modeling complex systems: Taming the complexity of expert models to improve decision making. *ACM Transactions on Interactive Intelligent Systems*, *11*(2), 1–49. https://doi.org/10.1145/3453172 (accessed on 8th July 2025).

Gill, S. S., Xu, M., Ottaviani, C., Patros, P., Bahsoon, R., Shaghaghi, A., Golec, M., Stankovski, V., Wu, H., & Abraham, A. (2022). AI for next generation computing: Emerging trends and future directions. *Internet of Things*, *19*, 100514. https://doi.org/10.1016/j.iot.2022.100514 (accessed on 8th July 2025).

Grundner, L., & Neuhofer, B. (2021). The bright and dark sides of artificial intelligence: A futures perspective on tourist destination experiences. *Journal of Destination Marketing & Management*, *19*, 100511. https://doi.org/10.1016/j.jdmm.2020.100511 (accessed on 8th July 2025).

Gupta, S., Modgil, S., Bhattacharyya, S., & Bose, I. (2022). Artificial intelligence for decision support systems in the field of operations research: Review and future scope of research. *Annals of Operations Research*, 1–60.

Huang, D., Chen, Q., Huang, J., Kong, S., & Li, Z. (2021). Customer-robot interactions: Understanding customer experience with service robots. *International Journal of Hospitality Management*, *99*, 103078. https://doi.org/10.1016/j.ijhm.2021.103078 (accessed on 8th July 2025).

Huang, M. H., & Rust, R. T. (2021). A strategic framework for artificial intelligence in marketing. *Journal of the Academy of Marketing Science*, *49*, 30–50.

Jabeen, F., Al Zaidi, S., & Al Dhaheri, M. H. (2022). Automation and artificial intelligence in hospitality and tourism. *Tourism Review*, *77*(4), 1043–1061. https://doi.org/10.1108/TR-09-2019-0360 (accessed on 8th July 2025).

Jaime, F. J., Muñoz, A., Rodríguez-Gómez, F., & Jerez-Calero, A. (2023). Strengthening privacy and data security in biomedical microelectromechanical systems by IoT communication security and protection in smart healthcare. *Sensors*, *23*(21), 8944.

Javaid, M., Haleem, A., Singh, R. P., Suman, R., & Gonzalez, E. S. (2022). Understanding the adoption of Industry 4.0 technologies in improving environmental sustainability.

Sustainable Operations and Computers, *3*, 203–217. https://doi.org/10.1016/j.susoc.2022.01.008 (accessed on 8th July 2025).

Kaur, D., Uslu, S., Rittichier, K. J., & Durresi, A. (2022). Trustworthy artificial intelligence: A review. *ACM Computing Surveys (CSUR)*, *55*(2), 1–38. https://doi.org/10.1145/3491209 (accessed on 8th July 2025).

Kumar, I., Rawat, J., Mohd, N., & Husain, S. (2021). Opportunities of artificial intelligence and machine learning in the food industry. *Journal of Food Quality*, 1–10. https://doi.org/10.1155/2021/4535567 (accessed on 8th July 2025).

Li, J. P. O., Liu, H., Ting, D. S. J., Jeon, S., Chan, R. V. P., Kim, J. E., Sim, D. A., Thomas, P. B. M., Lin, H., & Chen, Y. (2021). Digital technology, tele-medicine and artificial intelligence in ophthalmology: A global perspective. *Progress in Retinal and Eye Research*, *82*, 100900. https://doi.org/10.1016/j.preteyeres.2020.100900 (accessed on 8th July 2025).

Li, M., Yin, D., Qiu, H., & Bai, B. (2021). A systematic review of AI technology-based service encounters: Implications for hospitality and tourism operations. *International Journal of Hospitality Management*, *95*, 102930. https://doi.org/10.1016/j.ijhm.2021.102930 (accessed on 8th July 2025).

Lv, X., Yang, Y., Qin, D., Cao, X., & Xu, H. (2022). Artificial intelligence service recovery: The role of empathic response in hospitality customers' continuous usage intention. *Computers in Human Behavior*, *126*, 106993. https://doi.org/10.1016/j.chb.2021.106993 (accessed on 8th July 2025).

Mannuru, N. R., Shahriar, S., Teel, Z. A., Wang, T., Lund, B. D., Tijani, S., Pohboon, C. O., Agbaji, D., Alhassan, J., & Galley, J. (2023). Artificial intelligence in developing countries: The impact of generative artificial intelligence (AI) technologies for development. *Information Development*, 02666669231200628. https://doi.org/10.1177/02666669231200628 (accessed on 8th July 2025).

Marchant, G. E., & Gutierrez, C. I. (2022). Soft Law 2.0: An agile and effective governance approach for artificial intelligence. *Minn. JL Sci. & Tech.*, *24*, 375.

Mercan, S., Cain, L., Akkaya, K., Cebe, M., Uluagac, S., Alonso, M., & Cobanoglu, C. (2021). Improving the service industry with hyper-connectivity: IoT in hospitality. *International Journal of Contemporary Hospitality Management*, *33*(1), 243–262. https://doi.org/10.1108/IJCHM-06-2020-0621 (accessed on 8th July 2025).

Muheidat, F., & Tawalbeh, L. (2021). Artificial intelligence and blockchain for cybersecurity applications. In *Artificial Intelligence and Blockchain for Future Cybersecurity Applications* (pp. 3–29). Springer.

Nam, K., Dutt, C. S., Chathoth, P., Daghfous, A., & Khan, M. S. (2021). The adoption of artificial intelligence and robotics in the hotel industry: Prospects and challenges. *Electronic Markets*, *31*, 553–574.

Pan, Y., & Zhang, L. (2021). Roles of artificial intelligence in construction engineering and management: A critical review and future trends. *Automation in Construction*, *122*, 103517. https://doi.org/10.1016/j.autcon.2020.103517 (accessed on 8th July 2025).

Rahmadian, E., Feitosa, D., & Zwitter, A. (2022). A systematic literature review on the use of big data for sustainable tourism. *Current Issues in Tourism*, *25*(11), 1711–1730. https://doi.org/10.1080/13683500.2021.1974358 (accessed on 8th July 2025).

Rakova, B., Yang, J., Cramer, H., & Chowdhury, R. (2021). Where responsible AI meets reality: Practitioner perspectives on enablers for shifting organizational practices.

Proceedings of the ACM on Human-Computer Interaction, *5*(CSCW1), 1–23. https://doi.org/10.1145/3449081 (accessed on 8th July 2025).
Ray, P. P. (2023). ChatGPT: A comprehensive review on background, applications, key challenges, bias, ethics, limitations and future scope. *Internet of Things and Cyber-Physical Systems*. https://doi.org/10.1016/j.iotcps.2023.04.003 (accessed on 8th July 2025).
Roslan, F. A. B. M., & Ahmad, N. B. (2023). The rise of AI-powered voice assistants: Analyzing their transformative impact on modern customer service paradigms and consumer expectations. *Quarterly Journal of Emerging Technologies and Innovations*, *8*(3), 33–64.
Ruel, H., & Njoku, E. (2021). AI redefining the hospitality industry. *Journal of Tourism Futures*, *7*(1), 53–66. https://doi.org/10.1108/JTF-03-2020-0032 (accessed on 8th July 2025).
Salazar, E. J., Samper, M. E., & Patiño, H. D. (2023). Dynamic customer demand management: A reinforcement learning model based on real-time pricing and incentives. *Renewable Energy Focus*, *46*, 39–56. https://doi.org/10.1108/JTF-03-2020-0032 (accessed on 8th July 2025).
Shvaikovskaia, E. (2022). *Artificial Intelligence for Innovative Service Design and Guest Experience: Case: Hotels in Germany.* Thesis. LAB University of Applied Sciences.
Srivastava, V., Kishore, S., & Dhingra, D. (2021). Technology and the future of customer experience. In *Crafting Customer Experience Strategy: Lessons from Asia* (pp. 91–116). Emerald Publishing Limited. https://doi.org/10.1108/978-1-83909-710-220211006 (accessed on 8th July 2025).
Stone, P., Brooks, R., Brynjolfsson, E., Calo, R., Etzioni, O., Hager, G., Hirschberg, J., Kalyanakrishnan, S., Kamar, E., & Kraus, S. (2022). Artificial intelligence and life in 2030: The one-hundred-year study on artificial intelligence. *ArXiv Preprint ArXiv*, 2211.06318. https://doi.org/10.48550/arXiv.2211.06318 (accessed on 8th July 2025).
Swink, M., Hu, K., & Zhao, X. (2022). Analytics applications, limitations, and opportunities in restaurant supply chains. *Production and Operations Management*, *31*(10), 3710–3726. https://doi.org/10.1111/poms.13704 (accessed on 8th July 2025).
Trinchini, L. (2021). *Value Creation in Smart Destinations. The Case of Manchester*. Manchester Metropolitan University.
Tschang, F. T., & Almirall, E. (2021). Artificial intelligence as augmenting automation: Implications for employment. *Academy of Management Perspectives*, *35*(4), 642–659.
Tuo, Y., Ning, L., & Zhu, A. (2021). How artificial intelligence will change the future of tourism industry: The practice in China. In *Information and Communication Technologies in Tourism 2021: Proceedings of the ENTER 2021 eTourism Conference, January 19–22, 2021* (pp. 83–94).
Venkatesan, R., & Lecinski, J. (2021). *The AI Marketing Canvas: A Five-Stage Road Map to Implementing Artificial Intelligence in Marketing*. Stanford University Press.
Vinnakota, S., Mohan, M. D., Boda, M. J., Askarzai, W., Devkota, M. P., Shetty, M. S., Wangmo, M. T., & Choden, M. T. (2023). Venturing into virtuality: Exploring the evolution, technological underpinnings, and forward pathways of virtual tourism. *Educational Research (IJMCER)*, *5*(6), 8–49.
Vuong, N. A., & Mai, T. T. (2023). Unveiling the synergy: Exploring the intersection of AI and NLP in redefining modern marketing for enhanced consumer engagement and strategy optimization. *Quarterly Journal of Emerging Technologies and Innovations*, *8*(3), 103–118.

Walters, R., & Novak, M. (2021). *Cyber Security, Artificial Intelligence, Data Protection & the Law*. Springer.

Wang, F., & Aviles, J. (2023). Enhancing operational efficiency: Integrating machine learning predictive capabilities in business intelligence for informed decision-making. *Frontiers in Business, Economics and Management*, *9*(1), 282–286. https://doi.org/10.54097/fbem.v9i1.8694 (accessed on 8th July 2025).

Xu, Z. (2022). *Application & Security of Smart Contract in Financial Industries*.

Yadav, J. K., Verma, D. C., Jangirala, S., & Srivastava, S. K. (2021). An IAD type framework for blockchain enabled smart tourism ecosystem. *The Journal of High Technology Management Research*, *32*(1), 100404. https://doi.org/10.1016/j.hitech.2021.100404 (accessed on 8th July 2025).

Zhanbayev, R. A., Irfan, M., Shutaleva, A. V., Maksimov, D. G., Abdykadyrkyzy, R., & Filiz, Ş. (2023). Demoethical model of sustainable development of society: A roadmap towards digital transformation. *Sustainability*, *15*(16), 12478.

CHAPTER 9

EFFECTS OF SMART TOURISM ON DESTINATION DEVELOPMENT: A SUSTAINABLE APPROACH

SHOHEL MD. NAFI[1*] and MARCO VALERI[2]

[1]*Department of Tourism and Hospitality Management, Noakhali Science and Technology University, Noakhali, Bangladesh*

[2]*Faculty of Economics, Niccolò Cusano University, Rome, Italy*

*Corresponding author

ABSTRACT

Smart tourism has been gaining a flourishing development direction in recent decades. This study will investigate the effects of smart tourism on destination development. The object of this study is to investigate the existing development, experiences, and challenges of smart tourism in constructing a destination. In the previous literature, numerous aspects of smart tourism are mentioned, but the outcomes of smart tourism for sustainable tourism destinations are still under investigation. This study will focus on the major components of sustainable tourism with a focus on smart tourism to overcome the current limitations. This study is qualitative research in nature. The information was acquired from secondary sources, including journal articles, relevant literature, websites, and various published works. It is expected that results will show a positive correlation between smart tourism and sustainable destination development. This study outlines the concept of smart city, the tourism stakeholders' perspectives on smart tourism, the relationship of ecotourism with smart tourism, and the smart tourism effect. Findings will reveal the essential

Smart Tourism: Managing Technological Revolution in the Tourism Industry. Santus Kumar Deb (Ed.)

DOI: 10.1201/9781779641090-9

elements, benefits, current opportunities, and obstacles of smart tourism in destination development. It will contribute significantly to the future theoretical and practical fields of study.

9.1 INTRODUCTION

The actual meaning of "smart" is "having quick intelligence," which has become incredibly popular after being used to characterize the features and qualities of specific technologies (Ye et al., 2020). Alshamaila et al. (2023) described that technologies are reshaping human behavior, industries, societies, and overall economic activities. Because they enable prompt, automatic, natural, and efficient interactions between human and nonhuman actors, a variety of new technologies are regarded as intelligent. The tourism and travel industry is more dependent on technological innovation than any other industry. Researchers indicate that, before, during, and after travel, technologies have radically altered and impacted traveler activities (Deb et al., 2024). From the beginning of the 21st century, the name "smart tourism" has been introduced in the tourism industry. Gorden Philips defined smart tourism in 2000 for the first time (Li et al., 2017). Later, in 2009, UNWTO used the term "smart tourism" in Madrid at the UNWTO Tourism Resilience Committee. Recently, smart tourism has gained greater attention among academicians, researchers, and policymakers.

The tourism industry is known for its vast social and economic impacts on the local people. Richardson (2010) mentioned that it is increasingly thought that tourism is essential to fostering economic growth and combating poverty. According to UNWTO (2023), around 975 million tourists traveled globally in the first nine months of 2023, which is almost 38% more than the previous year. Following the devastating pandemic, the tourism industry contributed 7.4% of global GDP in 2022 (WTTC, 2023), which is almost 23% below the pre-pandemic situation. In the pandemic period, the entire world witnessed the significance of technological innovation and advancement in the tourism sector. At that time, the use of technological tools helped the tourism industry to survive and avert all odds situations. After that, the urge for smart tourism applications is more intensified in destination areas all over the world. According to a number of studies, it helps to develop sustainable and technologically sound cities

(Gössling, 2021; Kajan and Saarinen, 2013) and an income-generating way for developing nations (Pang et al., 2013; Rahman and Saima, 2021).

Bastidas-Manzano (2021) stated that the radical growth of tourism has created some serious problems at the destination level, which are: (i) Its exponential rise in urban waste (Deb and Nafi, 2023; González and Ruiz, 2006); (ii) mass production of tourism has a detrimental consequence on natural destinations (Jovicic, 2019); (iii) seasonality of tourism results in unstable job and financial activity for few months of the year (Capocchi et al., 2019; Gössling et al., 2008; Pou, 2012); (iv) it generates inflation because of a rise in demand over the supply (Amelung and Viner, 2006); (v) it causes a spike in pollution and noise due to an increase in vehicle use (Cai et al., 2019); and (vi) it causes devastation of landscapes and historical relies by uncontrolled construction in rural areas (Ballina, 2022; Castiblanco and Gisel, 2018, Mínguez and Jiménez, 2014). To address these problems, academicians and policymakers have made an effort to develop technologically enabled and sustainably designed tourist destinations for the future. As a result, smart tourism destination development has been directed by the policymakers, which are creative locations built on cutting-edge technology infrastructures that make use of the explosion of communication technologies to provide effective services.

According to Scott et al. (2016) and Fauzi (2023), regional sustainable development aims to ensure an accessible environment for everyone and encourage tourist involvement. Buhalis and Amaranggana (2013) were the first researchers who came to the conclusion that these locations are driven by creative cities, known as smart cities, that offer innovative features at the destination areas to enhance and long-lasting the visitor experience. However, this study aims to discover the importance of smart tourism on developing sustainable tourism destinations. The present study will cover the current practices of destinations to develop smart tourism to forecast the future.

9.2 LITERATURE REVIEW

Smart tourism is known as the dynamic fusion of human experiences with intelligent technologies. An advanced level of tourist informationization is called smart tourism. Celdran et al. (2018) stated that the concept of smart is a significant foundation of the tourism and travel industry in recent time. Although the terms "intelligence" and "smart tourism" may be synonymous in the literal sense, the key distinction is in anticipating tourists' wants through information (Jovicic, 2019). Buhalis (2015), a prominent researcher in smart tourism, describes a smart destination as "*a smart tourism destination successfully implements smartness which is fostered by open innovation, supported by investments in human and social capital, and sustained by participatory governance in order to develop the collective competitiveness of tourism destinations to enhance social, economic, and environmental prosperity for all stakeholders*." By utilizing advanced technologies, smart tourism seeks to increase sustainability, optimize competitiveness, and manage resources more effectively. Consequently, a large number of destinations are modernizing every aspect of their business, from interactive activities to transaction systems. Novel approaches to gathering and utilizing information from social networks, physical infrastructure, governmental and non-governmental organizations, and users in order to assist integrated operations at destinations.

Based on digital and virtual technologies, it comprises digital and virtual tourism. The concept of smart tourism is linked to three main components: information, exchange, and processing (Gretzel et al., 2015; Ballina, 2022). Sigala and Chalkati (2014) mentioned three main links between smart tourism and ICTs that are given below:

- In the first smart destination, integrating ICTs and innovation into the infrastructure for tourists is a crucial component (Capadevila and Zarlenga, 2015).
- According to Cantino et al. (2019), the creation of collaborative environments involving both public and private tourist collaborators constitutes the second component of smart business. It generates the greatest experiences, greater dialogue, and personalization (Vanolo, 2014).
- The last element is the community and social transformation brought about by the aggregation of ICTs and the traveler experience: the "added intelligence" or "smart tourist" (González-Reverté, 2019).
- Achieving sustainability through smart tourism in the tourism operation and development is crucial (Saravanan, 2010).

Figure 9.1 shows the different dimensions of smart tourism destinations. Gretzel (2015) and Wise and Heidari (2019) mentioned that the smart tourism concept has the possibility to expand upon ideas of smart

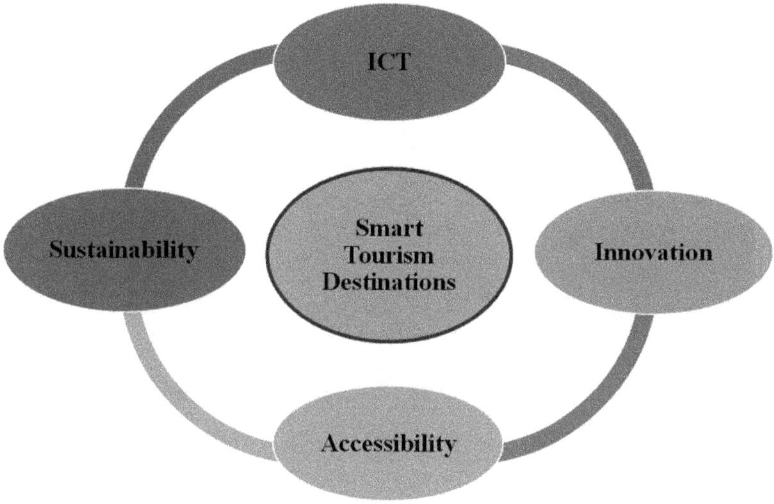

FIGURE 9.1 Dimensions of smart tourism destinations.

Source: Adapted from Saravanan (2010).

cities since it pertains not only to improving the living standard in urban areas but also to enlarge the quality, knowledge, and experience of the destination through the formation, exchange, and consumption of value. The terms "smart tourism destination" and "smart city" are frequently used by researchers. Smart cities are a center of solutions to socio-economic, cultural, and environmental issues (Koo et al., 2016), which are the major concern for tourism development. A city can be considered as smart when effective investment is made on human capital development programs along with the development of sociocultural resources and modern communication infrastructure development (Caragliu et al., 2011). That results in long-term economic growth and a good standard of living for the local population. This must be accomplished through participation in governance and the prudent natural resource management. Some have investigated the features of smart cities and the cities that fall under the category of intelligence (Xiang et al., 2015). The European Commission (2023) has described the importance of smart tourism from six different perspectives (Figure 9.2).

In urban areas, smart cities are working as a guideline for smart tourist destinations. Smart tourist destinations are measured against smart cities, which are mainly situated in urban regions. According to the majority of

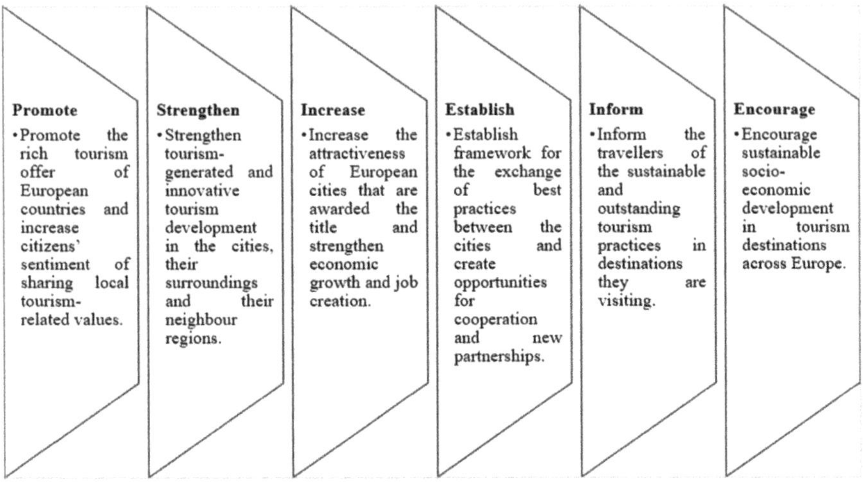

FIGURE 9.2 Characteristics of smart tourism.

Source: Adapted from European Commission (2023).

authors, smart cities increase their sustainability standards by using well-defined technologies. A portion of the literature on smart cities primarily addresses environmental and technical factors that impact urban efficiency (Lombardi, et al., 2011). Therefore, the idea of a smart city entails technological advancements in urban development that bolster natural and knowledge resources, as well as the physical and social infrastructure, to revitalize the urban environment and economy and enhance the delivery of public and social services (Buonincontri and Micera, 2016; González-Reverté, 2019). It is important to consider if smart tourist destinations can produce sustainability "as a result of several intertwined social, urban, and technological shifts" (Höjer and Wangel, 2014), given that they are a component of the techno-urban phenomenon that affects smart cities today. Figure 9.3 shows the destination development model through smart tourism. Cavalheiro et al. (2020) developed four layers for smart destination development ranging from 0–4. At the basic level, there should be a tourism destination with required tourism facilities. After that, smart ICT infrastructure and tourism applications need to support the overall smart tourism destinations. In this stage, smart network connection, infrastructure, smart service, and smart management should be introduced to support smart destinations. Lastly, desired tourist services, required investment, and other socio-economic factors should be developed for smart tourism destination development.

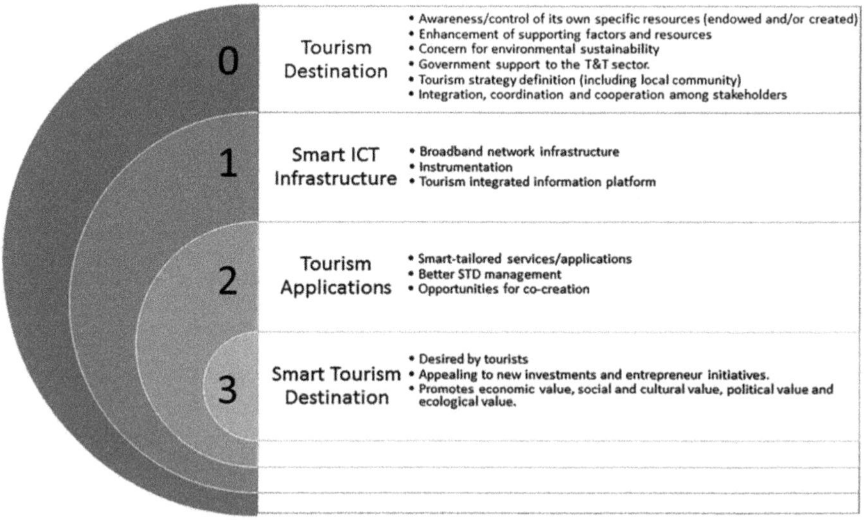

FIGURE 9.3 Smart tourism destination development model.

Source: Cavalheiro et al. (2020).

Several more realistic perspectives on integrating sustainability into tourist destinations (Bramwell et al., 1996; Sharple, 2000; Swarbrooke, 1999) state that smart destinations must take action in the areas of commitment, transparency, planning, growth limitation, and equity in their deployment and governance. However, previous studies have supported that the theme of sustainability is aligned with the perception of smart tourism. Dobrica and Jovicic (2019) stated the evolution of the smart tourism concept, which is shown in Figure 9.4. Figure 9.4 shows three phases of tourism concept development, which are the traditional concept, systematic approach, and smart tourism destination concepts. All three concepts focus on the geographic dimensions along with other factors. The traditional concept links with attractions and services at the destination level and ignores the cooperation among different parties at the destination area and the responsibility of travelers. The systemic dimension combines travelers, services, residents, and the macro-environment.

In the smart tourism domain, the integration of digital and physical domains within spatial dimensions, in conjunction with public-private-consumer partnerships, promotes participating governance and enables value co-creation. This setting fosters the growth of creative and knowledgeable people, which makes it possible for the delivery of individualized services.

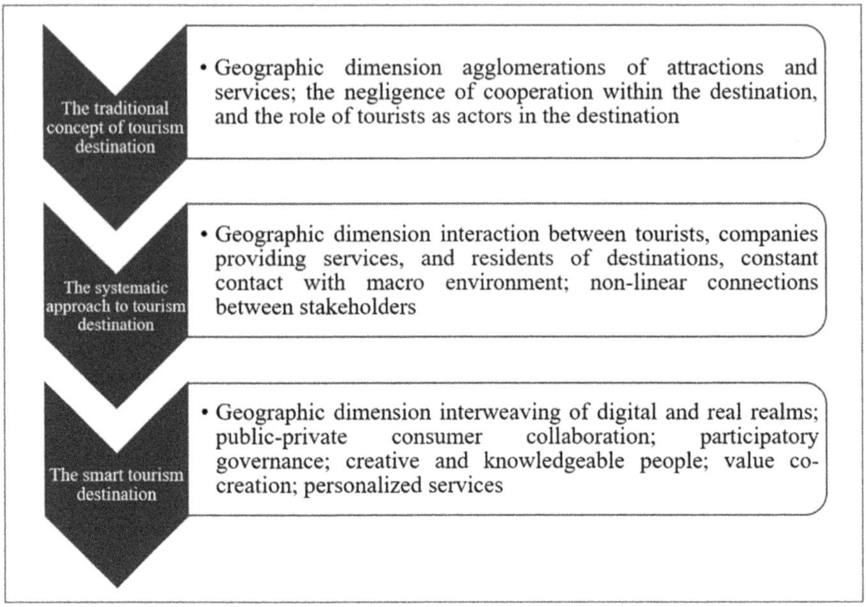

FIGURE 9.4 The evolution of tourism destination concept.

Source: Adapted from Dobrica and Jovicic (2019).

9.3 METHODOLOGY

The purpose of this study is to identify the impact of smart tourism on the development of destinations. For this reason, the enhancement of tourism services associated with smart tourism has been studied and mentioned. This study is qualitative in nature. The information was acquired from secondary sources, including journal papers, relevant literature, websites, and various published research. Boslaugh (2007) and Clark (2013) mentioned that secondary data was obtained by someone else without the author's involvement and used as information in other studies. Secondary data can also include information that has already been obtained but is being considered for use to solve new problems and gives new insights for which it was not intended (Vartanian, 2010). To find the relevant data, this study uses some specific keywords like smart tourism, smart destination, tourism in the 21st century, smart management, and smart city.

9.4 SMART CITY: A NEW ERA FOR TOURISM

Recently, the whole world has experienced two major trends: global urbanization and the digital revolution. Rapid urbanization makes it more difficult to provide services including housing infrastructure, transportation, education, health care, and security in a sustainable and high-quality manner (Armagan, 2018). On the other hand, technological innovation also helps to lessen these problems of urbanization. In this context, the smart city or smart destination concept has evolved. Through improved connection, smart cities employ data to provide individuals with information about themselves and to spur innovation, competitiveness, and overall living standard (Albino et al., 2015; Celikyay, 2017).

The concept of a smart city is defined in various ways, including "digital" or "numerical" cities (Pop and Prostean, 2018), "information cities," "cable cities," "information-based cities," "electronic communities," "electronic community areas," and "cyber areas" (Della Corte et al., 2017; Lee et al., 2013). The idea of a smart city is most likely described as one that seeks to address issues through the use of multi-stakeholder ICT. However, it should address the concept of sustainability, technology, and smart design in the development process. Bilici and Babahanoglu (2018) and Lombardi et al. (2012) stated some indicators and their importance to evaluate the performance of a smart city, such as "infrastructure, e-services, energy, greenhouse gas emissions, transportation, resource management, public participation, competitiveness, economy, environment, and city quality." These all indicators specify the sustainability measures for a destination. Doğan and Keleş (2022) mentioned that within 2025, at least 26 smart cities will be developed around the world. These cities have some unique characteristics. These characteristics are given in Figure 9.5.

Smart cities have six elements that are essential for the development of a smart city. They should consider smart economy management, smart people development, developing smart living systems, smart environment management, and smart government activities. Smart cities are helpful for tourism because tourists stay a short time in a destination and enjoy the features of a smart city (Giffinger et al., 2007). However, a number of research studies conclude that the development of smart tourism has mostly occurred within city initiatives that have already employed intelligent tactics aimed at sustainability, e-governance, residence, and circular economy (Akgul, 2013; Li et al., 2017). Thus, a smart city will help in tourism to ensure smart management, smart governance, smart development, smart service application, smart resource management, and, most importantly, increase tourist experiences.

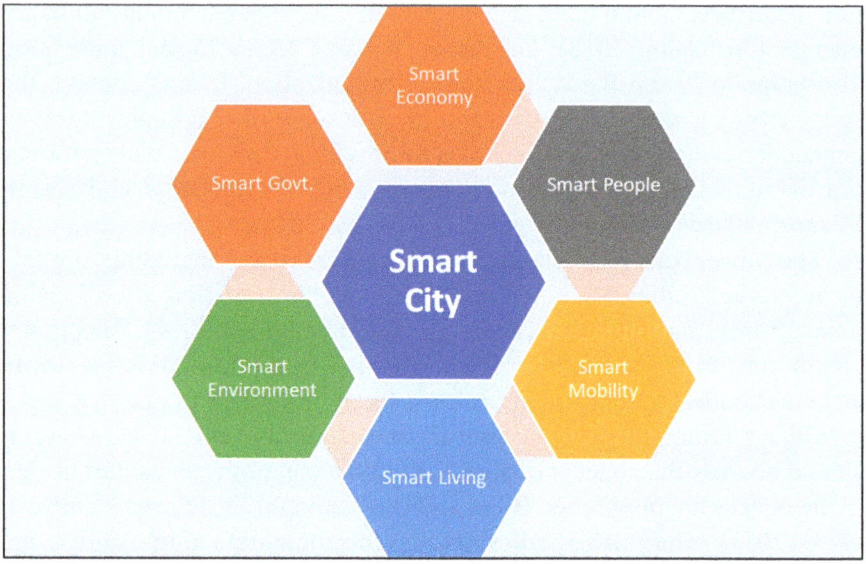

FIGURE 9.5 Elements of smart city.

Source: Adapted from Della Corte et al. (2017); Pop and Prostean (2018).

9.5 TOURISM STAKEHOLDERS' PERSPECTIVE IN SMART TOURISM DESTINATION DEVELOPMENT (BUSINESS AND TOURIST)

Smart tourism has formed a new model in the travel and tourism industry. It incorporates networked ICT and provides opportunities for tourists and business organizations (Li et al., 2017). They also make mechanisms available that enable stakeholders to engage as much as feasible in the innovation procedure (Boes et al., 2015; Koo et al., 2017). These innovations enable tourism stakeholders to use sophisticated products and services for improving their knowledge and experiences. According to Femenia-Serra et al. (2018), smart tourism strengthens the sense of shared purpose among locals and destinations and maintains their edge over competitors. To co-generate value for one another, all participants—including tourism organizations, ICT firms, governments, industry, locals, and individual travelers—become deeply involved (Koo et al., 2017). Dominguez et al. (2017) compiled five major stakeholders' characteristics in smart destinations. Table 9.1 shows the tourism stakeholders' characteristics in a smart destination environment.

TABLE 9.1 Tourism Stakeholders Characteristics in Smart Destinations Environment

Stakeholders	Characteristics in Smart Environment
Tourism Organizations	• Serve as a hub that combines pertinent data and facilitates consumers' easy access to up-to-date information.
Government	• Digitalization in core business processes.
	• Engage the local people and tourists in the service co-creation process for increasing tourism experiences.
Residents	• Adaptability inside the organization, quick decision-taking, and responsiveness to client need on real time perceptions.
Tourists	• Accurate segmentation and customized assistance.
	• Governance in information management that protects data privacy. and
	• Encourages data transparency.
Environment	• Establish private-public partnerships.
	• Aggressively engaged in evolving smart heritage/e-culture systems.
	• Innovation ecosystem.
	• Sensor network throughout the environment.
	• Integrate social circumstances with digital data to enhance geophysical reality.

Source: Dominguez et al. (2017).

Santos-Júnior et al. (2023) studied the local resident's perception of smart tourism. This study stated that smart tourism destinations are defined as creative areas built on up-to-date and sophisticated technology that can enhance the quality of life for locals while offering noteworthy tourist experiences. It is implied that applying ICT to smart tourist destinations influences residents' opinions of the positive or negative aspects of tourism. ICTs are recognized as influencing these views and general resident happiness in a cross-cutting way, which can help smart tourism destinations continue to grow. Lastly, because locals are the community's social capital, they ought to be heavily involved in the governance and planning of smart tourism destinations.

9.6 SMART TOURISM AND ECO-TOURISM DESTINATION

Eco-tourism is known as the sustainable, responsible, and efficient practice of tourism resource management at the destination area. Promoting the right practices, improving eco-tourism experiences, and operating tourism resources sustainably are ways to achieve the goals of ecotourism (Chai-Arayalert, 2020; Deb et al., 2022; Katsoni and Dologlou 2017). The term "smart ecotourism application" describes new methods and technologies for enhancing ecotourism activities and experiences. Digital media can leverage ICTs to educate individuals and foster eco-tourism knowledge. The use of digital media and smart technologies also help to reshape tourism activities in natural areas and reduce the destructiveness of tourists. Hunter et al. (2015) mentioned some main features of ICTs in ecotourism that are: (i) preserving nature-based destinations and reducing the detrimental effects of tourism on nearby communities; (ii) raising public awareness and providing educational opportunities; and (iii) encouraging stakeholders to contribute to conservation.

To preserve the natural environment and encourage ecotourism, many destinations globally have been declared as protected areas. These areas have some unique characteristics like unique nature, wildlife, and habitat (Hasan et al., 2024; Nafi and Ahmed, 2017). A variety of tools and smart technologies are used to protect ecotourism destinations. Chai-Arayalert (2020) mentioned some of the technologies for smart ecotourism development that are given below:

1. **Wireless Environmental Monitoring** can assist ecotourism management by providing accurate and useful data about the maintenance and improvement of ecosystems.
2. **Participatory Sensing** can increase and develop awareness about the ecosystem to the stakeholders.
3. **Geocaching** can enhance and improve interaction with nature.
4. **Web Presence** contains information about ecotourism destinations and creates awareness.
5. **Smart Buildings** can be achieved through high energy savings and developed new infrastructure in ecotourism destinations.
6. **E-guides** include do's and don'ts for visitors and create awareness about natural areas.

Eco-tourism and smart tourism locations provide technology-driven, sustainable experiences that enhance visitor enjoyment and encourage environmental conservation. These locations guarantee a healthy ecology for both wildlife and visitors by combining creativity, preservation, and responsible tourism practices in a balanced way. The merging of eco-tourism and smart tourism marks the beginning of a new phase in sustainable travel. Through the use of technology and a focus on conservation initiatives, these locations preserve natural resources while providing engaging experiences. They encourage a sense of stewardship in travelers and build a closer bond with the environment through ethical methods and creative solutions. Smart and eco-tourism locations act as beacons of development, paving the way for a brighter future for international travel as we continue to promote sustainability and harmony with nature.

9.7 EFFECT OF SMART DESTINATION DEVELOPMENT

The way that technology has affected tourism destinations and cities raises the importance of new management strategies to modify planning procedures to meet the opportunities and difficulties presented by the smart scenario. Smart tourism initiatives are shaping tourism development activities and considering technological characteristics. The goals of smart city and smart tourism destination strategies are to enhance visitor experiences, resident quality of life, and administrative efficiency. The smart tourism concept has a double effect on the destinations, which is known as improving the overall competitiveness and improving the different tourism stakeholders' performance in the destination area (Buhalis, 2020; Collado-Agudo et al., 2023 Gelter et al., 2021; Ortega and Malcolm, 2020).

9.7.1 SMART TOURISM DESTINATIONS IN DIFFERENT REGIONS

1. **North America:** It is estimated that North America would account for a sizeable portion of the global tourism industry because tourism service providers have made large expenditures to provide visitors with an increased experience. A significant percentage of the smart infrastructure market is based in North America. New York, San Francisco, and Boston are the top three cities that have adopted smart infrastructure initiatives. These programs include smart homes, smart transportation, smart agriculture, and smart cities, among many other initiatives. The main source of the

expansion of the smart tourist industry in North America is the rapid development of smart infrastructure projects.

2. **Europe:** Because of the related government's desire to expand tourism, Europe is expected to demonstrate a substantial contribution to the market's growth. The European Capital of Smart Tourism (ECST), announced in October 2019, will support the European Tourism Forum's commitment to promote Europe as a smart tourism destination. The European government's strong commitment to the growth of smart tourism has been shown by these declarations. Based on the theme of "smart tourism," the European Commission has organized two competitions aimed at strengthening and networking travel destinations, encouraging the sharing of best practices, and promoting sustainable and smart tourism throughout the EU. The extraordinary achievements were honored by the ECST in four areas related to European cities as travel destinations: digitalization, accessibility, sustainability, and cultural heritage and innovation. Subsequent to the European Destination of Excellence (EDEN) competition, the European Green Pioneer of Smart Tourism recognizes and honors smaller locations that have effectively employed tactics to enhance sustainable tourism by means of green transition methods.
3. **Asia-Pacific:** Because of the fast-growing tourism industry in several Asian nations, Asia-Pacific is presumed to have a substantial share in the worldwide smart tourism marketplace. As an illustration, the Indian city of Bhubaneswar created a foundation for a smart city by creating a number of smart applications, including an intelligent traffic management system, an integrated public service center, and a smart park. There are a number of destinations in the world that are trying to achieve the goal of smart destinations. Table 9.2 shows some examples of smart tourism destinations all over the world. Thus, technology and innovation are the leading themes of the smart destination development process.

9.8 CONCLUSION

Smart tourism has a significant impact on destination development because it provides a sustainable strategy that combines environmental and social concerns with technical improvements. In addition to improving the traveler

TABLE 9.2 Features of Smart Destinations

Name of the Smart Destination/City	Country	Specific Features
Neom City	Saudi Arabia	• Future mega city combines sustainable design and eco-friendly accommodation. • Advanced transportation like autonomous electric vehicles. • Up-to-date technology is used to enhance tourist experiences. • Emphasize on environmental conservation and sustainable use of resources.
El Hierro	Spain	• First smart island in the world to accomplish self-independence in energy. • Reduce pollution level. • Generate electricity from waste and substitute brick with volcanic stone in construction.
Taquila	Mexico	• Provides free Wi-Fi in its historic center. • Designed an app to provide real-time traffic and business activity updates to tourists. • Developed applications with related information on local or regional goods and services.
Gothenburg	Sweden	• Develop a smart navigation/map that promotes residents and travelers involvement to exchange and share in the city. • Utilizing a 3D model to initiate discussions with community people to make better decisions, foresee the effects of future development, and have public discussions. • One of the greenest cities in Europe because 60% of its constituency central heating comes from waste.

*Source: C*ompiled by author.

experience, smart tourism helps places become smart cities and smart destinations overall. The adoption of innovative technologies that optimize resource management, improve infrastructure, and enhance connection facilitates the evolution towards smart cities. These developments provide an environment that is favorable for sustainable growth and development, which benefits both locals and visitors. The way stakeholders see a project is crucial in determining how it develops. Effective stakeholder engagement promotes collaboration and undertakings that development projects are in line with the goals and values of all parties involved, including local communities, companies, and governmental organizations. Smart tourism can facilitate collaboration among the stakeholders in the digital environment. Incorporating varied perspectives and resolving concerns will enable smart tourism initiatives to acquire greater support and ultimately achieve long-term success. Furthermore, the idea of smart destinations represents a comprehensive strategy for development that gives sustainability, inclusivity, and resilience precedence over merely integrating technology. Responsible use of technology combined with an embracement of sustainability principles can help destinations become more competitive. These types of competitive advantages help to bring investment, protect natural resources, and preserve cultural heritage for future generations. Essentially, the shift towards smart tourism signifies a paradigm modification in destination development, with the aim of fostering more inclusive, resilient, and sustainable communities while providing visitors with unforgettable experiences. Adopting this strategy will not only help shape the future of tourism in a world that is changing quickly, but it is also essential. This kind of shift will help to achieve environmental sustainability along with visitor satisfaction.

KEYWORDS

- **smart city**
- **smart destination**
- **smart economy**
- **smart tourism**
- **stakeholders**
- **sustainable tourism**
- **technology**

REFERENCES

Akgul, M. K. (2013). eTransformation of the cities: Smart cities. http://anahtar.sanayi.gov.tr/tr/news/kentlerin-e-donusumu-akilli-kentler/416 (accessed on 8th July 2025).

Albino, V., Berardi, U., & Dangelico, R. M. (2015). Smart cities: Definitions, dimensions, performance, and initiatives. *Journal of Urban Technology*, *22*(1), 3–21.

Alshamaila, Y., Papagiannidis, S., Alsawalqah, H., & Aljarah, I. (2023). Effective use of smart cities in crisis cases: A systematic review of the literature. *International Journal of Disaster Risk Reduction*, *85*. https://doi.org/10.1016/j.ijdrr.2023.103521 (accessed on 8th July 2025).

Amelung, B., & Viner, D. (2006). Mediterranean tourism: Exploring the future with the Tourism Climatic Index. *Journal of Sustainable Tourism*, *14*(4), 349–366. https://doi.org/10.2167/jost549.0 (accessed on 8th July 2025).

Armagan, V. (2018). Smart cities and e-government in the digital transformation process. *İletisim Kuram ve Arastirma Dergisi*, *46*, 387–413.

Ballina, F. J. (2022). Smart business: The element of delay in the future of smart tourism. *Journal of Tourism Futures*, *8*(1), 37–54. https://doi.org/10.1108/JTF-02-2020-0018 (accessed on 8th July 2025).

Bastidas-Manzano, A. B., Sánchez-Fernández, J., & Casado-Aranda, L. A. (2021). The past, present, and future of smart tourism destinations: A bibliometric analysis. *Journal of Hospitality & Tourism Research*, *45*(3), 529–552.

Bilici, Z., & Babahanoglu, V. (2018). Smart city applications and sample of Konya. *Akademik Yaklasimlar Dergisi*, *9*(2), 124–139.

Boes, K., Buhalis, D., & Inversini, A. (2015). Conceptualizing smart tourism destination dimensions. In Tussyadiah, I. & Inversini, A. (Eds.), *Information and Communication Technologies in Tourism* (pp. 391–404). Springer, Cham.

Boslaugh, S. (2007). *Secondary Data Sources for Public Health: A Practical Guide*. Cambridge University Press.

Bramwell, B., Richards, G., Henry, I., Jackson, G., van der Straaten, J., van't Zelfde, J., & Evans, D. M. (1996). *Sustainable Tourism Management: Principles and Practice*. Tilburg University Press.

Buhalis, D. (2015). Working definitions of smartness and smart tourism destination. *Buhalis Blog*. http://t.co/xrLRpGipvu (accessed on 8th July 2025).

Buhalis, D. (2020). Technology in tourism—from information communication technologies to eTourism and smart tourism towards ambient intelligence tourism: A perspective article. *Tourism Review*, *75*(1), 267–272.

Buhalis, D., & Amaranggana, A. (2013). Smart tourism destinations. In Z. Xiang & I. Tussyadiah (Eds.), *Information and Communication Technologies in Tourism 2014* (pp. 553–564). Springer International.

Buonincontri, P., & Micera, R. (2016). The experience co-creation in smart tourism destinations: A multiple case analysis of European destinations. *Information Technology & Tourism*, *16*(3), 285–315.

Cai, W., Richter, S., & McKenna, B. (2019). Progress on technology use in tourism. *Journal of Hospitality and Tourism Technology*, *10*(4), 651–672. https://doi.org/10.1108/JHTT-07-2018-0068 (accessed on 8th July 2025).

Cantino, V., Giacosa, E., Alfiero, S., Shams, R., & Ferraris, A. (2019). Introduction: Smart tourism businesses (Sustainability, Measurability, Awareness, Recognition, and Technology). *Tourism Analysis*, *24*(3), 261–263.

Capdevila, I., & Zarlenga, M. I. (2015). Smart city or smart citizens? The Barcelona case. *Journal of Strategy and Management*, *8*(3), 266–282. https://doi.org/10.1108/JSMA-03-2015-0030 (accessed on 8th July 2025).

Capocchi, A., Vallone, C., Amaduzzi, A., & Pierotti, M. (2019). Is "overtourism" a new issue in tourism development or just a new term for an already known phenomenon? *Current Issues in Tourism*, *23*(18), 2235–2239. https://doi.org/10.1080/13683500.2019.1638353 (accessed on 8th July 2025).

Caragliu, A., Bo, C. D., & Nijkamp, P. (2011). Smart cities in Europe. *Journal of Urban Technology*, *18*(2), 65–82. https://doi.org/10.1080/10630732.2011.601117 (accessed on 8th July 2025).

Castiblanco, C., & Gisel, N. (2018). Analysis of the touristic sustainability. *Gestión y Ambiente*, *21*(1), 99–109.

Cavalheiro, M. B., Joia, L. A., & Cavalheiro, G. M. D. C. (2020). Towards a smart tourism destination development model: Promoting environmental, economic, socio-cultural and political values. *Tourism Planning & Development*, *17*(3), 237–259.

Celdran, M. A., Mazon, J. N., Ivars-Baidal, J. A., & Vera-Rebollo, J. F. (2018). Smart tourism. Un studio de mapeo sistematico. *Cuadernos de Turismo*, *41*, 107–138.

Celikyay, H. H. (2017). The view of smart cities through Istanbul perspective: Is it only technology makes cities smart? *Van Yuzuncu Yil Universitesi Sosyal Bilimler Enstitusu Dergisi*, *1*(4), 505–512.

Chai-Arayalert, S. (2020). Smart application of learning ecotourism for young eco-tourists. *Cogent Social Sciences*, *6*(1), 1772558. https://doi.org/10.1080/23311886.2020.1772558 (accessed on 8th July 2025).

Clark, G. (2013). Secondary data. In *Methods in Human Geography* (pp. 57–73).

Collado-Agudo, J., Herrero-Crespo, Á., & San Martín-Gutiérrez, H. (2023). The adoption of a smart destination model by tourism companies: An ecosystem approach. *Journal of Destination Marketing & Management*, *28*, 100783. https://doi.org/10.1016/j.jdmm.2023.100783 (accessed on 8th July 2025).

Deb, S. K., & Nafi, S. (2023). Impact of pandemic, crises, and challenges on event tourism. In *Event Tourism and Sustainable Community Development* (pp. 231–245). Apple Academic Press.

Deb, S. K., Kuri, B. C., & Nafi, S. M. (2024). Application of knowledge management in tourism and hospitality industry: A sustainable approach. In *Knowledge Management and Knowledge Sharing: Business Strategies and An Emerging Theoretical Field* (pp. 99–116). Cham: Springer Nature Switzerland. https://doi.org/10.1007/978-3-031-37868-3_7 (accessed on 8th July 2025).

Deb, S. K., Nafi, S. M., & Valeri, M. (2022). Promoting tourism business through digital marketing in the new normal era: A sustainable approach. *European Journal of Innovation Management*.

Della Corte, V., D'Andrea, C., Savastano, I., & Zamparelli, P. (2017). Smart cities and destination management: Impacts and opportunities for tourism competitiveness. *European Journal of Tourism Research*, *17*, 7–27.

Doğan, S., & Keleş, Y. (2022). Smart city applications in Asian capital cities. In *Handbook of Technology Application in Tourism in Asia* (pp. 507–532). Singapore: Springer Nature Singapore.

Dominguez, C. D., Hernández, M. R., Talavera, A. S., & López, E. P. (2017). Smart island tourism and strategic marketing: The case of the island of El Hierro. *Actas del Seminario Internacional Destinos Turísticos Inteligentes: Nuevos Horizontes en la Investigación y Gestión del Turismo*, Alicante, 236.

European Commission. (2023). European Capital & Green Pioneer of Smart Tourism. Available at: https://smart-tourism-capital.ec.europa.eu/index_en (accessed on 8th July 2025).

Fauzi, M. A. (2023). Sustainable tourism and sustainable development goals (SDGs): A state-of-the-art review of past, present, and future trends. *Environment, Development and Sustainability*, 1–22.

Femenia-Serra, F., Perles-Ribes, J. F., & Ivars-Baidal, J. A. (2018). Smart destinations and tech-savvy millennial tourists: Hype versus reality. *Tourism Review*, 1–19. https://doi.org/10.1108/TR-02-2018-0018 (accessed on 8th July 2025).

Gelter, J., Lexhagen, M., & Fuchs, M. (2021). A meta-narrative analysis of smart tourism destinations: Implications for tourism destination management. *Current Issues in Tourism*, *24*(20), 2860–2874.

Giffinger, R., Fertner, C., Kramar, H., Kalasek, R., Pichler-Milanovic, N., & Meijers, E. (2007). *Smart Cities Ranking of European Medium-Sized Cities*. Vienna University of Technology.

González, M. de la O. B., & Ruiz, D. F. (2006). The international competitiveness of tourist destinations: From the macroeconomic approach to the strategic approach. *Cuadernos de Turismo*, *17*, 7–24.

González-Reverté, F. (2019). Building sustainable smart destinations: An approach based on the development of Spanish smart tourism plans. *Sustainability*, *11*(23), 6874. https://doi.org/10.3390/su11236874 (accessed on 8th July 2025).

Gössling, S. (2021). Tourism, technology and ICT: A critical review of affordances and concessions. *Journal of Sustainable Tourism*, *29*(5), 733–750. https://doi.org/10.1080/09669582.2021.1873353 (accessed on 8th July 2025).

Gössling, S., Peeters, P., & Scott, D. (2008). Consequences of climate policy for international tourist arrivals in developing countries. *Third World Quarterly*, *29*(5), 873–901.

Gretzel, U., Werthner, H., Koo, C., & Lamsfus, C. (2015). Conceptual foundations for understanding smart tourism ecosystems. *Computers in Human Behavior*, *50*, 558–563.

Hasan, S., Halim, M. A., Md. Nafi, S., & Amin, M. R. (2024). Perception and attitude towards wildlife-based tourism: Bangladesh perspective. *Global Sustainability Research*, *3*(1), 61–76. https://doi.org/10.56556/gssr.v3i1.733 (accessed on 8th July 2025).

Höjer, M., & Wangel, J. (2014). Smart sustainable cities: Definition and challenges. In L. M. Hilty & B. Aebischer (Eds.), *ICT Innovations for Sustainability* (pp. 333–349). Springer: Berlin, Germany.

Hunter, W. C., Chung, N., Gretzel, U., & Koo, C. (2015). Constructivist research in smart tourism. *Asia Pacific Journal of Information Systems*, *25*(1), 105–120. https://doi.org/10.14329/apjis.2015.25.1.105 (accessed on 8th July 2025).

Jovicic, D. Z. (2019). From the traditional understanding of tourism destination to the smart tourism destination. *Current Issues in Tourism*, *22*(3), 276–282.

Jovicic, D. Z. (2022). From the traditional understanding of tourism destination to the smart tourism destination. *Current Issues in Tourism*, *22*(3), 276–282. https://doi.org/10.1080/13683500.2017.1313203 (accessed on 8th July 2025).

Kaján, E., & Saarinen, J. (2013). Tourism, climate change and adaptation: A review. *Current Issues in Tourism*, *16*(2), 167–195. https://doi.org/10.1080/13683500.2013.774323 (accessed on 8th July 2025).

Katsoni, V., & Dologlou, N. (2017). ICT applications in smart ecotourism environments. In A. Stratigea, E. Kyriakides, & C. Nicolaides (Eds.), *Smart Cities in the Mediterranean* (pp. 225–244). Springer: Cham.

Koo, C., Park, J., & Lee, J. (2017). Smart tourism: Traveler, business, and organizational perspectives. *Information and Management*, *54*(6), 683–686.

Koo, C., Shin, S., Gretzel, U., Hunter, W. C., & Chung, N. (2016). Conceptualization of smart tourism destination competitiveness. *Asia Pacific Journal of Information Systems*, *26*(4), 561–576.

Lee, J. H., Phaal, R., & Lee, S. H. (2013). An integrated service-device-technology roadmap for smart city development. *Technological Forecasting and Social Change*, *80*, 286–306.

Li, Y., Hu, C., Huang, C., & Duan, L. (2017). The concept of smart tourism in the context of tourism information services. *Tourism Management*, *58*, 293–300. https://doi.org/10.1016/j.tourman.2016.03.014 (accessed on 8th July 2025).

Lombardi, P., Giordano, S., Caragliou, A., del Bo, C., Deakin, M., Nijkamp, P., & Kourtit, K. (201). An advanced triple-helix network model for smart cities performance. *Vrije Universiteit Amsterdam, Research Memorandum.*

Lombardi, P., Giordano, S., Farouh, H., & Yousef, W. (2012). Modelling the smart city performance. *Innovation: The European Journal of Social Science*, *25*(2), 137–149.

Mínguez, M. del C., & Jiménez, P. R. (2014). Smart tourist destinations in Spain: An institutional project or the future of the sector? In Foronda-Robles (Ed.), *Espacios Turísticos E Inteligencia Nacional: Respuestas Ante La Crisis: Actas Del Coloquio* (pp. 65–78). Red de Impresión.

Nafi, S. M., & Ahmed, T. (2017). Sustainable tourism in Saint Martin Island: An observation on young tourist perception and awareness level. *Journal of Humanities and Social Science*, *22*(10), 73–80.

Ortega, J. L. C., & Malcolm, C. D. (2020). Touristic stakeholders' perceptions about the smart tourism destination concept in Puerto Vallarta, Jalisco, Mexico. *Sustainability*, *12*(5), 1–18.

Pang, S. F. H., McKercher, B., & Prideaux, B. (2013). Climate change and tourism: An overview. *Asia Pacific Journal of Tourism Research*, *18*(1, 2), 4–20. https://doi.org/10.1080/10941665.2012.688509 (accessed on 8th July 2025).

Pop, M. D., & Proştean, O. (2018). A comparison between smart city approaches in road traffic management. *Procedia–Social and Behavioral Sciences*, *238*, 29–36.

Pou, L. (2012). Tourism and employment: A realistic look. *P3T. Journal of Public Policies and Territory*, *1*(3), 39–44.

Rahman, M. H. A., & Saima, F. N. (2021). Technology application in the tourism industry for socio-economic development in Bangladesh. In *Technology Application in the Tourism and Hospitality Industry of Bangladesh* (pp. 347–365). https://doi.org/10.1007/978-981-16-2434-6_21 (accessed on 8th July 2025).

Richardson, R. B. (2010). The contribution of tourism to economic growth and food security (Food Security Collaborative Working Papers 97140). Michigan State University, Department of Agricultural, Food, and Resource Economics. https://ideas.repec.org/p/ags/midcwp/97140.html (accessed on 8th July 2025).

Santos-Júnior, A., Almeida-García, F., Mendes-Filho, L., & Simões, J. M. (2023). Residents' perspectives in smart tourism destinations development: A theoretical approach. In *Handbook of Tourism and Quality-Of-Life Research ii: Enhancing the Lives of Tourists, Residents of Host Communities and Service Providers* (pp. 519–531). Cham: Springer International Publishing.

Saravanan, R. (2010). *ICTs for Agricultural Extension: Global Experiments, Innovations and Experiences*. New India Publishing.

Scott, D., Hall, C. M., & Gössling, S. (2016). A review of the IPCC fifth assessment and implications for tourism sector climate resilience and decarbonization. *Journal of Sustainable Tourism*, *24*(1), 8–30.

Sharpley, R. (2000). Tourism and sustainable development: Exploring the theoretical divide. *Journal of Sustainable Tourism*, *8*, 1–19.

Sigala, M., & Chalkati, K. (2014). Investigating the exploitation of Web 2.0 for knowledge management in Greek tourism industry: A utilization-importance analysis. *Computers in Human Behavior*, *30*, 800–812.

Swarbrooke, J. (1999). *Sustainable Tourism Management*. CABI: London, UK.

UNWTO. (2023). International tourism to end 2023 close to 90% of pre-pandemic levels. https://www.unwto.org/news/international-tourism-to-end-2023-close-to-90-of-pre-pandemic-levels (accessed on 8th July 2025).

Vanolo, A. (2014). Smart mentality: The smart city as disciplinary strategy. *Urban Studies*, *51*(5), 883–898. https://doi.org/10.1177/0042098013494427 (accessed on 8th July 2025).

Vartanian, T. P. (2010). *Secondary Data Analysis*. Oxford University Press.

Wise, N., & Heidari, H. (2019). Developing smart tourism destinations with the Internet of Things. In *Big Data and Innovation in Tourism, Travel, and Hospitality: Managerial Approaches, Techniques, and Applications* (pp. 21–29).

WTTC. (2023). Economic impact research. https://wttc.org/research/economic-impact (accessed on 8th July 2025).

Xiang, Z., Tussyadiah, I., & Buhalis, D. (2015). Smart destinations. *Journal of Destination Marketing & Management*, *4*(3), 143–201.

Ye, B. H., Ye, H., & Law, R. (2020). Systematic review of smart tourism research. *Sustainability*, *12*(8), 3401.

CHAPTER 10

EMERGENCE OF A CONSTRUCTED ENVIRONMENT FOR SMARTER TOURISM: A CRITICAL REFLECTION

A. S. SOCHIPEM ZIMIK[1*] and ARUP BARMAN[2]

[1]*Department of Management, SRM University, Sikkim, India*

[2]*Department of Business Administration, Assam University, Silchar, Assam, India*

*Corresponding author

ABSTRACT

The chapter spotlights the widespread embrace of smart tourism, underlining the imperative of meticulous planning and resource deployment. Subsequently, the Constructed Environment (CE) concept has emerged, embodying strategies designed to harmonize man-made settings with the diverse expectations and imaginations of tourists. A key focus of the chapter is to transform smart tourism to smarter industries through CE. The implementation of smarter strategies through CE helps in the development of destinations housing iconic landmarks, contemporary structures, and immersive 3D theaters that transport visitors across historical epochs and spatial dimensions, fostering attraction, contentment, and encouraging repeat visits, thereby promoting the destination as a recognizable brand. Understanding tourist psychology through the lens of the Maslowian triangle, coupled with stakeholder perspectives and strategic investments, serves as a fundamental cornerstone in the transformation of a smarter industry. The principal essence of the study lies in fostering a sense of

Smart Tourism: Managing Technological Revolution in the Tourism Industry. Santus Kumar Deb (Ed.)

DOI: 10.1201/9781779641090-10

security, adventure, leisure, wellbeing, and inviting tourists to take part in policymaking significantly in a smart nation. The chapter underscores the critical role and involvement of policymakers, tourists, local communities, and society at large in shaping policies, investments, and innovations concerning the CE, thereby fortifying the tourism industry for sustained economic prosperity.

10.1 INTRODUCTION

The phenomenon of smart tourism has emerged as a distinctive hallmark, substantiated by empirical studies (Gretzel et al., 2015; Huertas et al., 2021; Mehraliyev et al., 2020; Tiago et al., 2019). This paradigm holds significant implications for optimizing resource utilization, fostering sustainability, enhancing tourist experiences, and augmenting industry competitiveness (Gidumal, 2022; Lasisi et al., 2023; Shen et al., 2020; Sonuç and Süer, 2023; Yang et al., 2022; Zhang and Li, 2022). The transformation of a destination into a smart one necessitates unwavering commitment and the enhancement of local life quality, complemented by technological advancements (Bettini, 2018; Fonseca, 2021; Goo et al., 2022; Signes et al., 2020; Sorokina et al., 2022; Sustacha et al., 2023; Torabi et al., 2023). Lee and Co (2020) posit that urbanization has propelled destinations towards smart tourism, driven by the evolving demands and dynamism within the industry (Matos et al., 2019; Zhang et al., 2022). Pai et al. (2020) identify the components of smart tourism as infrastructures, services, and the 4 A's (Accommodation, Access, Amenities, and Attractions), marking a contemporary trend in the tourism sector. Further accentuating these concerns, Lazarevic and Pavlović (2018) underscore the pivotal role of technology, human capital, and innovation in smart tourism development. Within the technological and digitalization framework, Kumar (2020) advocates for a more astute and intelligent approach in the tourism industry. Smart tourism emergess as a pressing concern, serving as an ideal means for sustenance and economic viability. The escalating demand from tourists and the influence of information technology communication (ITC) necessitate a reevaluation of resource management strategies by developing nations, urging them to adopt the most efficient and reliable practices to compete with developed nations.

10.1.1 SIGNIFICANCE OF FOURTH INDUSTRIAL REVOLUTION ON SMART AND TOURISM ENVIRONMENT

The 21st century presents a significant opportunity for transformative changes, akin to the economic impact of the 19th-century industrial revolutions. This phenomenon aligns seamlessly with the concept of the Fourth Industrial Revolution (4IR), which is integral to discussions within the context of constructed environment (CE) as an optimal platform for facilitating the implementation of 4IR in the tourism industry. Envisaging limitless possibilities, the imagination and expectations of tourists extend infinitely, achievable through the synergistic collaboration of key elements such as cloud computing, artificial intelligence, the internet of things, big data, and cyber security. To meet tourist expectations, smart destinations must inherently reshape, with the safety and integration of cloud computing serving as critical criteria for establishing platforms that enhance the tourist experience. Practical implications of a smart environment must be simulated and experienced by tourists exploring destinations. Redefining a smart environment through artificial intelligence, in alignment with updated criteria, becomes an intriguing aspect. Hence, the significance of contemplating and embracing 4IR as imperative for developing smart destinations is undeniable. Considering the pervasive influence of globalization, the potential for universal redefinition of tourism industries is evident. Comprehending destinations as smart environments necessitates the adoption of sustainable principles, renewable energy, pollution-free policies, artificial intelligence interventions, and effective management. Emphasizing human hospitality and widespread social acceptance emerges as pivotal in shaping a smart external tourism environment, drawing insights from the Maslowian triangle in subsequent discussions. Internally, involving policymakers, constructors, and tourists, the focus on wealth maximization becomes paramount. Strategies oriented towards a "*win-win*" scenario, addressing both monetary and non-monetary aspects such as high tourist ratings, return visits, and legal support, underline the importance of this approach. The concept of investment and reinvestment, particularly from the margin, emerges as the ideal approach. The assessment of return on investment challenges investors to reconsider their involvement in constructed environment, presenting it as an eminent opportunity according to Zimik and Barman (2023). The chapter's intentions evoke a bold stance in advocating for philosophical practices of investing and reinvesting in the pursuit of sustainable and enriched tourism experiences.

10.1.2 STUDY BACKGROUND

The term CE refers to the human-made surroundings encompassing infrastructure, shops, amenities, and nature, crucial for meeting tourist satisfaction and expectations (Zimik and Barman, 2021, 2023). This chapter underscores that CE significantly influences tourism development (Zimik and Barman, 2024). Despite CE being traditionally viewed as a means for fostering tourism industry development, our study seeks to pioneer the application of CE in enhancing the intelligence of tourism industries. This approach stems from a thorough examination of existing secondary data on smart tourism, revealing a conspicuous absence of substantial research in this domain. The model proposed by Zimik et al. (2022) advocates constructing environments that cater to tourist satisfaction, psychology, and the awareness of various industry stakeholders. Importantly, our research establishes that no prior contributions have investigated the role of CE in elevating the intelligence of tourism industries, apart from acknowledging its impact on the sector.

The tabulated data in Table 10.1 strongly indicates a shift in the tourism industry from smart to smarter practices. This transition is influenced by in-depth research on smart tourism, emphasizing the role of CE in advancing towards smarter tourism. The analysis of this transformation is substantiated by diverse elements of Customer Experience.

10.1.3 AIMS AND SCOPE

Recognizing the pivotal role of tourism industries in driving economic growth, there is a compelling need to enhance their intelligence for sustained economic reliance. Despite abundant literature on smart tourism, it is evident that the evolving and increasingly intricate expectations of tourists require a more dynamic approach. This study focuses on delving deeper into the built environment and its components, which present formidable challenges in the transition of tourism industries into smarter entities. The central argument posits that, while tourism is acknowledged as "smart industries," there is a notable dearth of substantial research demonstrating the transformation of these industries from smart to even more intelligent enterprises.

TABLE 10.1 Transforming Smart to Smarter Tourism through Constructed Environment (CE)

Smart Tourism	Smarter Tourism
• Infrastructure	• Competence
• Hospitability	• Innovations
• ITC	• Strategies (Policies)
• Technology	• Satellite
• Facilities Offers	• Sustainability
• Strategies	• Dexterity
• Investment	• Consciousness
• Tourists Demand	• Professionalism (Educations)
• Tourists Return Tendency	• Collaborations and ITC (Communications)
• Various Models	• Maslowian Triangulation
• Offers	• Re designing of Models

Source: Author Compilation.

10.1.4 OBJECTIVES

- To comprehend the transformation of smart to smarter tourism industries through constructed environment.
- To analyze the roles of tourists and stakeholders in shaping smarter tourism industries with the help of constructed environment.

10.2 LITERATURE REVIEW

The literature review undertaken involves an exploration of diverse contributions within the main sub-sections (Nature, Smart Tourism, and Sustainability), as depicted in the Venn diagram above, highlighting their interconnections. Recognizing the urgency to address these contributions triggers the identification of research gaps and areas for improvement. This refinement aims to align the diagram more elegantly with the study's objectives, facilitating discussion on a broader platform.

Mansilla et al. (2023) delve into a holistic approach involving stakeholders, researchers, local communities, and tourists to define a smart destination and its corresponding perception. However, the integration of nature into smart tourism remains theoretically driven, lacking practical implications (Agudo et al., 2023). The genesis of smart tourism from smart city projects, as noted by Gretzel et al. (2015), raises concerns about potential compromises to nature. In addition, scheduling peak tourist visits during optimal seasons is emphasized by Corluka (2019), Corluka et al. (2016), and Sæþórsdóttir et al. (2019) as a collective management effort within smart tourism. Lohmann and Hübner's (2013) research reveals tourists' adaptability to dynamic psychological processes based on perceptions and weather-related activities. Addressing environmental concerns necessitates a collective stakeholder effort, encompassing control of tourist density (Sun and Liu, 2020), rainwater treatment, sewage management (Chen et al., 2021), and the establishment of carbon-neutral destinations (Gössling, 2009; Gössling, 2020; Zhang et al., 2021) to maintain destination attractiveness (Figure 10.1).

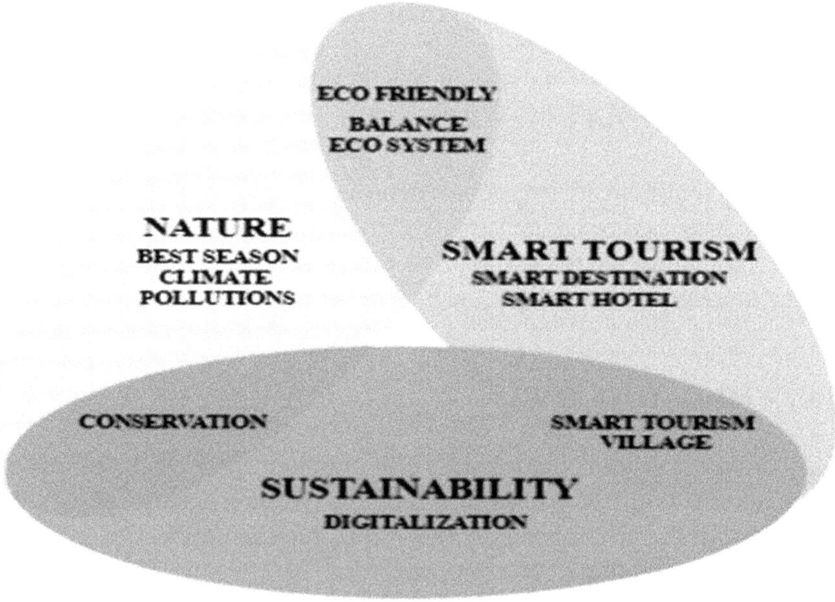

FIGURE 10.1 Literature contribution in smart tourism.

Source: Author Compilation.

Baloch et al. (2022) assert that a healthy natural environment and biodiversity sustainability stimulate tourists, overcoming negative experiences. Tourist behaviors, influenced by climatic conditions and seasons, emphasize the importance of favoring tourist satisfaction. Tiago et al. (2020) employ scientometric methods, revealing the pivotal roles of smart ecosystems and technology in smart tourism. Innovative tools and technology's significant role in destination transformation is highlighted by Yang et al. (2022). Buhalis and Amaranggana (2014) suggest industries can leverage smart city development, and Mehraliyev et al. (2019) find a sudden rise in smart tourism industries through text mining and network analysis. Liberato et al. (2018) define smart destinations based on information and communications technology accessibility. The emergence of sense-making and social systems as a new paradigm for smart accommodations and destinations is posited by Gelter et al. (2022), emphasizing the roles of professionals and academics in smart destination transformation (Manzano et al., 2020). Sustainability, encompassing resource enhancement through recycling, reuse, and sustainable development (Azarmi et al., 2019), is crucial. Digitalization plays a pivotal role in destination transformations, with a focus on making nights more attractive. Balancing market demands, rivalries, and long-term development under the UN Sustainable Development Goals (Abouzeid, 2022) necessitates quick adoption of advanced technology, especially for underdeveloped nations (Shahzad et al., 2022). Digitalization acts as a catalyst for attracting tourist destinations (Busulwa et al., 2022; Zhang et al., 2023). To enrich the study, the review will further delve into the intersection of the sets, exploring topics such as ecosystem, eco-friendly practices, smart villages, and conservation. Zorba's (2023) proposal of eco-city tourism as an ideal approach for sustaining industries emphasizes protection from exploitation and promotion of local products. Bhuiyan et al. (2022) suggests enhancing sustainability values through smart ecotourism destinations. Logistic regression models reveal artificial intelligence as a key element for smart heritage and rural development (Li et al., 2022).

The existing body of literature reveals a notable dearth of substantial research and concerted efforts beyond the domain of smart tourism. This observation underscores the identification of a research gap, justifying the rationale for the proposed study. In order to bridge this gap, a suggested approach involves the implementation of a resource management mechanism that extends beyond the operational scope of smart tourism. This strategic shift is deemed essential for advancing the current understanding and contributing meaningfully to the evolving discourse in the field.

10.3 METHODOLOGY

Due to a lack of existing literature and identified research gaps, we employ a descriptive methodology to be more feasible for this study. Our focus lies in introducing a groundbreaking concept: the establishment of a constructed environment to enhance the intelligence of tourism industries. This necessitates a breakthrough theoretical construction achieved through an in-depth examination of pertinent literature addressing smarter tourism. The study addresses serious research problems and relies on secondary data from diverse sources, including journals, books, and websites, to synthesize insights into how a constructed environment contributes to the smartening of tourism industries through depth interviews.

In this research, collaboration with an expert in the field of smart tourism was undertaken to leverage their expertise in enhancing the efficacy of smart tourism mechanisms. The study is underpinned by a series of 17 interviews conducted to align with the defined objectives (Table 10.2). The interviewees encompassed three experts with official representation from the tourism department, two bureaucrats actively engaged in policy framing, nine individuals affiliated with contractor and worker associations, and three representatives from civil society. The diverse perspectives of these groups are systematically presented in Table 10.2. The investigation delves into the nascent realm of smarter tourism, recognizing that while the concepts may appear to be under-implemented, they warrant scrutiny due to the transitional nature of smart destinations. The burgeoning demand and technological advancements impel researchers and policymakers to extend their focus beyond conventional smart destinations. Consequently, the study solicits collective opinions to foster a collaborative effort, resulting in the emergence of the "constructed environment" concept as a compelling and holistic strategy for transforming smart destinations into smarter ones. This approach aims to encapsulate the multifaceted dimensions essential for the evolution of smarter tourism destinations (Figure 10.2).

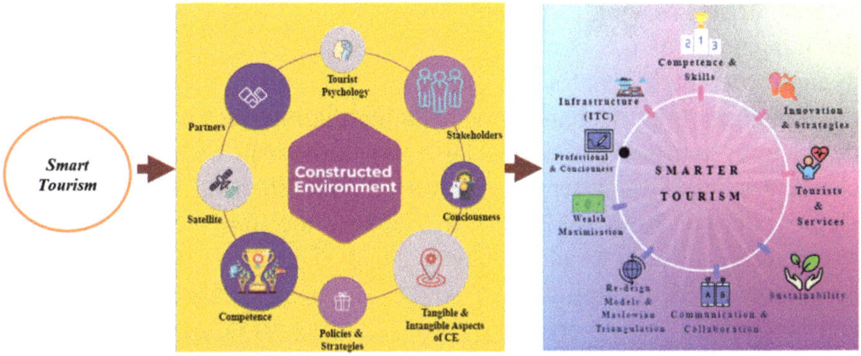

FIGURE 10.2 The transformation of smart to smarter tourism industries through constructed environment (CE).

Source: Author Compilation.

TABLE 10.2 Interview Groups

Category	No.	Opinions
Tourism Experts	3	Investment in infrastructure and FDI key to a smarter development of the industries
Bureaucrats	2	Collaborative efforts in making policies within department and region, job creation and regional sustenance, Promote culture and resources, and investment in infrastructure enhance the tourism status
Contractors	4	Budget allocation and sanction play a vital role, Logistic cost need to minimized by strategically choosing the destinations, and timely completion of the project is the strategic motion for smart destinations
Workers	5	Workers are well competent and willing to take challenge task
Civil Society	3	CE need to improve along with the cultural conservation, protection from all kind of exploitations, and improve the wellbeing of the local community
Total	**17**	

Source: Author Compilation.

10.4 SIGNIFICANCE OF CE FOR SMARTER TOURISM INDUSTRIES

The focal point of discussions lies in the transition of smart tourism industries towards a more intelligent form, prompting an exploration of the associated intricacies. Understanding the gravity of this inquiry necessitates a meticulous examination of diverse literature related to smart tourism, serving as a foundational step. This research emphasizes the integration of CE concepts to catalyze the evolution towards more dynamic and intelligent tourism industries. Figure 10.1 visually delineates the transformation mechanism, elucidating the interplay between CE and the smarter tourism paradigm. The study primarily aims to unveil the characteristics of a smarter tourism, with a comprehensive analysis of each element derived from Table 10.1. To enhance the study's vibrancy, CE is strategically tailored to address every facet of smart tourism, fostering a heightened level of intelligence through alignment with the competencies and awareness of diverse stakeholders. The subsequent subsections systematically justify the imperative for transformation and elucidate how CE emerges as the paramount approach, demonstrating its adaptability to changes and applications.

10.4.1 SMART TOURISM

The study primarily focuses on understanding the literature surrounding smart tourism, extensively evaluating diverse research contributions in this domain. Integral to tourism communication, ICT serves as a key component through the internet, embodying the identity of smart tourism destinations (Chakrabarty, 2023; Limna, 2022; Nguyen and Duy, 2021; Sun, 2023). Technological advancements, including Artificial Intelligence (AI), Big Data, Social and Internet platforms, and 5G, establish a robust infrastructure for promoting destinations and enhancing accessibility (Ahammed et al., 2023; Roy, 2018; TRAI, 2023). Intriguingly, the study delves into the realm of accessibility, exploring both virtual accessibility (Fernandes and Bernier, 2021; Fernández-Díaz et al., 2023; Liberato et al., 2018; Sambhanthan and Good, 2013) and physical transportation systems (Darcy et al., 2020; Kovačić, 2022; Rucci and Porto, 2022) to enhance the appeal of smart tourist destinations. Smart technologies play a crucial role in conjunction with the 4 A's (Chuang, 2023), emphasizing their significant impact on community building (Gil, 2021). These technologies vary based on destinations and tourist expectations (Consultants, 2023; Shen et al., 2020) and have emerged as catalysts for tourist satisfaction, experience, and loyalty (Sziva et al., 2023). Moreover, the integration of smart tourism technology with sustainability principles covering economic, environmental, and social aspects is highlighted (Sonuç and Süer, 2023). Green technology is presented as an alternative avenue for advancing smart technology (Wallace, 2024). Public awareness, policy, and local interventions are underscored as significant factors alongside proposed projects for smart tourism (Vardopoulos et al., 2023).

Tourists exhibit greater contentment when destinations facilitate interaction and integration with the environment, necessitating the enhancement of facilities related to natural sights and discussed amenities. Gardens by the Bay serves as an exemplary case, transforming natural resources into an environment-admired destination and providing strategies to augment vibrant tourist spots. Moving beyond conventional push and pull strategies, the chapter recommends creating a tourist persona, market familiarization, and implementing a comprehensive marketing mix to enhance industry competitiveness (Steeves, 2021). Strategies extending beyond marketing are deemed pivotal for the transformation of tourism, with branding involving influencers and ITC support gaining prominence (Rappensperger, 2023; Zimik et al., 2022). Situational analysis, emphasizing safety, education, satisfaction, and efficient resource mobilization, is considered integral for smart tourism sustenance (Tomaso, 2022). Artificial intelligence is recognized for its significant contribution to facilitating smart tourism industries (Xu, 2023), underscoring that technological innovations enhance tourist facilities alongside hospitality and offers (Wang, 2015; Yang et al., 2022).

The chapter posits that foresight-enhancing facilities in tourist attractions add value to smart tourism (Singh and Duhan, 2022). Zhang and Li (2022) illuminate the inseparable connection between tourism activities or facilities and economic development, emphasizing their impact on improving the standard of living. Strategizing emerges as a key element for comprehending smart tourism, with business volume generated by tourism necessitating situational strategies or models for balance and sustenance (Archi et al., 2023; Chen, 2022; Guo and Gu, 2022; Zimik et al., 2022). Manzano et al. (2020) advocate for understanding the past, present, and future of tourism destinations through academic and professional approaches, underscoring the importance of research and development. The role of marketing strategies such as "push, pull, and marketing mix" resurfaces in comprehending tourist psychology (Zimik and Goenka, 2021). Highlighting the crucial focus on investment for smart destinations, Zimik and Barman (2023) stress the significance of investing in the CE for tourism development.

The global significance of the tourism sector, contributing to one-fourth of global employment (WTTC, 2022) and 10% of global GDP (WTTC, 2019), mitigates risks for investors. Globalization and liberalization policies further encourage investment in the sector, with an emphasis on green tourism (Lu et al., 2021) and intelligent management (Choudhury and Dixit, 2018; Gelter et al., 2020; Habsi et al., 2022). Smart tourism is characterized by the evolution of traditional and e-tourism services, incorporating safety measures, robot guides, online ticketing, and virtual tours (PTI, 2022). From a background perspective, smart tourism's essence is discerned through tourist demand and satisfaction levels in smart destinations. Zheng (2023) argues that smart tourism technology, including AI, virtual reality, smartphones, and other technologies, enhances positive emotions and motivates revisit tendencies (Suanpang et al., 2021). Social and managerial perspectives are acknowledged for their vital roles in perceiving benefits and psychological well-being (Gani et al., 2023). Numerous smart tourism models are proposed, encompassing effective policy-making, technology-oriented approaches, economic and environmental considerations, and strategic planning (Cavalheiro et al., 2020; Huda et al., 2021; Shafiee et al., 2019). Smart tourism has become a new brand tag, embodying the expectations of tourists seeking innovative and technologically advanced experiences.

Upon closely examining the discourse on smart tourism, it becomes apparent that various facets of smart tourism face challenges due to emerging issues. Consequently, CE concepts emerge as a paramount strategy, not only fostering a global shift toward smarter tourism but also addressing the challenges associated with smart tourism elements. CE provides a strategic approach to resource management, effectively bridging any expectation gaps in tourist experiences within the context of smart tourism.

10.4.2 CONSTRUCTED ENVIRONMENT (CE)

The escalating tourist demand, market potential, and competence standards underscore the necessity for a transformative mechanism addressing the sixth sense sensation. Commonly, CE denotes the human-made environment catering to needs; however, its true purpose lies in satisfying tourists. Within the research's objectives and scope, CE signifies the experience of exploring smarter destinations, evident in the dearth of literature.

The shift from smart to smarter tourism lacks substantial research. Acknowledging the intricate operations, intellectual aspects, and synchronization in tourism, constructing a smarter tourism industry necessitates stakeholder roles, conscious and unconscious tourist expectations, competence, and policy considerations.

10.4.3 THE STAKEHOLDERS

The advancement of development activities is intricately linked to the involvement of various stakeholders across different governance structures. The consciousness and competency levels of these stakeholders play a crucial role in addressing challenges faced by the tourism industry. The formulation of policies and the approval of diverse strategies rest within the purview of stakeholders, emphasizing the importance of their roles. Effective collaboration among organizations, communities, and the optimal utilization of human resources contribute to the seamless functioning of the tourism sector. A noteworthy aspect to explore is the feasibility of funding, which has witnessed a significant upswing in developing and underdeveloped nations compared to their developed counterparts. This surge in development is attributed to the availability of funding options coupled with the increasing trend of globalization, redefining and transforming destinations.

The complexity of defining "smarter tourism" becomes apparent, with variations in definitions for developing and developed nations. The transformation of infrastructure, the provision of facilities, hospitality, and leisure requires a concerted effort from partnerships and elements of the CE, fostering smarter tourism. The success of tourism industries hinges on the active involvement of communities and social organizations. Policy decisions play a pivotal role in determining the intensity of tourism development, underscoring the need for resource utilization through collaborations among professional departments. Financial aid emerges as a key driving force, influencing industries and controlling stakeholders. The significance of investment in smart tourism becomes evident, reflecting the initiative and commitment of stakeholders. The transformation of destinations serves as a testament to the dedication of key role players, including bureaucrats, entrepreneurs, and constructors, in enhancing destinations. The credibility and willingness of participants are tested in the pursuit of smarter destinations, gauged by the satisfaction levels of tourists. This prompts critical questions regarding the commitment to implementing the latest policies and appropriate measures. Understanding tourist psychology emerges as a cornerstone in presenting smart destinations favorably and ensuring satisfaction for tourists.

10.4.4 THE CONSCIOUSNESS

In the realm of CE, the scrutiny of consciousness levels revolves around interactive analyses, considering perspectives from both tourists and stakeholders. Primarily, an examination of the awareness among policymakers and constructors is essential. Destinations must be equipped to offer heightened convenience with minimal effort from tourists, ensuring an optimal experience. The construction of destinations should not only embody visual allure but also align with the principles of smart destinations, encompassing aspects from initial planning and execution to necessary corrections or supplements. Policymakers, beyond catering to local demands, must establish a benchmark for work output comparable to smart destinations in developed nations, focusing on sustainability. It is imperative to conscientiously educate the local community about the concept of serving and entertaining visitors. Various research and development reports have prompted bureaucrats to contribute to pragmatic implementations, utilizing CE's rationality and sensibility to foster an allegiance approach.

The formulation and approval of any development project should incorporate a nuanced understanding of push and pull strategies. Empowering smart tourism experts and academicians is crucial to infuse meaningful knowledge into policymakers' consciousness, thereby enhancing their effectiveness. Case study analyses serve as a valuable tool for making informed decisions and refining intuitions. Both conscious and unconscious elements in policymakers' decision-making processes can be enhanced. Media plays a pivotal role in conveying the benefits and consequences of tourism management decisions. As decisions and intuitions intertwine, a balanced assessment of situations and associated risks is imperative. Drawing inspiration from "The Elephant Whisperers" exemplifies how visitors can partake in globally recognized experiences, reflecting an intention to fulfill their wishes and bucket lists. This underscores the significance of ensuring that visitors leave with a profound sense of satisfaction. This gives ask research questions (***RQx***) as:

1. **RQ1:** To what extent do policymakers prioritize fulfilling desires, reaching the pinnacle, and making a mark on individuals' bucket lists?
2. **RQ2:** Is the management of the destination's environment consistent with expectations and prior communications?
3. **RQ3:** Can the destination be deemed superior to conventional smart tourism destinations by surpassing established criteria for intelligent and innovative tourism?

Examining the aforementioned research questions (***RQs***) and aligning them with Maslow's hierarchy of self-actualization and esteem needs, it can be deduced that the level of consciousness among individuals engaged in tourism significantly impacts destination development. Enhancing destination intelligence goes beyond conventional smart tourism, relying heavily on decision-making, competence, and expertise in the field.

10.4.5 TANGIBLE AND INTANGIBILITY ASPECTS OF CONSTRUCTED ENVIRONMENT (CE)

Numerous scholarly works underscore the pivotal role of infrastructure in fostering tourism development. Embedding the concept of physical evidence within the framework of CE is crucial for imbuing the study with practicality and facilitating real-world applications. Within the context of CE stakeholders and awareness, particular emphasis should be placed on enhancing visibility. This is substantiated by the Maslowian Triangle, revealing a diminishing role of tangibility as hierarchical levels ascend and intangibility becomes more prominent. The pursuit of ideal objectives involves leveraging virtual reality and sensory experiences. Engaging with construction and workforce associations becomes imperative for executing construction projects and facilities that align with tourist satisfaction and yield creative outputs. Developing nations stand to benefit from their cost-effective labor and resources, coupled with technology imports, to enhance efficiency. The focal point remains the coordination among various departments through collaborative projects. However, joint collaboration projects between different entities are observed as underutilized.

Tangible aspects are frequently outsourced or managed with minimal resources, raising concerns about the longevity of output and management issues related to repairs, reproduction, or skill transfer opportunity loss.

The intangible aspect of CE, as elucidated by Zimik and Barman (2023), can be effectively managed through the application of methodologies such as Six Sigma (6σ), 5S (Sort, Set, Shine, Standardize, and Sustain), project management, and Total Quality Management (TQM). Implementing academically standardized practices is essential and should be accompanied by active involvement to enhance the intangible aspects of tourist destinations. Facilitating opportunities for tourists to engage in cultural events, experience social acceptance, savor gastronomy, ensure safety, receive hospitality, and avail cordial guidance is paramount. The presence of CCTV surveillance and access to medical assistance enhances confidence, contributing to the development of ancillary services. Providing leisure options like hologram adventures and other recreational facilities encourages tourists to extend their stays. Construction projects should address challenges such as increased footfall and offer multiple options for tourists to choose from. The aspiration for a heavenly experience demands commitment from all stakeholders involved.

10.4.6 POLICIES AND STRATEGIES

The alignment of policies and strategies is a captivating aspect within the context of CE. In examining consciousness and stakeholder roles, the implementation of policies necessitates collaborative efforts among various stakeholders. Unfortunately, the lack of shared platforms between departments or organizations and tourists results in policy lags. Emulating successful strategies becomes a logical approach for developing nations aiming to promote tourism. Exemplary projects like Singapore's Garden of Bays and the creation of destination wonders evoke a more appealing and attractive ambiance. Developing and underdeveloped nations, often skeptical and resistant to changes in policies, design, or initiatives, face challenges in effectively implementing approved policies. This skepticism becomes a significant impediment, hindering the realization of research contributions. The overarching concept of wealth maximization underpins CE, utilizing local products, fostering local employment, generating revenue, and supporting educational institutions. CE primarily focuses on leveraging local resources and sustainability to minimize costs in terms of time, finances, and employment. The synergy between strategies and policies steers destinations toward the path of becoming smarter destinations. Self-reliance emerges as a hallmark of smarter destinations, emphasizing the need for destinations to be self-sufficient. Corporate social responsibility stands out as a key criterion that CE strives to fulfill, underscoring its commitment to ethical and socially conscious practices.

10.4.7 COMPETENCE

Competence is a pivotal factor influencing the sustainability of tourism industries, encompassing a fusion of skills and behavior aligned with established benchmarks. This necessitates a focus on educational proficiency within the tourism sector. Competence serves as the powerhouse driving CE, linking stakeholder commitment to result-oriented outputs. Policymakers must adopt a conceptual approach attuned to both tourist psychology and destination dynamics. Smart tourism industries require an examination of stakeholders' cognitive levels to ensure a sophisticated understanding. Constructors and workers should leverage acquired cognitions to create well-sensational infrastructures, etching destination images into visitors' minds. Management within the sector must actively pursue self-actualization feelings in tourists, fostering dynamic and rapid transformations during each visit. The infusion of delight and enthusiasm by organizations contributes to destinations' competence, aligning them with the smart destinations benchmark. Resource and time management emerge as critical discussion points, pivotal for completing projects and generating vibrant outputs. Continuous training and education, a focus of CE, enhance worker dexterity, ensuring engaging work replications. Consequently, CE becomes imperative for fostering flexibility and aptitude that propels destinations beyond the smart destinations benchmark, making them more appealing and sustainable.

10.4.8 THE SATELLITE

Besides the conventional methods used to identify satellites based on their purposes, a crucial aspect of CE is its ability to gather information, exhibit vigilance, be service-oriented, and maintain reliability. Given that the role of CE is dynamic and involves constant change, innovative variations are communicated and implemented. A comprehensive and valuable report is compiled from various corners of the world, offering the most efficient strategies to stakeholders. Research and development institutions serve as a driving force, providing a positive vibrancy that gives an accurate view and an opportunity for SWOT (Strengths, Weaknesses, Opportunities, and Threats) analysis in every situation. Reports from organizations such as UNWTO (United Nations World Tourism Organization), WTO (World Tourism Organization), official reports, research contributions, expert opinions, social media, and magazines serve as sources to provide key updated information essential for creating survey reports for stakeholders. CE offers the opportunity to visualize reports and findings essential for transforming and adopting situational strategies from various sources. This concept emphasizes the need for policymakers to rely on their sixth sense and intuition for forecasting and fostering economic generations. Such an approach signifies minimal cost and maximum information collection for all global tourism activities.

10.4.9 PARTNERS

A separate analysis of the nature of partnerships is essential to make destinations resonate globally and be recognized as more vibrant destinations. Considering tourists as partners represents a new concept emerging in smarter tourism practices. Moving beyond competence in work output and tourist satisfaction, trust appears to be the key criterion for establishing confidence among various stakeholders and fostering healthy relations. Increasing the hierarchical level in the Maslowian triangle places more emphasis on safety and security needs. Studying this aspect is the need of the hour for transforming destinations into smarter ones. "Win-win" strategies suggest overcoming challenges with reward sharing, once again proving what smarter destinations should resemble currently. Corporate social responsibility (CSR) emphasizes improving the workplace environment and creating a positive ecosystem among tourists, stakeholders, nature, and economic sustainability. Tourist roles in partnerships for smarter destinations can be understood by examining past experiences of tourist visits and suggesting improvements. These contributions are noted as the best ways tourists can participate as partners, not only aiding in transformation but also reducing the likelihood of negative feedback. Providing tourists with the opportunity to rate destinations in terms of smartness and attractiveness, through ratings or grades, enhances their involvement in the process. This practice not only facilitates transformation but also increases the likelihood of receiving positive feedback and reducing negative impressions.

10.4.10 PSYCHOLOGY

Understanding tourist psychology is the primary strategy for boosting the tourism industry. This aspect can be explored through the Maslowian triangle, as described by Zimik et al. (2022). The triangle is based purely on the psychology of tourists and stakeholders' perceptions. Smart tourism often emphasizes technology's role in attracting tourists, but it leaves a gap in comprehending tourist psychology accurately. This challenges policymakers' awareness and the competence of constructors in aligning with tourist psychology. The ultimate goal is to perceive and connect with tourist imagination, transforming it into reality. Higher levels of hierarchy tend to focus more on the intangible aspects of the CE, while tangible aspects, such as basic safety and infrastructure, are more prevalent in lower hierarchies. These pose challenges for constructors and worker associations to be competent enough to overcome disturbances and ensure tourists feel comfortable.

The sections demand more inspiration, involvement, and commitment from stakeholders in response to tourist expectations, hopes, and experiences. Higher hierarchies rely on perceived intangible aspects, working on friendliness, hospitality, and social acceptance to make tourists feel at home and safe. Issues like CCTVs and active medical facilities are part of the CE's focus. Esteem and self-actualization are not just about accomplishing goals; they are also influenced by tourists' perceptions of destinations. Stakeholders are examining ways to counter negative feelings towards visits. The CE's main intention is to achieve synergistic efforts from all discussed aspects, enhancing smart destinations' functionality with positive collective effects. Any room for improvement in smart tourism destinations is achievable through CE, making stakeholders aware and prompting the adoption of optimal strategies. This transformation enhances areas by incorporating the best alternative strategies, ensuring a holistic and integrated approach for continuous improvement in the tourism sector.

10.4.11 ENCOUNTERING SMARTER TOURISM INDUSTRIES

It is evident from the current study that there is a significant literature gap, necessitating a thorough discussion on the scope to enhance the performance of tourism industries as an improved version of smart tourism. In the continuation of the process from smart to smarter tourism, it cannot only be comprehended as a simulated case but also as a pragmatic approach to showcase information and knowledge flow for efficient and well-kept contemporary practices. Much research has been conducted for smart tourism, raising the question of whether there are limitations to room for improvement. This contemplation on transformation demands dynamic improvements to define smart tourism. A smarter tourism destination involves unique and situational infrastructural transformations. A substantial improvement is needed in the domains of 5As and technology, with discussions required on sustainability, accessibility, and digitalization for rural development. Tourist destinations require more options, emphasizing the role of operational research disciplines. Issues like traffic can be managed by improving parking spaces, utilizing areas such as ocean beaches, riverbanks, and other spaces. Road transportation can be managed by considering parameters such as vehicle wheels, public transportation, and technological backup. Smarter destinations symbolize 24/7 operations, contingent on the legality and social cooperation of the host. Safety and social needs hinge on the destination's welcome address, necessitating strong infrastructural support in training and education. Leisure and adventure remain top priorities for a smarter tourist experience.

A robust Information and Communication Technology (ICT) system is essential to counter any challenges and support tourists and policymakers in responding to and controlling both positive and negative situations. An organization consisting of government departments, communities, and organizations should exist to enhance collaborations at destinations. While smart destinations exist, proposed infrastructure consisting of such a body does not exist in reality. It is high time to demand such initiatives become a reality. Policies and suggestions should be infrastructural to enhance the performance of tourism activities. In the transition to a green economy, infrastructure should encourage the use of renewable energy sources, such as solar power stations, CCTVs, well-guided route maps, grid LEDs, and lawns for walking. Collective efforts from host hospitality, technological service support, and tourist initiatives would elevate destinations towards esteem and attractiveness. Education and training play vital roles, where tourist suggestions and problems can be analyzed. Smarter destinations should encourage tourists to interact with guides or stakeholders to enhance tourist satisfaction. Infrastructure should respond dynamically to the dynamism and complexity of tourist expectations, with roles and responsibilities deeply dependent on service sectors.

Tourists are more inclined to ask for available facilities, posing a Herculean task for management. A sound and vibrant infrastructure approach becomes subjective and scientific, accommodating every suggestion and satisfaction of tourists. Accessibility and attraction depend on infrastructure, and enhanced virtual reality should not be optional but rather made compulsory. A holistic approach to the tourism industry needs practical implementation with the existing effort of infrastructure benefits to reflect tourist imaginations.

Innovations and strategies remain another area to witness in smarter destinations. Innovations go beyond imagination and appreciation, questioning the competence and professional capability of stakeholders. Historical facts can be brought alive and observed through immersive 3D theaters, revolutionizing experiences and educational purposes. The seven wonders of the world can be brought to life in a theme park. Innovative taglines and marketing promotions through newsletters and social media can enhance promotion and attract more tourists. Approaches to attract tourists involve not only infrastructure but also adopting more efficient strategies such as using available resources like ICT, apps, and online booking options to reduce queues and time. Innovations should involve artificial intelligence, cloud technology, and contactless travel to maintain the definition of smarter tourism. Tourism demands more innovation with updated technology and ideas, as suggested by the survey report from the conceptual satellite proposed by the CE. Any ideas for improving leisure, attractions, accessibility, and technology in tourism may result in amplified ripple effects, such as increased revenue generation, satisfaction, and spending nature. Resource management involving academic students, researchers, and free volunteers with proper training can improve cost-cutting and enhance opportunities for encounters and experiences. Innovations should focus on reducing risk and liability, not just for tourist satisfaction but also for a safer identity for smarter destinations. Service sectors act as supplementary agents, making tourists conscious enough to experience contentment, a key aspect in Maslow's hierarchy. To achieve such levels, tourist opinion assessments and comprehensive responsibility need correlation. Sustainability signifies more than just environmental, social, and economic benefits; it works towards growth rate improvements and adaptations. Smarter tourism is a holistic approach from each stakeholder, synchronizing every effort beyond expectations. This points toward steady improvement while working on existing tourism phenomena for effective performance. Communications and collaborations happen not only for tourist convenience but also symbolize smarter destinations.

Database maintenance and reports from reputed institutions need highlighting for more insightful knowledge extraction. Communication between various departments and collective efforts need to be a reality, remaining a high note to be achieved. Another domain of communication is network accessibility and physical transportation comfort. Situations arise where excellent network services become idle with the increase in tourist footfall, especially at hill stations, questioning the feasibility of cashless operations and social media updates. Any lag in social media networks acts as a blind spot for tourists. Social media acts as the next best world, which tourists wish to exist in, making connections a dream and solving problems of security and safety through cashless transactions. The auto-conversion of current digital platforms would surely make destinations proactive, demanding more cooperation and integrative effort from the banking service. App creations would certainly make destinations informed and well-assisted. To date, numerous literature and models speak about smart tourism and show potential to perform as smarter destinations. This mounts the amount of research and exploration work demanded to improve the researched recommendations and redesign existing models. Redefining each element considered in various proposed models needs updating based on perspectives of smarter tourism industries and the dynamic demands of tourists. Such adaptation helps in generalizing to modify and encourage change to improve tourism activities.

Furthermore, basic needs may vary depending on tourists and destinations over time, leading to more complex models comprising an infinite series of elements. Delimiting models of tourist satisfaction is recommended to counter any flaws and create room for improvement. The study findings suggest various conceptual models contributing and seeking experts to intervene through simulation and practical applications of designs. Thus, it acts as a means of feedback for communicating with the CE to modify its strategies and approach. Tourism is all about tourist satisfaction for sales and spending, but a smarter tourism seeks to go beyond tourist exploitation to wealth maximization schemes. "Win-Win" strategies seem to be the right tune to comprehend wealth maximization. The concept tries to move away from the conventional profit-making to a new dimension that goes beyond self-centered profit maximization to adopting an approach of revering competitors and making suppliers, stakeholders, and local vendors find gains while skimming the market. Sharing dominance of the market together with other enterprises does not only mark profit-sharing but also signifies the healthy nature of the market. Wealth maximization, from the financial perspective, specifically speaks about considering shareholders and other stakeholders. In the tourism industry, more triggering features occur when stakeholders are not definable due to constant involvement of various other enterprises, starting from transportation facilities to investors. Thus, a strong policy is needed where investors find means to benefit from the investment, promote local products and resources, and consider measures to improve the standard of living. Wealth maximization can be interpreted as the likability and happiness index of the tourist and local host community, which strongly bases on the ease of doing business and transparency toward work efficiency from worker associations to top bureaucracy, creating an ideal work environment for a healthy work environment. Availability of investment in the region for development projects is another intriguing feature for a positive wave of sustainable growth.

The practical implications and examinations of the consciousness of policymakers should be visible in smarter tourism destinations. Counting on efficiency and realistic benchmarks would definitely mark how deep and clear the projection was made. Any destination should be called a better and improving mode with better proficiency of work deliverance, or even while policies were formulated. A destination needs to reflect attractiveness, management, sound technology, and an image that emerges from the effective professionalism and consciousness of tourist psychology. The forecasted and pre-decisions made at CE need to be examined critically, which may be relevant not only from the standpoint of profit or tourist satisfaction but also from the capacity and ability to manage tourist destinations. Smarter tourism would always yield results beyond presumed and expected results, as a thorough assessment was made during CE applications. Ironically, it may be presumed that any irregularity or huge contrast and deviation from the expected benchmark can be considered as incompetence of professionals and consciousness. Nevertheless, smart tourism receives special attention for smarter tourism destinations through tourist opinions and serious efforts, coupled with advanced technology, seek to make the industries function in each fully sustainable function. Lastly, the competence level and skills benchmark output were analyzed to counter all the discussed criteria in the CE as a means of inspirations and passion for work involvement. Qualitatively, it can be considered that smarter tourism refers to a way above the outstanding performance of the industries through CE.

10.5 CONCLUSIONS

The exposition delves into the realm of intelligent tourism, offering profound insights and emphasizing the imperative for future research and earnest discussions to integrate the CE concept. Acting as a potent catalyst, CE propels a positive metamorphosis, translating tourists' imaginative ideals into tangible realities. It introduces a paradigm shift, prompting a reconsideration of smart tourism and its management to foster the evolution of smarter tourism destinations. The proposal advocates a comprehensive overhaul of the tourism industry's management and operational processes, envisioning a radical transformation through the assimilation of CE principles. The involvement of the state and various stakeholders is deemed crucial for crafting intelligent industries. In contrast to the historical era where substantial funds facilitated construction, contemporary complexities with intricate chains of command necessitate the professional and competent application of CE.

The study highlights CE concepts as the focal point driving transformative changes, concurrently establishing a compelling connection with the envisioned global industrial revolutions, particularly the Fifth Industrial Revolution. This synthesis creates an optimistic outlook on enhancing industries within a smart nation, underscored by strategic sustainability and eco-balancing initiatives. The role of CE extends beyond mere attraction, actively contributing to shaping a mindset conducive to niche markets (such as catering, wellness tourism, or cultural heritage), employing a data-driven approach for comprehensive information dissemination, and affording tourists the opportunity to participate in policymaking as responsible stakeholders in the tourism ecosystem. In conclusion, it becomes evident that the formulation of tourism policies requires a nuanced understanding of the consciousness and psychology of the visitors, taking into account their past experiences. It is imperative to bridge the gap between policymaking and the intricate aspects of tourist perception, ensuring that the strategies implemented align with the expectations and values of responsible tourism. The integration of CE principles not only enhances the industries within the smart nation but also fosters a symbiotic relationship between sustainable development and the diverse needs of the evolving tourism landscape.

Overcoming the intricacies and time constraints, CE proves instrumental in navigating the contemporary landscape of construction. With escalating tourist demands, a surge in research contributions, and evolving industry trends, there arises a compelling need for destinations to transition from being merely smart to becoming truly intelligent. This shift poses challenges to both public and academic realms, prompting a reevaluation of differences. Consequently, CE emerges as a recent beacon, providing a comprehensive perspective on meeting tourist demands while aligning with stakeholders' perceptions, fostering the emergence of smarter tourism destinations derived from the foundational principles of smart tourism. A key takeaway from the discourse underscores that CE thrives on collaborative efforts from diverse stakeholders, demonstrating competence in enhancing fairness—a facet conspicuously absent elsewhere.

KEYWORDS

- **constructed environment (CE)**
- **fourth industrial revolution (4IR)**
- **local communities**
- **Maslowian triangle**
- **policymakers**
- **smart strategies**
- **smart tourism**
- **stakeholders**

REFERENCES

Abouzeid, N. (2022). The impact of digitalization on tourism sustainability: Comparative study between selected developed and developing countries. In *2nd International Conference on Advanced Research in Management, Economics, and Accounting* (pp. 30–50). Milam: NA.

Agudo, J., Crespo, Á., & Gutiérrez, H. (2023). The adoption of a smart destination model by tourism companies: An ecosystem approach. *Journal of Destination Marketing & Management*, *28*, 1–12.

Ahammed, T., Patgiri, R., & Nayak, S. (2023). A vision on the artificial intelligence for 6G communication. *ICT Express*, *9*(2), 197–210.

Archi, Y., Benbba, B., Nizamatdinova, Z., Issakov, Y., Vargáné, G., & Dávid, L. (2023). Systematic literature review analyzing smart tourism destinations in context of sustainable development: Current applications and future directions. *Sustainability*, *15*(6), 50–86.

Azarmi, S., Vaziri, R., & Oladipo, A. (2019). Environmental impact of the ugly face of tourism: Pollution and management perspectives. In *The Ugly Face of Tourism* (pp. 312–326). Taylor & Francis.

Baloch, Q. B., Shah, S. N., Iqbal, N., Sheeraz, M., Asadullah, M., Mahar, S., & Khan, A. U. (2022). Impact of tourism development upon environmental sustainability: A suggested framework for sustainable ecotourism. *Environmental Science and Pollution Research*, https://doi.org/10.1007/s11356-022-22496-w (accessed on 8th July 2025).

Bettini, J. (2018). IDB improving life. https://blogs.iadb.org/sostenibilidad/en/smart-destinations-quality-tourism-for-visitors-and-sustainable-development-for-residents/ (accessed on 8th July 2025).

Bhuiyan, K., Jahan, I., Zayed, N., Islam, K., Suyaiya, S., Tkachenko, O., & Nitsenko, V. (2022). Smart tourism ecosystem: A new dimension toward sustainable value co-creation. *Sustainability*, *14*(22), https://doi.org/10.3390/su142215043 (accessed on 8th July 2025).

Buhalis, D., & Amaranggana, A. (2014). Smart tourism destinations. In *Information and Communication Technologies in Tourism 2014* (pp. 553–564). NA: Springer Nature.

Busulwa, R., Pickering, M., & Mao, I. (2022). Digital transformation and hospitality management competencies: Toward an integrative framework. *International Journal of Hospitality Management*, *102*, https://doi.org/10.1016/j.ijhm.2021.103132 (accessed on 8th July 2025).

Cavalheiro, M., Joia, L., & Cavalheiro, G. (2020). Towards a smart tourism destination development model: Promoting environmental, economic, socio-cultural and political values. *Tourism Planning & Development*, *17*(3), 237–259.

Chakrabarty, A. (2023). Smart tourism: Conceptual aspects and related issues. *International Journal of Research Publication and Reviews*, *4*(8), 3230–3234.

Chen, G. (2022). Tourism management strategies under the intelligent tourism IoT service platform. *Computational Intelligence and Neuroscience*, 1–11.

Chen, P., Zhao, Y., Zuo, D., & Kong, X. (2021). Tourism, water pollution, and waterway landscape changes in a traditional village in the Huizhou region, China. *Land*, *10*(8), 2–22.

Choudhury, R., & Dixit, S. (2018). Prospects and challenges in smart tourism in India: Case study of smart city Bhubaneswar. *International Journal of Creative Research Thoughts*, *6*(1), 242–248.

Chuang, C. (2023). The conceptualization of smart tourism service platforms on tourist value co-creation behaviors: An integrative perspective of smart tourism services. *Humanities and Social Sciences Communications*, *10*, https://doi.org/10.1057/s41599-023-01867-9 (accessed on 8th July 2025).

Consultants. (2023). Visa Index. Available at: https://visaindex.com/blog/what-is-smart-tourism/ (accessed on 8th July 2025).

Corluka, G. (2019). Tourism seasonality–An overview. *Journal of Business Paradigms, 4*(1), 22–43.
Corluka, G., Mikinac, K., & Milenkovska, A. (2016). Classification of tourist season in coastal tourism. *UTMS Journal of Economics, 7*(1), 71–83.
Darcy, S., McKercher, B., & Schweinsberg, S. (2020). From tourism and disability to accessible tourism: A perspective article. *Tourism Review, 75*(1), 140–144.
Fernandes, D., & Bernier, E. (2014). Virtual accessibility in smart tourism destinations: The cases of Curitiba and Malaga. *Smart Tourism, 2*(1), 1–12. https://doi.org/10.54517/st.v2i1.1717 (accessed on 8th July 2025).
Fernández-Díaz, E., Jambrino-Maldonado, C., Iglesias-Sánchez, P., & Heras-Pedrosa, C. (2023). Digital accessibility of smart cities–tourism for all and reducing inequalities: Tourism Agenda 2030. *Tourism Review, 8*(2), 361–380.
Fonseca, F. (2021). Orioly. Available at: https://orioly.com/8-solutions-that-are-turning-cities-into-smart-destinations/ (accessed on 8th July 2025).
Gani, M., Roy, H., Faroque, A., Rahman, M., & Munawara, M. (2023). Smart tourism technologies for the psychological well-being of tourists: A Bangladesh perspective. *Journal of Hospitality and Tourism Insights*, https://doi.org/10.1108/JHTI-06-2022-0239 (accessed on 8th July 2025).
Gelter, J., Fuchs, M., & Lexhagen, M. (2022). Making sense of smart tourism destinations: A qualitative text analysis from Sweden. *Journal of Destination Marketing & Management, 23*, 1–12.
Gelter, J., Lexhagen, M., & Fuchs, M. (2020). A meta-narrative analysis of smart tourism destinations: Implications for tourism destination management. *Current Issues in Tourism*, 1–16. https://doi.org/10.1080/13683500.2020.1849048 (accessed on 8th July 2025).
Gidumal, J. (2022). Post-COVID-19 recovery of island tourism using a smart tourism destination framework. *Journal of Destination Marketing & Management, 23*, https://doi.org/10.1016/j.jdmm.2022.100689 (accessed on 8th July 2025).
Gil, K. (2021). *Smart Tourism, The Bet of Smart Cities*. BBVA. Available at: https://www.bbva.ch/en/news/smart-tourism-the-bet-of-smart-cities/ (accessed on 8th July 2025).
Goo, J., Huang, C., Yoo, C., & Koo, C. (2022). Smart tourism technologies' ambidexterity: Balancing tourist's worries and novelty seeking for travel satisfaction. *Information Systems Frontiers, 24*, 2139–2158.
Gössling, S. (2009). Carbon neutral destinations: A conceptual analysis. *Journal of Sustainable Tourism, 17*(1), 17–37.
Gössling, S. (2020). The low-carbon imperative: Destination management under urgent climate change. *Journal of Travel Research, 60*(6), 1167–1179.
Gretzel, U., Sigala, M., Xiang, Z., & Koo, C. (2015). Smart destinations: Foundations, analytics, and applications. *Journal of Destination Marketing & Management, 4*(3), 179–188.
Gretzel, U., Sigala, M., Xiang, Z., & Koo, C. (2015). Smart tourism: Foundations and developments. *Electronic Markets, 25*(3), https://doi.org/10.1007/s12525-015-0196-8 (accessed on 8th July 2025).
Guo, K., & Gu, Y. (2022). The construction of smart tourism city and digital marketing of cultural tourism industry under network propaganda strategy. *Security and Communication Networks, 2022*, https://doi.org/10.1155/2022/4932415 (accessed on 8th July 2025).
Habsi, R., Agarwal, A., & Naidu, V. (2022). Design and develop a mobile application for smart tourism in Oman. *5th Middle East College Student Research*, 1–7. Oman: JSR.

Huda, C., Ramadhan, A., Trisetyarso, A., Abdurachman, E., & Heryadi, Y. (2021). Smart tourism recommendation model: A systematic literature review. *International Journal of Advanced Computer Science and Applications*, *12*(12), 167–174.

Huertas, A., Moreno, A., & Fontanilles, J. (2021). Place branding for smart cities and smart tourism destinations: Do they communicate their smartness? *Sustainability*, *13*(19), 2–18.

Kovačić, N. (2022). The acceptance of smart technologies in tourist regions with respect to mobility preferences of their visitors. *Transportation Research Procedia*, *64*, 257–269.

Kumar, A. (2020). Smart destination and technology driven tourism for sustainable development. *AVAHAN: A Journal on Hospitality and Tourism*, *8*(1), 23–31.

Lasisi, T., Odei, S., & Eluwole, K. (2023). Smart destination competitiveness: Underscoring its impact on economic growth. *Journal of Tourism Futures*, 1–21.

Lazarevic, J., & Pavlović, G. (2018). Analysis of factors of smart tourism development in Serbia. *Menadzment u Hotelijerstvu i Turizmu*, *6*(1), 81–91.

Lee, P., Hunter, W., & Chung, N. (2020). Smart tourism city: Developments and transformations. *Sustainability*, *12*(10), 39–58.

Li, D., Du, P., & He, H. (2022). Artificial intelligence-based sustainable development of smart heritage tourism. *Wireless Communications and Mobile Computing*, 1–13.

Liberato, P., Alen, E., & Liberato, D. (2018). Smart tourism destination triggers consumer experience: The case of Porto. *European Journal of Management and Business Economics*, *27*(1), 6–25.

Limna, P. (2022). Artificial intelligence (AI) in the hospitality industry: A review article. *International Journal of Computing Sciences Research*, *6*, 1–12.

Lohmann, M., & Hübner, A. (2013). Tourist behavior and weather. *Open Edition Journals*, 44–59.

Lu, C., Huang, J., Chen, C., Shu, M., Hsu, C., & Bapu, B. (2021). An energy-efficient smart city for sustainable green tourism industry. *Sustainable Energy Technologies and Assessments*, *47*, https://doi.org/10.1016/j.seta.2021.101494 (accessed on 8th July 2025).

Mansilla, E., Tussyadiah, I., Campo, S., & Rubio, N. (2023). Smart destinations: A holistic view from researchers and managers to tourists and locals. *Tourism Management Perspectives*, *51*, https://doi.org/10.1016/j.tmp.2024.101223 (accessed on 8th July 2025).

Manzano, A., Fernández, J., & Aranda, L. (2020). The past, present, and future of smart tourism destinations: A bibliometric analysis. *Journal of Hospitality & Tourism Research*, *45*(3), 529–552.

Matos, A., Pinto, B., Barros, F., Martins, S., Martins, J., & Oliveira, M. (2019). Smart cities and smart tourism: What future do they bring? In Á. R. et al., *New Knowledge in Information Systems and Technologies* (pp. 358–370). Switzerland: Springer Nature.

Mehraliyev, F., Chan, I., Choi, Y., Koseoglu, M., & Law, R. (2020). A state-of-the-art review of smart tourism research. *Journal of Travel & Tourism Marketing*, *37*(1), 78–91.

Mehraliyev, F., Choi, Y., & Köseoglu, M. (2019). Progress on smart tourism research. *Journal of Hospitality and Tourism Technology*, *10*(4), 522–538. https://doi.org/10.1108/JHTT-08-2018-0076 (accessed on 8th July 2025).

Nguyen, H., & Duy, T. (2021). Information technology infrastructure for smart tourism in Da Nang City. *International Journal of Hyperconnectivity and the Internet of Things*, *5*(1), 98–108.

Pai, C., Liu, Y., Kang, S., & Dai, A. (2020). The role of perceived smart tourism technology experience for tourist satisfaction, happiness and revisit intention. *Sustainability*, *20*(16), 1–14.

PTI. (2022). *The Economic Times*. https://travel.economictimes.indiatimes.com/news/technology/embrace-smart-tourism-for-adopting-futuristic-trends-in-the-industry/93949395 (accessed on 8th July 2025).

Rappensperger, T. (2023). *Regiondo*. https://pro.regiondo.com/blog/tourism-marketing-strategies/ (accessed on 8th July 2025).

Roy, A. (2018). *National Strategy for Artificial Intelligence*. New Delhi: NITI Aayog.

Rucci, A., & Porto, N. (2022). Accessibility in tourist sites in Spain: Does it really matter when choosing a destination? *European Journal of Tourism Research*, 1–30.

Sæþórsdóttir, A., Hall, C., & Stefánsson, P. (2019). Senses by seasons: Tourists' perceptions depending on seasonality in popular nature destinations in Iceland. *Sustainability*, *11*(11), 2–18.

Sambhanthan, A., & Good, A. (2013). A second life based virtual community model for enhancing tourism destination accessibility in developing countries. *International Journal of Collaborative Enterprise*, *3*(4), 269–286.

Shafiee, S., Ghatari, A., Hasanzadeh, A., & Jahanyan, S. (2019). Developing a model for sustainable smart tourism destinations: A systematic review. *Tourism Management Perspectives*, *31*, 287–300.

Shahzad, M., Qu, Y., Rehman, S., & Zafar, A. (2022). Adoption of green innovation technology to accelerate sustainable development among manufacturing industry. *Journal of Innovation & Knowledge*, *7*(4), https://doi.org/10.1016/j.jik.2022.100231 (accessed on 8th July 2025).

Shen, S., Sotiriadis, M., & Zhou, Q. (2020). Could smart tourists be sustainable and responsible as well? The contribution of social networking sites to improving their sustainable and responsible behavior. *Sustainability*, *12*(4), 2–21.

Signes, E., Merino, R. P., & Adán, I. (2020). Transition towards a tourist innovation model: The smart tourism destination: Reality or territorial marketing? *Journal of Innovation & Knowledge*, *5*(2), 96–104.

Singh, R., & Duhan, A. (2022). Smart tourism and tourist attractions. *PARIPEX–Indian Journal of Research*, *11*(6), 39, 40.

Sonuç, N., & Süer, S. (2023). Smart tourism destinations and digitalization of cultural heritage for sustainability. *Smart Cities for Sustainability*, *32*, 151–168.

Sorokina, E., Wang, Y., Fall, A., & Lugosi, P. (2022). Constructing a smart destination framework: A destination marketing organization perspective. *Journal of Destination Marketing & Management*, 1–29.

Steeves, K. (2021). *Checkfront*. https://www.checkfront.com/blog/tourism-marketing-strategy/ (accessed on 8th July 2025).

Suanpang, P., Netwong, T., & Chunhapataragul, T. (2021). Smart tourism destinations influence a tourist's satisfaction and intention to revisit. *Journal of Management Information and Decision Sciences*, *24*(1S), 1–10.

Sun, Q., & Liu, Z. (2020). Impact of tourism activities on water pollution in the West Lake Basin (Hangzhou, China). *Open Geosciences*, *12*(1), https://doi.org/10.1515/geo-2020-0119 (accessed on 8th July 2025).

Sun, W. (2023). Research on the construction of smart tourism system based on wireless sensor network. *Mathematical Problems in Engineering, 2023,* 1–8.

Sustacha, I., Pino, J., & Valle, E. (2023). The role of technology in enhancing the tourism experience in smart destinations: A meta-analysis. *Journal of Destination Marketing & Management, 30,* https://doi.org/10.1016/j.jdmm.2023.100817 (accessed on 8th July 2025).

Sziva, I., Keller, K., & Kovács, L. (2023). Smart positioning: How smart technologies can increase the attractiveness of heritage tourism destinations? The case of a small-scale Hungarian heritage city. *Journal of Heritage Tourism,* https://doi.org/10.1080/1743873X.2023.2276271 (accessed on 8th July 2025).

Thiagarajan, K. (2023). *PNR.* https://www.npr.org/sections/goatsandsoda/2023/03/05/1160659634/the-elephant-whisperers-an-oscar-nominated-love-story-about-people-and-pachyderm (accessed on 8th July 2025).

Tiago, F., Borges-Tiago, T., & Veríssimo, J. M. C. (2019). A review of smart tourism studies in the tourism. *Travel and Tourism Research Association's*, Bournemouth: Bournemouth University.

Tiago, T., Verissimo, J., & Tiago, F. (2020). Smart tourism: A scient metric review (2008–2020). *European Journal of Tourism Research, 30,* 1–15.

Tomaso, M. (2022). *Solimar International.* https://www.solimarinternational.com/how-to-conduct-a-situation-analysis-to-develop-a-tourism-strategy/ (accessed on 8th July 2025).

Torabi, Z., Pourtaheri, M., Hall, C., Sharifi, A., & Javidi, F. (2023). Smart tourism technologies, revisit intention, and word-of-mouth in emerging and smart rural destinations. *Sustainability, 15*(14), 2–21.

TRAI. (2023). *Digital Transformation Through 5G Ecosystem.* New Delhi: TRAI: Telecom Regulatory Authority of India.

Vardopoulos, I., Evangelou, M., Nosova, B., & Salvati, L. (2023). Smart 'tourist cities' revisited: Culture-led urban sustainability and the global real estate market. *Sustainability, 15*(5), https://doi.org/10.3390/su15054313 (accessed on 8th July 2025).

Wallace, M. (2024). *Sea Going Green Organization.* https://www.seagoinggreen.org/blog/5-green-technologies-that-support-sustainable-tourism (accessed on 8th July 2025).

Wang, N. (2015). Research on construction of smart tourism perception system and management platform. In *International Conference on Education, Management and Computing Technology (ICEMCT 2015)* (pp. 89–92). Atlantis Press.

WTTC. (2019). *WTTC Research Reveals Global Travel & Tourism Sector Suffered a Loss of Almost US$4.5 Trillion in 2020 due to the Impact of COVID-19.* London: WTTC: World Travel and Tourism Council.

WTTC. (2022). *Travel and Tourism Economic Impact 2022.* London: WTTC: World Travel and Tourism Council.

Xu, M. (2023). Research on smart tourism system based on artificial intelligence. In *3rd International Conference on Information Technology, Big Data and Artificial Intelligence (ICIBA 2023)* (pp. 201–205). IEEE.

Yang, S., Yumeng, L., & Ziqi, Y. (2022). Tourists' risk perception of smart tourism impact on tourism experience. *2022 International Conference on Social Sciences and Humanities and Arts (SSHA 2022)* (pp. 368–375). Atlantis Press.

Zhang, C., Liu, Y., & Pu, Z. (2023). How digital financial inclusion boosts tourism: Evidence from Chinese cities. *Journal of Theoretical and Applied Electronic Commerce Research, 18*(3), 1619–1636.

Zhang, L., Ma, D., & Hu, J. (2021). Research on the sustainable operation of low-carbon tourism supply chain under sudden crisis prediction. *Sustainability*, *13*(15), 2–25.
Zhang, Y., & Li, Q. (2022). The development status and trend of urban smart tourism based on Internet of Things technology. *Scientific Programming*, *2022*, 1–11.
Zhang, Y., Sotiriadis, M., & Shen, S. (2022). Investigating the impact of smart tourism technologies on tourists' experiences. *Sustainability*, *14*(5), 30–48.
Zheng, Y. (2023). An investigation of how perceived smart tourism technologies affect tourists' well-being in marine tourism. *PLOS ONE*, *18*(8), https://doi.org/10.1371/journal.pone.0290539 (accessed on 8th July 2025).
Zimik, S., & Barman, A. (2021). Constructed environment as unkempt issue in North East Indian tourism: A review-based triangulation for research. *International Journal of Multidisciplinary Educational Research*, *10*(5), 137–153.
Zimik, S., & Barman, A. (2023). Emergence of constructed environment concept for tourism: A study based on triangulation review. *10th International Baskent Congress on Humanities and Social Sciences* (pp. 381–386). Baskent: Baskent Congress.
Zimik, S., & Barman, A. (2023). Global trend and roles of constructed environment in tourism development. *European Economic Letters*, *13*(3), 783–793.
Zimik, S., & Barman, A. (2023). Investment in constructed environment for tourism development. *Formosa Journal of Multidisciplinary Research*, *2*(11), 1799–1814.
Zimik, S., & Barman, A. (2023). Tourist destination innovation: The role of constructed and. *Mukt Shabd Journal*, *12*(9), 807–826.
Zimik, S., & Barman, A. (2023). Tourist destination innovation: The role of constructed and deconstructed environments. *Mukt Shabd Journal*, *12*(9), 807–826.
Zimik, S., & Barman, A. (2024). Role of constructed environment for tourism development: A global trend analysis based on triangulated review. *Journal of Business Management and Economic Development*, *2*(1), 159–190.
Zimik, S., & Goenka, H. (2021). Investigating tourism industries through the lenses of functional management: A concept note. In *Exploring the Paradigms for North East Development* (pp. 193–257). Guwahati: A M Publication.
Zimik, S., Barman, A., & Dey, R. (2022). Re-designing tourists' satisfaction models through review triangulations. *Manager–The British Journal of Administrative Management*, *58*(154), 196–225.
Zorba, B. (2023). Eco-city tourism in smart cities for sustainability. In A. S. E. Abdelli (Ed.), *Smart Cities for Sustainability* (*Advanced Series in Management*, Vol. 32) (pp. 139–150). Leeds: Emerald Publishing Limited.

CHAPTER 11

SMART TOURISM: A NEXT LEVEL OF EMPOWERING CITIZEN ENGAGEMENT

SADIA AFRIN ANANYA*

Department of Tourism and Hospitality Management, Gopalganj Science and Technology University, Gopalganj, Bangladesh; and University of South Australia, Adelaide, South Australia, Australia

*Corresponding author

ABSTRACT

In the contemporary digital age, the proliferation of smart technologies has revolutionized the way tourism is experienced and managed. This chapter is an initiative to find out the pathways for citizen engagement with the manifold dynamics of smart tourism. By examining the relationships between smart tourism and citizen participation, the study fills a notable gap, drawing on a thorough review of the literature. Using a secondary data analysis approach, insights from various sources focus on the potential of smart tourism in shaping inclusive and participatory experiences for residents. Content analysis is employed to identify trends, themes, and connections relevant to citizen participation and smart tourism. A proposed model is included in the findings, highlighting the foundational components crucial for sustainable community engagement in smart tourism. Findings emphasize a three-dimensional structure encompassing foundational infrastructure, active engagement elements, and overall impact assessment. The foundation dimension prioritizes resilient digital infrastructure, ensuring secure engagement. The engagement dimension focuses on information dissemination, feedback

Smart Tourism: Managing Technological Revolution in the Tourism Industry. Santus Kumar Deb (Ed.)

DOI: 10.1201/9781779641090-11

mechanisms, and technology integration for a responsive tourism system. The implications broaden the theoretical conversation and provide a thorough framework, and the practical insights may guide stakeholders to create beneficial smart tourism initiatives that will strengthen citizens' capacity.

11.1 INTRODUCTION

Smart city strategies have been an integral part of urban planning wherein communities leverage technologies to increase collaboration and standard of living (Iqbal and Olariu, 2020). At the forefront of this evolution, the fusion of Information and Communication Technologies (ICTs) with the tourism sector heralds a new era, implying enhanced accessibility, information dissemination, and participatory frameworks for decision-making (Januszewska et al., 2015). Smart tourism is the result of a confluence of interrelated frameworks that leverage the potential of state-of-the-art technologies such as the Internet of Things (IoT), extensive data analytics, Artificial Intelligence (AI), and Augmented Reality (AR) for the enhancement of the satisfied tourist experience (Jeong and Shin, 2020). This evolution transcends mere technological assimilation, encapsulating a more expansive perspective wherein local inhabitants are empowered to actively participate and influence the tourism system (Shafiee et al., 2023). The role of residents within a destination takes on a new dimension, emphasizing the potential for technology-driven initiatives to reshape citizen involvement within tourism landscapes.

The growth of smart cities involves the utilization of resources, the use of smart technologies, and the application of citizen-centric approaches so that the collaboration among residents, tourism businesses, and concerned authorities could be fostered (Boes et al., 2016; Cardullo and Kitchin, 2019). As smart cities strive for increased efficiency, sustainability, and improved governance, the role of smart tourism emerges as a dynamic component that redefines citizen engagement within urban landscapes. Building on a comprehensive review of the literature, this chapter addresses the lack of research on the relationship between smart tourism and citizen participation. Also, through the use of secondary data analysis, connections between these domains are demonstrated.

Citizen engagement is a concept that goes beyond physical coexistence in communities; it becomes an impetus for shared responsibility, ownership, and sustainable development. In this context, smart tourism appears as a revolutionary force, presenting a fusion of data-driven insights, digital innovations, and citizen-centric approaches that hold the capability to completely transform the tourism industry. The first section of the chapter examines the various elements that make up smart tourism, using previous research to identify and categorize the three main components: smart destinations, smart experiences, and smart businesses (Gretzel et al., 2015). These elements form the basis for comprehending how community involvement and innovative technology combine to improve the total tourism experience provided by the destination's tourism businesses.

The study outlines different approaches as its research progresses, including the Human Driven Method (HDM), Technology Driven Method (TDM), and Smart City Framework. By acknowledging the mutually beneficial relationship between technological innovation and human capital in the growth of smart cities, these strategies open the door to a balanced integration. The study offers practical insights into citizen engagement in the context of smart tourism, going beyond the theoretical foundations. The use of ICTs and smart systems in tourism is emphasized, with particular attention given to how these technologies affect organizational and macro system management. The conversation includes how citizen involvement can be included into the early phases of smart city planning, shifting the emphasis to plans that are just, rooted in the community, and equitable. Smart tourism initiatives could ensure sustainable tourism practices that benefit both visitors and host communities, preserve cultural heritage, and harness local knowledge by involving citizens as stakeholders.

This chapter examines different aspects of citizen involvement in smart tourism within the context of enabling citizen engagement. Furthermore, it envisioned a time when technology is seamlessly integrated with inclusive and community-focused tourism experiences, offering a model for the successful fusion of technology and community-driven initiatives. The exploration aims to guide stakeholders in smart tourism initiatives and promote a thorough understanding of the potential for transformation that comes from the dynamic connection of technology, tourism, and citizen engagement by bridging the theoretical foundations with practical insights and adding to the body of academic literature.

11.2 LITERATURE REVIEW

There is a deep relationship between smart city and smart tourism (Shafiee et al., 2021). The integration of technology and tourism is transforming conventional tourism paradigms into smart tourism, which is the subtype of the smart city concept (Matos et al., 2019). Therefore, it is essential to understand the concept of smart city. A smart city is an integrated community that uses shared resources and cutting-edge technologies to facilitate the collaboration between residents, businesses, academic institutions, and municipal authorities, and continuously increasing the quality of life (Khan et al., 2017). Supporters argue that leveraging ICTs, improved governance, and citizen human capital would lead to beneficial social changes in smart cities (Chen and Chan, 2023). On the other hand, smart tourism, an evolution from traditional and e-tourism, leverages ICTs to create innovative experiences (Gretzel et al., 2015). It includes three key components: smart destinations, smart experience, and smart business (Figure 11.1).

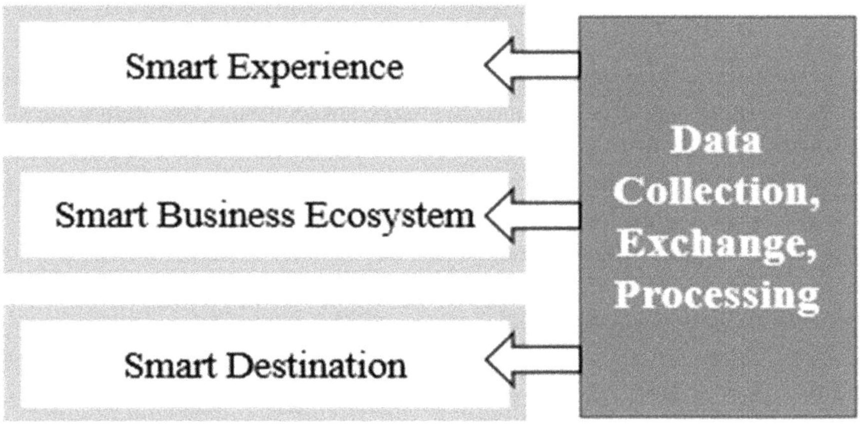

FIGURE 11.1 Components of smart tourism.
Source: Adapted from Gretzel et al. (2015).

Based on the definition of Gretzel et al. (2015), smart tourism involves using advanced technology to collect and analyze data from various sources at a destination, transforming it into an advanced destination with smart on-site experiences and business value, with a focus on efficiency and sustainability. To improve the entire tourism experience, it integrates social networking sites, government resources, and physical infrastructure (Lee et al., 2021). This nascent concept compresses a dynamic blend of digital innovations, data-driven insights, and citizen-centric approaches, fostering a sincere revolution in the tourism landscape (Akdu, 2020; Lung-Amam et al., 2021; Sarkar et al., 2020). Amidst this evolution, the amplification of citizen engagement stands as a fundamental cornerstone, promising to redefine interactions between tourists, destinations, and local communities (Lung-Amam et al., 2021).

11.2.1 CITIZEN ENGAGEMENT

The concept of community or citizen has transcended the mere physical coexistence of individuals in the time of societal interconnectedness. Smith et al. (2011) illuminate these communities of practice as informal hubs where like-minded individuals gather to share, discuss, and propel their common interests forward. These spontaneous gatherings serve as crucibles for the evolution and advancement of shared practices, forming the essence of community beyond geographical boundaries. However, what defines a community is that the term *community* suggests the existence of a single, homogeneous identity and places an emphasis on coherence over conflict, unity over variety, spontaneity over mediation, emotions over thinking, and stability over change (Mannarini and Fedi, 2009). This depiction, while emphasizing emotional connections and stability, often sidelines the richness brought by diversity and the need for adaptability within evolving citizens. Empowering citizens fosters human ecosystems, driving holistic growth by addressing financial, human, and natural resource development (Lee and Jan, 2019).

Citizen engagement yields multifaceted benefits that resonate across various spheres of societal development. Primarily, it fosters a sense of shared ownership and responsibility within communities, establishing a framework where individuals actively contribute to addressing local needs and initiatives (Loeffler and Martin, 2015). Nichols and Ralston (2012) highlight the intrinsic value of volunteering as a form of community engagement, elucidating how individuals, by investing their time and skills, play an instrumental role in enhancing community well-being. This active participation not only addresses immediate concerns but also nurtures a sense of solidarity and cohesion among community members. Moreover, citizen engagement transcends individual actions to collective decision-making processes through mechanisms like community meetings and consultations, as expounded by Loeffler and Martin (2015). These open forums provide a platform for residents to exchange ideas, deliberate on issues, and contribute feedback, thereby fostering a more inclusive and participatory environment (Giampiccoli et al., 2021). Furthermore, citizen engagement extends to structured avenues such as citizen panels and direct democracy, where individuals are empowered to deliberate on complex matters or make decisions that directly impact their communities (Loeffler and Martin, 2015; Tseng et al., 2019). It also serves as a catalyst for social and cultural enrichment within societies. It encapsulates the views and beliefs of individuals impacted by developmental processes, shaping the socio-cultural fabric of destination communities in the context of tourism (Lee and Jan, 2019). This engagement harnesses the diverse perspectives of stakeholders and fosters an environment conducive to sustainable tourism practices, nurturing a balance between economic development and cultural preservation.

Citizen engagement plays a focal role in shaping smart tourism initiatives, as highlighted by Lung-Amam et al. (2021). Involving locals in the early phases of planning shifts the emphasis away from technology-centric solutions and towards more locally grounded, fair, and community-focused smart city designs and investments (Cardullo and Kitchin, 2019; Lung-Amam et al., 2021). Furthermore, consideration of community's resident perspectives regarding tourism's impact on economic development and quality of life influence enhancing visitor experiences. Mandić and Kennell (2021) assert that technology-driven innovations hold the potential to assist Destination Management Organizations (DMOs) in optimizing tourism development, addressing critical aspects such as carrying capacity, stakeholder management, and community involvement. By enhancing digital literacy and cultivating a culture of active participation among community members, smart tourism can facilitate decision-making processes and fostering community engagement (Mandić and Kennell, 2021).

11.2.2 INFLUENCE OF SMART SYSTEMS ON TOURISM AND CITIZEN ENGAGEMENT

The growing ubiquity of smart systems, which are meant to adapt to user needs, is having a significant impact on organizational and macro system management in the perspective of smart tourism and citizen involvement. The term smart refers to a notion that may be applied to a city or country at a macro level, as well as a product or service at a micro level (Angelakoglou et al., 2019; Khan et al., 2017). This interconnected influence is further exemplified in the concept of smart tourism, where technological innovations enhance accessibility, communication, and decision-making for users (Akdu, 2020; Buhalis and Amaranggana, 2013). It represents the merge of ICT with travel experiences, offering increased, systemic, and ample services, meeting user demands for swift, wide information transmission in the sphere of tourism (Akdu, 2020). ICT-driven destinations that provide knowledge-driven environments qualify in the definition of smart tourism (Jovicic, 2019).

Technologies like cloud computing and IoT enable instantaneous interchange of tourism-related experiences and information at these sites through ICT platforms. These technologies are contributing to improved visitor experiences, stakeholder collaboration, and the integration of non-tourism industries into the destination network are their main goals (Jovicic, 2019). Therefore, the integration of smart systems with ICT plays a pivotal role in creating rich and effective smart tourism experiences which endeavors to construct a smart city by seamlessly integrating technologies, fostering stakeholder collaboration, and actively engaging citizens in the development and offerings of the destination (Rayyan et al., 2019; Shafiee et al., 2021). So, again it can be affirmed that there is a substantial connection between the concepts of smart tourism and smart cities.

However, there are no specific theories in the literature detailing how smart tourism directly enhances or involves local communities. Yet, different approaches in the literature support people as a strong component of smart city and suggesting that smart cities inherently require the integration of smart tourism (Cavalheiro et al., 2020; Chen and Chan, 2023; Finch and Tene, 2018). Therefore, it is worthwhile to look some of these approaches where this idea is advocated.

11.2.3 TECHNOLOGY-DRIVEN METHOD (TDM) AND HUMAN-DRIVEN METHOD (HDM)

There are two approaches in discussing smart cities called TDM and HDM, which could assist us in understanding how these interconnected theories are supporting technology-based tourism with its inhabitants' participation. A smart city, according to TDM, is a networked area where the use of ICT substantially improves the standard of living for its citizens (Kummitha and Crutzen, 2017). Within the context of smart tourism, by incorporating the latest innovations, For example, IoT, data analytics, and AI, into the infrastructure of the tourism industry, TDM can promote smart tourism. This improves the effectiveness and allure of tourism activities and includes real-time information, smart navigation, and personalized experiences. Furthermore, through the provision of digital platforms for communication and participation in urban governance, TDM promotes citizen engagement. Citizens can actively participate in the decision-making stage, voice concerns, and keep up with local developments through smart applications and online platforms.

On the other hand, HDM emphasizes how essential skill development and human capital are to the smart city paradigm (Kummitha and Crutzen, 2017; Sarkar et al., 2021). HDM places a strong emphasis on citizens' active participation in the field of citizen engagement. By emphasizing their abilities and expertise, it promotes cooperation between local governments and communities in order to achieve successful urban development. Thus, it could be expected that, by placing a strong emphasis on the human experience, HDM could enhance smart tourism. It can imply that technology improves rather than replaces human interactions and fosters the growth of the cultural and social components of tourism. This strategy enhances and adds to a more genuine travel experience. Moreover, with a focus on social justice, inclusivity, and community involvement, HDM can encourage citizen participation. By empowering people to actively shape their urban surroundings, it promotes a feeling of community and shared accountability. HDM can ensure that technological interventions are in line with the welfare of society and human values. Therefore, it is worth saying that there is a need for a balanced integration of both approaches for the holistic development of smart cities that prioritize technological innovation alongside community participation through tourism.

11.2.4 SMART CITY FRAMEWORK

Cavalheiro et al. (2020) attempted to build a model delineating the essentials of a smart city and methods to progress towards this vision by adopting shared services and overcoming the obstacles. Their framework categorized eight key elements into two tiers of impact, emphasizing the higher influence exerted by outer factors such as governance, community involvement, environment, infrastructure, and the economy over the important inner factors of technology, management, and policy in determining the success of smart city edges (Chourabi et al., 2012). This framework highlights the significance of citizen involvement (people and communities) as one of the major factors influencing the success of smart city initiatives, which is ultimately a reflex of smart tourism destination (Cavalheiro et al., 2020). Hence, in building a smart tourism destination, involving and collaborating with residents becomes essential to ensure a holistic and impactful destination experience (Cavalheiro et al., 2020).

Dubai is a prime example; it is a large metropolis in the Middle East and North Africa that has had tremendous economic expansion due to different sectors, including tourism and travel. 14.87 million visitors visited Dubai in 2016, placing it fourth in the region (Khan et al., 2017). Dubai's government and stakeholders recognize the value of tourism, as it contributed a major part to the economy. The Smart Dubai initiative, which uses the *Happiness* vision as the motivating force, has guided the emergence of smart technologies for building a future city. However, challenges remain due to Dubai's existing resource allocation and preparedness to meet the demands of growth. Government reports, publications, articles, and research statistics are used to analyze current advancements in the smart city idea and assess Dubai's reputation as a smart travel destination.

11.2.5 CITIZEN EMPOWERMENT WITH SMART TOURISM AND DIGITAL CHALLENGES

Smart tourism, with its technological integration, holds potential implications for residents. Involving residents in the creation and use of technological innovations has the potential to promote citizen involvement and improve the visitor experience in general (Buhalis and Amaranggana, 2013). The concept of Smart Tourism Destinations (STDs) was introduced with the foreword of *smartness* in the tourism framework, which is a result of the rapid growth of technology. This allows destinations to take advantage of the synergy between technology and social components. According to Buhalis and Amaranggana (2013), this integration not only improves the quality of tourist experiences but also enables destinations to efficiently meet the needs of travelers prior to, during, and following their travels. In addition, the integration of cutting-edge technologies such as IoT, Radio Frequency Identification (RFID), and Near Field Communication (NFC) in community-based tourism apps promotes the sharing of indigenous knowledge, illuminates obscure local treasures, and improves engagement with guests. Gretzel et al. (2015) pointed out that these kinds of programs additionally improve the experiences of tourists yet foster a feeling of community involvement and local pride.

Moreover, smart tourism, with its integration of intelligence into travel destinations, plays a pivotal role in fostering local economic development and citizen empowerment. The utilization of technology platforms facilitates the real-time exchange of tourism-related information among stakeholders, thereby creating opportunities for residents to actively engage in tourism-related businesses and services (Buhalis and Amaranggana, 2013). Citizen empowerment, essential for sustainable tourism development, hinges on collective action and engagement, aiming to meet community needs, manage local resources, and control environmental aspects. Smart tourism initiatives offer a pathway to citizen empowerment, fostering sustainable practices that benefit both residents and visitors (Cavalheiro et al., 2020). For many of the Caribbean's emerging governments, predominantly comprising small island nations, tourism stands as a pivotal economic driver, significantly bolstering employment, growth, and sustainable development. Within this landscape, community-based smart tourism holds profound importance, fostering a natural link with the interests of local stakeholders and enhancing inclusivity. The process is centered on human participation, paving the way for reinforcing community tourism. It achieves this by amplifying the visibility of small businesses and enhancing their service offerings. An essential outcome aligning with *Sustainable Development Goal (SDG) 11*, aimed at *sustainable cities and communities*, lies in the potential of these initiatives to substantially advance inclusive and sustainable development (Hussain et al., 2023; United Nations, 2024). Furthermore, they enhance the capacity for participatory, integrated, and sustainable local communities. Leveraging data curated by those deeply familiar with community resources, the utilization of map-based information and localized digital content tailors these efforts to specific geographical contexts. Barcelona's Smart Tourist Initiative is another example, which aims to promote eco-friendly travel while improving the overall experience of tourists. To mitigate the environmental effects of traffic, the city has introduced electric bike sharing programs and smart parking systems. Barcelona has also launched the *Barcelona Smart Destination* portal, which offers visitors personalized suggestions, tailored schedules, and up-to-date details on events and activities related to culture. Interestingly, the website includes a *Citizen Participation* section wherein locals can offer their opinions, criticisms, and suggestions to improve the city's tourism offerings.

On the contrary, a convergence of challenges comes with the application of smart tourism initiatives. Primary among these challenges is the persistence of a digital divide, marking discrepancies in technological access and proficiency among community members (Warschauer, 2004). This disparity could impede equal participation and inhibit the inclusive engagement of all segments within the community. In addition, the ethical dimensions surrounding data privacy emerge as a critical concern, necessitating a delicate balance between leveraging data-driven insights for tourism enhancement and safeguarding residents' privacy rights (Finch and Tene, 2018). Tensions between the preservation of cultural authenticity and the commodification of local heritage in the digital realm further pose challenges, requiring astute management to prevent cultural dilution or exploitation within smart tourism frameworks (Chhabra, 2021). Addressing these challenges demands a nuanced approach that amalgamates technological innovation with ethical considerations and community inclusivity.

11.3 METHODS

In this section, the methodology employed to investigate the role of smart tourism as the next level of citizen engagement is outlined. This study's methodology builds upon the insights gleaned from the literature review, highlighting the need for secondary data analysis in examining the intricate link between citizen engagement and smart tourism, as there were no relevant studies that clearly conceptualize the connection of both. Thus, to understand the connection between smart tourism and citizen participation, the study uses a method of secondary data analysis that makes use of already-existing data sources.

Secondary data analysis involves the examination and interpretation of existing data that was originally collected for a different research study, rather than gathering data directly from participants (Irwin, 2013). Researchers generally analyze and make sense of existing data that has already been collected and made accessible. Opting for secondary data analysis in the research on smart tourism and citizen engagement is justified by its cost and time efficiency, providing a broad scope of information and facilitating literature analysis. The approach enables the validation of findings across sources, making it appropriate for research. It also offers insights into how smart tourism initiatives affect citizen engagement in real-world scenarios. While acknowledging potential limitations, like data quality and relevance, we take consideration to ensure that secondary data effectively contributes to addressing our research objectives.

A comprehensive study of the literature is done to pinpoint important ideas, hypotheses, and research findings around smart tourism and citizen participation. This comprises an analysis of journal articles, academic books, papers from conferences, essays, and reports that further knowledge on the topic. Content analysis is employed to systematically examine and interpret textual information from the selected literature, reports, and case studies (Hermann, 2008; Irwin, 2013). This approach helps to find trends, themes, and connections that are pertinent to citizen participation and smart tourism. Analyzing the content also involves identifying relationships between different concepts, theories, or studies. This can lead to valuable insights and a greater understanding of the subject. Based on the results and suggestions of various papers, a proposed model has been established, which is exhibited in Figure 11.2. This model likely aims to represent the relationships, frameworks, or processes relevant to smart tourism and citizen engagement.

11.4 FINDINGS AND DISCUSSIONS

11.4.1 A FUTURISTIC APPROACH TO CITIZEN ENGAGEMENT IN SMART TOURISM

A blueprint of citizen involvement through smart tourism can promote inclusive and participatory tourism experiences and allow locals to enthusiastically engage in shaping the development of their communities. It would also help tackle the obstacles for involving citizens and could ensure that tourism initiatives align with their needs, values, and aspirations, encouraging a sense of ownership and pride.

The presented comprehensive model (Figure 11.2) delineates fundamental components crucial for citizen engagement in smart tourism. It envisions smart tourism and citizen engagement as a multifaceted system. The model's conceptualization aligns with a three-dimensioned structure, illustrating its interrelation. Each dimension embodies distinct yet interdependent aspects necessary for sustainable and efficient community engagement in smart tourism. This approach emphasizes the integration of cutting-edge tools, cohesive teamwork, and meaningful local involvement as foundational principles shaping the model's efficacy and impact within the realm of smart tourism initiatives.

The **foundation** dimension comprises the core infrastructure and resources needed for smart tourism. A resilient *digital infrastructure* within smart tourism ensures citizens' secure and seamless engagement with initiatives. Without sacrificing data security, it makes interactive platforms, information retrieval, and feedback gathering easier (Gretzel et al., 2015). Reliability, ease of use, and security of technology are essential for successful citizen participation. This digital infrastructure in smart tourism generally includes mobile apps, websites for information and booking, and IoT devices, all supported by secure databases and high-speed internet for optimal user experiences (Benckendorff et al., 2019). In the context of citizen engagement in smart tourism, having *tourism data* in the foundational dimension implies that this data forms the bedrock for understanding citizens' preferences, behaviors, and needs regarding tourism experiences. It entails employing modern technologies like artificial intelligence and machine learning to gather real-time data about events, transportation, lodging, and attractions. This data-driven understanding makes it possible to customize tourism experiences so that they more effectively connect with the public and raise

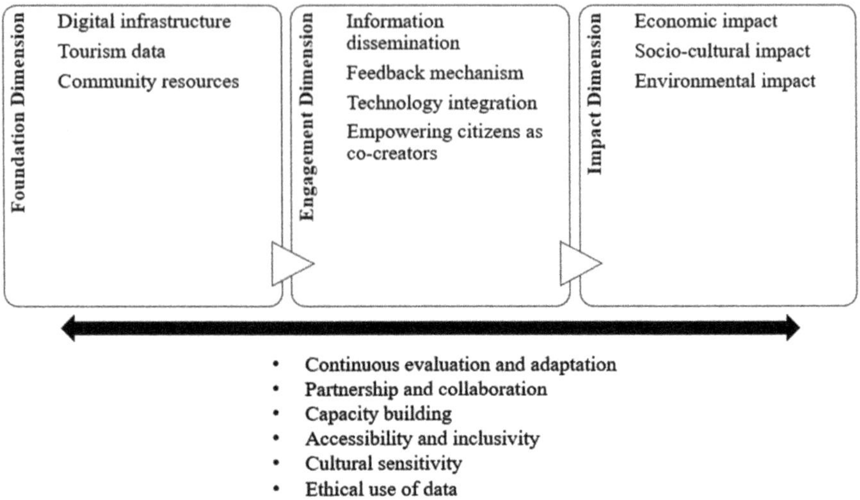

FIGURE 11.2 A proposed model of residents' involvement through smart tourism.

Source: Author.

their level of happiness and engagement. *Community resources* represent the local people, local businesses, cultural institutions, and community organizations that play a crucial role in shaping tourism experiences (Eslami et al., 2019). Engaging citizens in preserving their cultural heritage within the tourism framework raises a sense of pride and belonging. Forming community engagement through partnerships and forums endows citizens to actively contribute to modelling a more inclusive and community-driven approach to smart tourism. By ensuring that citizens' perspectives are heard, and their cultural values are respected in the tourism setting, this involvement generates an awareness of ownership.

The **engagement** dimension, here, refers to a stage or element of a system or model where people (including visitors and locals) actively interact, become involved, and participate. In the engagement dimension, *information dissemination* plays a pivotal role in providing accessible, relevant, and real-time information to both tourists and local citizens (Paul and Das, 2023). By providing travelers with instant access to essential details on attractions, events, and services, this information is disseminated across a variety of platforms, such as mobile apps, websites, and kiosks, thereby improving their experience. At the same time, providing residents with timely and correct information empowers them to interact with tourists in a proactive manner, offering advice, cultural insights, and local knowledge, which enhances the entire tourism experience. This stage can contribute to community awareness. The process of creating channels for citizens to express their opinions, ideas, and concerns is known as the *feedback mechanism*. Through two-way communication, a responsive tourism system where citizens' opinions are acknowledged and taken into account is fostered (Halme, 2001). This results in ongoing improvements and customized experiences that satisfy the needs of both residents and visitors (Eslami et al., 2019; Halme, 2001). Through the facilitation of this conversation, it raises an atmosphere of comprehensiveness, adaptability, and proactive engagement within the tourism industry, guaranteeing that the experiences provided are more closely aligned with the interests and requirements of all relevant parties. In the engagement dimension of smart tourism's citizen engagement model, *technology integration* refers to adopting innovative tools, such as AR, mobile apps, IoT devices, Virtual Reality (VR), geolocation services, and so forth. By offering interactive guides, tailored recommendations, and immersive experiences at tourist attractions, these technologies seek to improve the tourist's experience (Buhalis et al., 2019; Neuhofer et al., 2012).

Additionally, these tools also give individuals a platform to actively participate and engage, such as sharing local insights via smartphone apps or taking part in AR-enhanced historical excursions. By incorporating these technologies, citizens could participate in defining and improving the overall tourism experience, in addition to enjoying improved experiences. *Empowering citizens as co-creators* is positioned within the engagement dimension to actively involve locals in shaping and crafting tourism experiences. This approach enables the public to participate fully in the planning and creation of tourism-related projects (Bello et al., 2016; Lee and Jan, 2019). Through techniques such as innovation challenges, workshops, or collaborative sessions, residents can contribute their ideas and thoughts and become essential partners in the tourism services. Because their efforts are reflected in the tourism scene and local values and goals are authentically represented, locals feel a sense of pride and ownership as a result of their involvement. Furthermore, this involvement facilitates capacity building by integrating the community's skills, knowledge, and resources in smart tourism projects, providing them with the tools, resources, and expertise necessary for participation and initiative benefit.

Here, the **impact** dimension is concerned with assessing the benefits and outcomes that smart tourism programs offer to local communities. The financial benefits of tourism efforts are examined in the *economic impact*. This includes increased tourism revenue, job creation, and the positive influence on local businesses. Assessing the benefits and contributions that smart tourism brings to the local economy is a necessary step in involving people in this aspect (Boes et al., 2016). It secures that locals recognize and make use of the opportunities provided by smart tourism, promoting job creation, business expansion, and general economic progress within the community. This is accomplished by including the community in comprehending and contributing to these economic benefits. *Socio-cultural impact* assesses how tourism-associated initiatives influence the community's social and cultural aspects. This analysis involves understanding changes in local traditions, values, and community dynamics resulting from tourism activities. Involving citizens in this area is centered on protecting cultural legacy and making sure that tourism respects and enhances regional social systems (Bichler, 2021; Wu et al., 2022). Citizen involvement can include efforts to protect customs, historical landmarks, or cultural events to preserve the community's distinct identity. Through citizen involvement, smart tourism works to maintain and honor the community's unique qualities by fostering a feeling of pride and identity within the community and making sure that tourism activities are in line with and beneficial to local social structures. *Environmental impact* scrutinizes how tourism practices influence the natural surroundings. This evaluation involves measuring the ecological footprint of tourism initiatives, including aspects like energy consumption, waste generation, and the preservation or depletion of natural resources. Engaging citizens in this realm involves quantifying the environmental effects of tourism activities, seeking insights into their consequences. Furthermore, citizen involvement promotes eco-friendly practices within the community, advocating for measures that reduce waste, conserve energy, and support environmental conservation efforts (Murat Koçoğlu et al., 2023). By actively involving residents, smart tourism strives to mitigate negative environmental impacts caused by tourism, fostering a culture of sustainability and responsible resource management within the community's tourism framework.

To strengthen the framework for citizen engagement in the setting of smart tourism, it is essential to incorporate basic tenets like continuous evaluation and adaptation, partnership and collaboration, capacity building, accessibility and inclusivity, cultural sensitivity, and ethical use of data across all dimensions. Continuous evaluation and adaptation require a flexible and adaptable strategy to match changing citizen requirements with the smart tourism environment (Jovicic, 2019). In order to unite efforts towards sustained involvement, partnership and collaboration serve as the cornerstone (Boes et al., 2016; Jovicic, 2019). They encourage shared duties, synergistic interactions among stakeholders, and the amplification of varied perspectives. Building capacity becomes essential to achieving inclusive growth because it gives local people the tools and resources they need to actively engage in and contribute to the smart tourism ecosystem (Mandić and Kennell, 2021). For creating a fully participative environment, accessibility and inclusivity are essential components that promote fair access to opportunities, services, and information for all individuals, regardless of their differences (Chan and Agapito, 2022). A key aspect of the smart tourism paradigm is cultural sensitivity, which respects and preserves the history, customs, and values of the area while fostering a sense of identification and belonging (Cavalheiro et al., 2020; Mandić and Kennell, 2021). Additionally, ethical data use acts as a moral compass, supporting the safe and open management of information, protecting privacy, and building stakeholder trust—all of which strengthen the groundwork for citizen participation in smart tourism initiatives (Buhalis and Amaranggana, 2013).

11.5 IMPLICATIONS OF THE STUDY

From a theoretical standpoint, the study makes multiple theoretical contributions to academic literature. The study's theoretical foundations are based on the connection between citizen engagement, smart tourism, and the larger field of smart city development. The study expands and connects current theories about citizen engagement and smart tourism. It builds on established theories in both fields by highlighting the transformative potential of smart tourism in empowering citizens and demonstrating the synergies between technological advancements and citizen engagement. Also, the study offers a theoretical framework for comprehending how initiatives related to smart tourism may promote citizen engagement. A model describing the fundamental components, dynamics of engagement, and effects of smart tourism on nearby communities is presented.

By offering a methodical strategy for combining technology, community involvement, and sustainable development, this framework enhances the theoretical landscape. Again, it enhances the theoretical understanding of citizen engagement in smart tourism by emphasizing their active involvement in decision-making, co-creation of experiences, and contribution to the overall tourism ecosystem. On top of that, by emphasizing the crucial role that smart tourism plays within the larger framework of smart cities, the study advances the theoretical discourse on smart cities. It looks at how technological advancements in the travel industry can support the broader goal of smart cities. By acknowledging the importance of the tourism industry in influencing urban development, this extension widens the purview of current smart city theories. Finally, the study contributes to the debate on responsible and sustainable smart city development by incorporating these inclusive and ethical aspects into the theoretical framework. It offers theoretical foundations for further investigation and real-world application of smart tourism projects that give ethical issues, diversity, and cultural sensitivity top priority in the dynamic field of urban planning and development.

The study's practical implications include giving helpful guidance to develop initiatives concerning smart tourism. For stakeholders involved in the planning and implementation of smart tourism projects, the study provides a conceptual framework outlining key components, engagement dynamics, and impacts on local communities. Again, the study's emphasis on ethical issues has immediate application for people who gather and use data for smart tourism. To ensure that privacy rights are upheld, policymakers and practitioners can put policies and procedures for responsible data use into place. This encourages public participation and trust and has real effects on the development of safe and reliable smart tourism platforms. Additionally, the chapter encourages stakeholders to put into action doable steps for inclusive travel services. This includes making sure physical spaces are universally accessible, developing accessible digital platforms, and providing varied information in multiple languages. The goal is to make tourism experiences inclusive of a broad spectrum of tourists. It also highlights how cultural sensitivity can be practically incorporated into astute tourism planning. For doing this, practitioners must respect and celebrate the destination's distinctive heritage through digital platforms and experiences that preserve and promote local cultures. Here, it is suggested that stakeholders implement plans to involve locals in decision-making, experience co-creation, and tourism-related initiatives. Furthermore, the framework that has been suggested highlights a pragmatic method that stakeholders can use to consistently evaluate the success of smart tourism projects. To do this, feedback from visitors and locals is gathered, data on the effects on the economy, society, and environment is analyzed, and necessary adjustments are made to meet the needs of the evolving community. The study promotes practical community capacity building for proactive participation in smart tourism, as this will guarantee that communities are prepared to take part in and profit from smart tourism development.

11.6 CONCLUSION

The intricate interplay between smart tourism and local citizen empowerment unveils an evolving potential poised to redefine the contours of sustainable development within the tourism sector. Empowering local communities through smart tourism initiatives emerges as a fulcrum for achieving goals spanning resource optimization, competitive economic returns, and bolstering support for protected areas. Moreover, the integration of citizen engagement mechanisms such as volunteering, community consultations, and direct democratic participation cultivates a robust framework for inclusive tourism development, underscoring the pivotal role of communities in shaping their tourism landscape. Additionally, smart tourism combines technology into community development initiatives to improve citizen involvement, efficiency, and overall quality of life. By promoting community-based tourism and a more enriching visitor experience, the strategic combination of smart technology in tourism promotes economic inclusiveness and participatory involvement.

Crucially, the nexus between smart tourism and sustainable development aligns seamlessly with the aspirations delineated in global agendas such as the SDGs 11. Goal 11 is about sustainable communities and cities that find resonance in the symbiotic relationship between smart tourism and local community empowerment (United Nations, 2024). Citizen-based tourism can stand as a lighthouse, illuminating the path toward inclusivity and sustainability, underlining the imperative of a human-centered, participative framework for robust community engagement. Thus, the integration of smart tourism heralds a new era of transformative possibilities, forging a synergistic alliance between technological innovation and grassroots participation, thereby catalyzing a holistic and sustainable evolution within the realm of tourism.

The study makes several recommendations for future directions for citizen engagement and smart tourism research. Initially, scholars could look into how specific community segments are affected by smart tourism initiatives, as well as how various demographic groups view and profit from these efforts. Furthermore, examining how smart tourism affects destination communities over the long run would yield important information for sustainable development. Subsequent research endeavors may concentrate on enhancing and broadening the suggested theoretical structure. Further quantitative investigation could be carried out to extend the frame by identifying any additional determinants to be added in the framework from a larger pool of respondents.

KEYWORDS

- **citizen engagement**
- **community**
- **content analysis**
- **data analytics**
- **information dissemination**
- **smart tourism**
- **sustainability**
- **technology**

REFERENCES

Akdu, U. (2020). Smart tourism: Issues, challenges and opportunities. In *The Emerald Handbook of ICT in Tourism and Hospitality* (pp. 291–308).

Angelakoglou, K., Nikolopoulos, N., Giourka, P., Svensson, I. L., Tsarchopoulos, P., Tryferidis, A., & Tzovaras, D. (2019). A methodological framework for the selection of key performance indicators to assess smart city solutions. *Smart Cities*, *2*(2), 269–306.

Bello, F. G., Carr, N., & Lovelock, B. (2016). Community participation framework for protected area-based tourism planning. *Tourism Planning & Development*, *13*(4), 469–485.

Benckendorff, P. J., Xiang, Z., & Sheldon, P. J. (2019). *Tourism Information Technology*. CABI.

Bichler, B. F. (2021). Designing tourism governance: The role of local residents. *Journal of Destination Marketing & Management, 19*, 100389.

Boes, K., Buhalis, D., & Inversini, A. (2016). Smart tourism destinations: Ecosystems for tourism destination competitiveness. *International Journal of Tourism Cities*, *2*(2), 108–124.

Buhalis, D., & Amaranggana, A. (2014). Smart tourism destinations. In *Information and communication technologies in tourism 2014: Proceedings of the International Conference: Dublin, Ireland, January 21–24, 2013* (pp. 553–564). Springer.

Buhalis, D., Harwood, T., Bogicevic, V., Viglia, G., Beldona, S., & Hofacker, C. (2019). Technological disruptions in services: Lessons from tourism and hospitality. *Journal of Service Management, 30*(4), 484–506.

Cardullo, P., & Kitchin, R. (2019). Smart urbanism and smart citizenship: The neoliberal logic of 'citizen-focused' smart cities in Europe. *Environment and Planning C: Politics and Space*, *37*(5), 813–830.

Cavalheiro, M. B., Joia, L. A., & Cavalheiro, G. M. C. (2020). Towards a smart tourism destination development model: Promoting environmental, economic, socio-cultural and political values. *Tourism Planning & Development*, *17*(3), 237–259.

Chan, C. S., & Agapito, D. (2022). Managing destination experience design for tourists with disabilities: ICT and accessibility. In *Handbook on the Tourist Experience: Design, Marketing and Management*.

Chen, Z., & Chan, I. C. C. (2023). Smart cities and quality of life: A quantitative analysis of citizens' support for smart city development. *Information Technology & People*, *36*(1), 263–285.

Chhabra, D. (2021). *Resilience, Authenticity and Digital Heritage Tourism*. Routledge.

Chourabi, H., Nam, T., Walker, S., Gil-Garcia, J. R., Mellouli, S., Nahon, K., Pardo, T. A., & Scholl, H. J. (2012). Understanding smart cities: An integrative framework. *2012 45th Hawaii International Conference on System Sciences*.

Eslami, S., Khalifah, Z., Mardani, A., Streimikiene, D., & Han, H. (2019). Community attachment, tourism impacts, quality of life and residents' support for sustainable tourism development. *Journal of Travel & Tourism Marketing, 36*(9), 1061–1079.

Finch, K., & Tene, O. (2018). Smart cities: Privacy, transparency, and community. In *The Cambridge Handbook of Consumer Privacy* (pp. 125–148).

Giampiccoli, A., Mtapuri, O., & Nauright, J. (2021). Tourism development in the Seychelles: A proposal for a unique community-based tourism alternative. *Journal of Tourism and Cultural Change*, *19*(4), 444–457.

Gretzel, U., Sigala, M., Xiang, Z., & Koo, C. (2015). Smart tourism: Foundations and developments. *Electronic Markets*, *25*, 179–188.

Halme, M. (2001). Learning for sustainable development in tourism networks. *Business Strategy and the Environment*, *10*(2), 100–114.

Hermann, M. G. (2008). Content analysis. In *Qualitative Methods in International Relations: A Pluralist Guide* (pp. 151–167). Springer.

Hussain, S., Ahonen, V., Karasu, T., & Leviäkangas, P. (2023). Sustainability of smart rural mobility and tourism: A key performance indicators-based approach. *Technology in Society*, 102287.

Iqbal, A., & Olariu, S. (2020). A survey of enabling technologies for smart communities. *Smart Cities*, *4*(1), 54–77.

Irwin, S. (2013). Qualitative secondary data analysis: Ethics, epistemology and context. *Progress in Development Studies*, *13*(4), 295–306.

Januszewska, M., Jaremen, D., & Nawrocka, E. (2015). The effects of the use of ICT by tourism enterprises. *Zeszyty Naukowe Uniwersytetu Szczecińskiego. Service Management*, *16*, 65–73.

Jeong, M., & Shin, H. H. (2020). Tourists' experiences with smart tourism technology at smart destinations and their behavior intentions. *Journal of Travel Research*, *59*(8), 1464–1477.

Jovicic, D. Z. (2019). From the traditional understanding of tourism destination to the smart tourism destination. *Current Issues in Tourism*, *22*(3), 276–282.

Khan, M. S., Woo, M., Nam, K., & Chathoth, P. K. (2017). Smart city and smart tourism: A case of Dubai. *Sustainability*, *9*(12), 2279.

Kummitha, R. K. R., & Crutzen, N. (2017). How do we understand smart cities? An evolutionary perspective. *Cities*, *67*, 43–52.

Lee, M., Hong, J. H., Chung, S., & Back, K. J. (2021). Exploring the roles of DMO's social media efforts and information richness on customer engagement: Empirical analysis on Facebook event pages. *Journal of Travel Research*, *60*(3), 670–686.

Lee, T. H., & Jan, F. H. (2019). Can community-based tourism contribute to sustainable development? Evidence from residents' perceptions of the sustainability. *Tourism Management*, *70*, 368–380.

Loeffler, E., & Martin, S. (2015). Citizen engagement. In *Public Management and Governance* (pp. 301–318).

Lung-Amam, W., Bierbaum, A. H., Parks, S., Knaap, G. J., Sunderman, G., & Stamm, L. (2021). Toward engaged, equitable, and smart communities: Lessons from West Baltimore. *Housing Policy Debate*, *31*(1), 93–111.

Mandić, A., & Kennell, J. (2021). Smart governance for heritage tourism destinations: Contextual factors and destination management organization perspectives. *Tourism Management Perspectives*, *39*, 100862.

Mannarini, T., & Fedi, A. (2009). Multiple senses of community: The experience and meaning of community. *Journal of Community Psychology*, *37*(2), 211–227.

Matos, A., Pinto, B., Barros, F., Martins, S., Martins, J., & Au-Yong-Oliveira, M. (2019). Smart cities and smart tourism: What future do they bring? In *New Knowledge in Information Systems and Technologies: Volume 3*.

Murat Koçoğlu, C., Pinaroğlu, B., & Yaşarsoy, E. (2023). Smart destinations and eco-friendly practices. In *Smart Cities for Sustainability: Approaches and Solutions* (pp. 15–32). Emerald Publishing Limited.

Neuhofer, B., Buhalis, D., & Ladkin, A. (2012). Conceptualizing technology enhanced destination experiences. *Journal of Destination Marketing & Management*, *1*(1, 2), 36–46.

Nichols, G., & Ralston, R. (2012). The rewards of individual engagement in volunteering: A missing dimension of the Big Society. *Environment and Planning A*, *44*(12), 2974–2987.

Paul, S., & Das, S. (2023). Investigating information dissemination and citizen engagement through government social media during the COVID-19 crisis. *Online Information Review*, *47*(2), 316–332.

Rayyan, M. A., Octaviawan, A., & Billiyanto, R. (2019). From "smart city" to "smart tourism": Developing the role of tourism in Asia. *ASEAN/Asian Academic Society International Conference Proceeding Series*.

Sarkar, N. I., Iasepi, L. S., & Gul, S. (2021). Designing a smart city: Technologies, challenges and solutions. In *Green Computing in Smart Cities: Simulation and Techniques* (pp. 51–70).

Sarkar, S. K., Toanoglou, M., & George, B. (2020). The making of data-driven sustainable smart city communities in holiday destinations. In *Digital Transformation in Business and Society: Theory and Cases* (pp. 273–296).

Shafiee, S., Jahanyan, S., Ghatari, A. R., & Hasanzadeh, A. (2023). Developing sustainable tourism destinations through smart technologies: A system dynamics approach. *Journal of Simulation, 17*(4), 477–498.

Shafiee, S., Rajabzadeh Ghatari, A., Hasanzadeh, A., & Jahanyan, S. (2021). Smart tourism destinations: A systematic review. *Tourism Review, 76*(3), 505–528.

Smith, M., Muhajarine, N., Smith, J. A., & Delanoy, S. (2011). Community. In *The SPHERU KT Casebook*.

Tseng, M. L., Lin, C., Lin, C. W. R., Wu, K. J., & Sriphon, T. (2019). Ecotourism development in Thailand: Community participation leads to the value of attractions using linguistic preferences. *Journal of Cleaner Production, 231*, 1319–1329.

United Nations. (2024). https://sdgs.un.org/goals (accessed on 8th July 2025).

Warschauer, M. (2004). *Technology and Social Inclusion: Rethinking the Digital Divide*. MIT Press.

Wu, M. Y., Wu, X., Li, Q. C., & Tong, Y. (2022). Community citizenship behavior in rural tourism destinations: Scale development and validation. *Tourism Management, 89*, 104457.

CHAPTER 12

IMPACT OF GENERATIVE AI ON HOSPITALITY EMPLOYEES' MENTAL HEALTH: IDENTIFYING KEY ISSUES AND ADDRESSING CHALLENGES

PATITA PABAN MOHANTY* and SUSANTA RANJAN CHAINI

Faculty of Hospitality and Tourism Management, Siksha 'O' Anusandhan (Deemed to be University), Bhubaneswar, Odisha, India

*Corresponding author

ABSTRACT

In the hospitality industry, generative artificial intelligence (GAI) presents significant opportunities in all functional areas, ranging from operations, production, and guest services to marketing, human resources, revenue management, accounting, and finance. The adoption of GAI in the hospitality business brings both guest delight and employee happiness, simultaneously leading to misery. AI is gradually replacing human expertise and posing a challenge to the mental health of hospitality personnel. Most developed countries have already embraced generative AI, and significant results have been achieved. The hospitality industry in many developing nations is grappling with this complex issue and is at the crossroads of pondering over part of the problem or part of the solution. Hence, this study explores under-researched perspectives on mental health in the hospitality sector and how the advent of generative AI has influenced it, providing a clear picture of this complex issue. Thus, to address this research gap,

Smart Tourism: Managing Technological Revolution in the Tourism Industry. Santus Kumar Deb (Ed.)

DOI: 10.1201/9781779641090-12

this study examines whether technology-driven changes in the hospitality industry are part of the problem or part of the solution. Therefore, this study fills an important knowledge gap by investigating the impact of generative AI on the mental health of hospitality employees in developing nations such as India.

12.1 INTRODUCTION

In the 21st century, the hospitality industry has been navigating a dynamic landscape that encounters challenges and opportunities. Despite these complexities, the industry is open to innovation, collaboration, and growth. The steady growth of the industry generates USD 7.6 trillion and creates 292 million jobs globally, one in 10 jobs, and nearly 10% of global GDP (OECD, 2018; Statista Research Department, 2018; World Travel and Tourism Council, 2017). Recently, technological innovation within industries has been highlighted as a significant research focus (Kakatkar et al., 2020; Mustak et al., 2021). Technological innovation has leveraged organizations' potential to bring about competitive advantage. Therefore, a plethora of research has emphasized innovation-driven performance (Mariani et al., 2021). Recently, there has been an increase in interest in innovation scholarships in the hospitality and tourism fields (Orfila-Sintes et al., 2005; Shin and Kang, 2020).

The hotel industry has embraced artificial intelligence (AI) as a cutting-edge innovation for delivering exceptional guest experiences worldwide (Lin et al., 2019; Murphy et al., 2017). The exponential growth of the hotel sector is underpinned by AI-related technologies and applications (Ivanov et al., 2019). Specifically, AI technology is entering the hospitality business (Mariani and Borghi, 2019), creating and retaining value through service interactions (Huang and Rust, 2020; Wirtz et al., 2018).

Many aspects of the hotel sector are significantly affected by AI (Blöcher and Alt, 2020; Henkel et al., 2020; Tussyadiah, 2020). AI is increasingly employed in hotels worldwide to replace repetitive and time-consuming operations. Generative artificial intelligence (GAI), a subfield of artificial intelligence, has transformed business by allowing organizations to create new experiences through the integration of virtual and physical worlds. Researchers, academics, and the industrial community are exploring the boundless potential of GAI through the metaverse.

GAI is a key technology in both virtual and real commercial platforms, including Google Bard AI and OpenAI ChatGPT. Since its inception in November 2022, OpenAI, a research business funded by Microsoft, has received substantial attention from ChatGPT. OpenAI's ChatGPT offers consumers an intuitive and personalized experience by merging AI-based natural language processing (NLP) with machine learning (ML). ChatGPT can be used to improve customer assistance, simplify client inquiries, and facilitate interactive conversation.

Implementing GAI in the hotel business makes employees happy while delighting guests. It enables employees to work more intelligently and produce better business outcomes by developing new competencies and capabilities such as technological knowledge, social and emotional intelligence, and creative abilities (Ivanov and Webster, 2019). Technological advancements have enabled AI to replace human talent to streamline the organizational process and structure (Cain et al., 2019; Ivanov and Webster, 2017; OECD, 2018; Prentice et al., 2020).

According to experts, AI is predicted to replace 47% of employment in the US over the next several decades (Frey and Osborne, 2017–2020). Approximately 55% of front-line manufacturing employment in China can be readily replaced by AI (Fan and Tang, 2020). Manufacturing industries are gradually replacing labor with AI devices, leaving no hope for the hospitality industry. The impact of AI on employees' career growth is posing unforeseen challenges in today's human resources domain (Brougham and Haar, 2020).

AI, fueled by rapid technological breakthroughs, has transformed not only employment jobs but also the nature of the task. Employment has become more demanding, service-oriented, and cognitively complex. Poor mental health is currently the leading cause of low productivity in many developed countries (Black, 2008; Harvey et al., 2009; Whiteford et al., 2013). Generative AI has been implemented in most developed nations, resulting in both positive and negative outcomes. The hospitality sector in many developing nations struggles with this complicated issue and is in a doldrum position.

Hence, in this backdrop, the remainder of the chapter in the form of a review of literature explores mental health and how it has transformed over the years. Additionally, the penetration of AI into the hotel business and operation and how it advances by affecting mental health both positively and negatively.

12.2 THEORETICAL PERSPECTIVES

12.2.1 MENTAL HEALTH

The mental health of hospitality workers has become an area of growing concern after Covid-19 (Goforth, 2021). Mental health is an emerging concern declared by the World Health Organization (WHO) in many developing and developed nations. Mental health is not only limited to the hospitality service sector, but its presence has been understood in areas such as public health, organizational psychology, positive psychology, business, and tourism.

The WHO has provided one of the better definitions of mental health, which goes beyond simply being free from mental illnesses or deficiencies like depression or "a condition of wellness and good health in which a person reaches their full potential, is able to manage everyday stressful situations, is able to work with precision and effectiveness, and is capable of contributing back to their community" (WHO, 2018).

Mental illness and mental health have been used interchangeably in the literature on psychological well-being (Antaramian et al., 2010; Westerhof and Keyes, 2010). However, people's mental health without symptoms, such as anxiety, bipolar disorder, and depression, is debatable in psychological literature (American Psychiatric Association, 2000).

Hospitality is an extremely high-pressure field (Chang et al., 2020; Kim et al., 2022) with frequent shifts, heavy workloads, unfavorable working hours, close contact with customers, and intense emotional labor (Ahmad et al., 2021; Andrade et al., 2021). Front-line hotel employees, specifically those working long hours, are required to maintain a high level of customer service and often face difficult customer interactions (Chang et al., 2020). These factors contribute to the development of mental health issues among hotel workers. Depression is a mental health condition that frequently affects hotel employees. Hotel workers are vulnerable to depression because of financial instability, employment insecurity, and social isolation. Elevated levels of stress in the workplace may exacerbate depressive and anxiety symptoms.

According to the WHO report, organizations should foster work cultures that promote employees' mental health and well-being (WHO, 2021). Employees and organizations face challenges due to mental health issues, such as depression and anxiety (Thanem and Elraz, 2022), particularly in the tourism and hospitality industries (Saah et al., 2021). Therefore, hotel employees' mental health demands more attention from theory and practice (Zhang et al., 2020). Previous studies have found that increasing hotel employees' mental health can boost productivity and performance (Bayighomog and Arasli, 2022; Varga et al., 2021). Hotels can effectively manage the mental health challenges of their employees and attain commercial success by identifying underlying causes and problems (O'Neill and Davis, 2011; Sariisik et al., 2022).

According to the Royal Society of Public Health, "One in five hospitality workers suffer from work-related severe mental health issues, four out of five hospitality workers reported increased stress which was believed to be a direct consequence of their job." In addition, a 2019 study by Paychex also found that four out of five employees in the hospitality and food service industries reported feeling burned out by their workload.

Employees' mental health issues arise because of the wide gap in service delivery and service recovery in the hospitality business. According to a survey report, people working in the hospitality industry, casual workers, and young adults (18–24) have higher rates of mental health issues (SuperFriend, 2021). Mental health is defined as a significant disruption in behavior, emotional regulation, or cognition that causes a significant amount of distress in social, occupational, or other important activities (American Psychiatric Association, 2013). Mental health encompasses a wide range of illnesses including abuse, anxiety, despair, sleep disorders, and psychotic disorders.

Mental health poses serious issues at workplaces, creating imbalance among employees and diminishing the organization's output. Employees associated with jobs with high emotional, cognitive, and physical demands are more vulnerable to mental health issues. Workplace culture becomes very obnoxious because of feelings of job insecurity, depression, lack of stability and safety, job dissatisfaction, and lack of motivation among employees (Yan et al., 2021). The well-being of employees and the organization is significantly impacted by mental health issues in the workplace. There was a substantial correlation between employee productivity, mental health, and job performance.

Hospitality firms are now facing challenges of safety and security in the workplace, increased absenteeism, stress, and fatigue, leading to diminished staff turnover and decreased performance and productivity (Chi and Gursoy, 2009; Gamor et al., 2014; Koc and Bozkurt, 2017;

O'Neill and Davis, 2011; PwC, 2014). AI has the potential to enhance workers' productivity and similarly has the power to change the nature of the tasks and impact employee satisfaction (Bankins et al., 2022).

12.2.2 POSITIVE IMPACT OF GENERATIVE ARTIFICIAL INTELLIGENCE (GAI)

The hospitality industry is evolving constantly with the emergence of new technologies and trends. GAI is the largest invention of the 21st century (Mahroof, 2019; Rust and Huang, 2014). AI has a huge potential to transform society and humanity. GAI has been gradually advancing to improve the hospitality business through a range of innovative ideas (Carvalho and Ivanov, 2023; Dwivedi et al., 2023a; Dwivedi et al., 2023b; Mich and Garigliano, 2023). GAI has been accomplishing a variety of tasks, including personalized travel recommendations, exploring locations and hotel amenities, and creating virtual tours for hotels and tourist destinations (Lalli, 2023). The advent of GAI has transformed hospitality businesses into novel experiences in virtual and real-world environments (Mondal et al., 2023).

According to Dogru et al. (2023) and Lu et al. (2019), the power of GAI can improve workplace productivity, reduce expenses, and enhance customer experience. According to Ivanov and Webster (2017), the development of novel customer experiences by GAI has significantly transformed the traditional means of communication between hotel employees and visitors into a useful way.

From predicting tourist arrivals to allocating rooms to valued customers, offering delicious cuisine, room cleaning, and front-office concierge tasks, AI is now embedded in every nook and corner of hotels (Buhalis et al., 2023; Leung and Law, 2013; Ozdemir et al., 2023; Pandey and Srivastava, 2013). AI is also capable of identifying a property's service encounter, identifying potential for growth, improving online reputation management, and offering direction for marketing plans (Luo and Xu, 2021; Tran et al., 2019; Xia et al., 2019).

Although digital transformation has radically revolutionized the hotel and tourism industry (Buhalis, 2020), AI is now being used in a variety of other service industries. The hotel sector has attempted to reduce various problems on a regular basis by strategically implementing AI (Dwivedi et al., 2023). One of AI's most important and innovative products, chatbots, is doing an incredible job offering interactive, personalized customer services (Melián-González et al., 2021; Yun and Park, 2022). Moreover, this can expedite response times and enhance customer service, boosting satisfaction, loyalty, and guest engagement (Pereira et al., 2022) (Figure 12.1).

FIGURE 12.1 Artificial intelligence (AI): Game changer for the hotel industry.

Source: Hotelogix (2018). https://blog.hotelogix.com/how-artificial-intelligence-is-changing-the-game-for-the-hotel-industry/

12.2.3 NEGATIVE IMPACT OF GENERATIVE ARTIFICIAL INTELLIGENCE (GAI)

Recently, hotel organizations have partnered with Meta, Amazon, and Jasper to invest extensively in generative AI technology (Dwivedi et al., 2023a). It demonstrates the enormous potential of generative AI technologies such as ChatGPT for businesses and operations in the hotel industry. From the First Industrial Revolution onward, machines have progressively assumed the role of humans in labor. The development of AI and AI-enabled technologies has led to the emergence of a new generation of machines, such as service robots, which can now compete with humans and replace them in nearly every job (Brynjolfsson and McAfee, 2011). This suggests that the hospitality industry, which was previously unaffected by AI technology, is currently facing jeopardy. According to Bowen and Morosan (2018), who projected that robots will replace 25% of workers in the hospitality business over the next 10 years, the introduction of robotics technology represents a paradigm change (Bowen and Morosan, 2018). Some academics believe that artificial intelligence poses the greatest threat to humankind (Musk, 2014) (Figure 12.2).

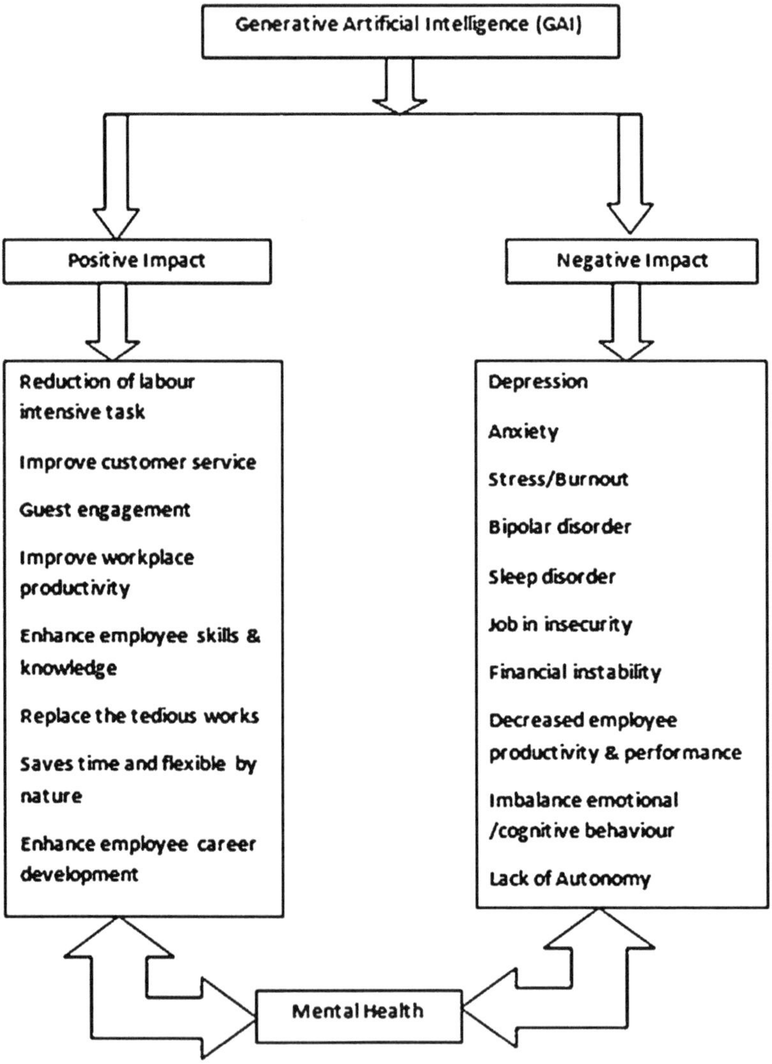

FIGURE 12.2 Framework showing impact of Generative artificial intelligence (GAI) on mental health of hospitality employees.

Source: Authors' own elaboration.

However, despite the fact that AI is becoming an unavoidable aspect of hotel industry innovation, it is undeniable that AI may pose a threat to human employment in the hotel industry. Consequently, it is possible that, in the future, AI will handle massive volumes of data and work in place of humans (McClure, 2018; Rust and Huang, 2014). AI influences staff and customers both negatively and positively (Huang and Rust, 2018; Li et al., 2019; Lu et al., 2019; Stock, 2018). While AI can replace analytical and mechanical intelligence, human employees' intuitive and emotional intelligence is still crucial. Workers in lower-skilled jobs, in particular, are aware of the high likelihood that AI will soon replace them in their current positions.

AI can stimulate employees' feelings of job insecurity. Job insecurity causes tension among hotel personnel (Nam, 2019). This affects employees' well-being and reduces business efficiency (Ashford et al., 1989; Greenhalgh and Rosenblatt, 2010; Nam, 2019). The hotel sector is a people-intensive industry, with people performing the majority of their daily duties (Prentice et al., 2019). Jobs in the hospitality business will eventually disappear because of AI penetration (Koo et al., 2021). This may lead to tension, decreased output, and a negative attitude toward the hotel and work (Kong et al., 2021; Li et al., 2019).

Although AI is considered beneficial to technological advancement, it has a detrimental impact on several issues, including employee engagement in the workplace, unemployment, job satisfaction, stress, and health. AI automation has resulted in a decline in employment for tasks that are "cognitively routine and manually routine" (Aaronson and Phelan, 2020). Researchers found that workers who suffered from "automation anxiety," or the concern that computers would take over their professions, had higher rates of insomnia, "decreased income," "being transferred to an uninteresting job," and "difficulty in using skills and ability" (Baek et al., 2022). Hotels typically employ a combination of highly trained and unskilled laborers. However, they also emphasize that "low-skilled workers" who undertake "routine-based tasks" and are most at risk of automated replacement are driving this "negative effect" (Schwabe and Castellacci, 2020).

Furthermore, the use of AI in the workplace weakens "relationships between human co-workers" and may have a detrimental effect on "trust and individuals' happiness." "Fear of change and the unknown," "fear of losing human relationships in the workplace," "lack of confidence in using AI," and "fear of being replaced by AI" are the main causes of the negative consequences of AI (Loureiro et al., 2022). These fears can negatively influence health (Nazareno and Shiff, 2021), sleep patterns (Baek et al., 2022), job satisfaction (Schwabe and Castellacci, 2020), and HRM employee behavior (Cappelli et al., 2019). AI is considered to pose a serious threat to human employment. Employee anxiety, worry, depression, and job burnout are exacerbated by unpredictability and threat (Ivanov, 2020).

12.3 METHODOLOGY

Narrative synthesis is the process of performing a systematic review and data synthesis from multiple studies that primarily use language and text to describe and summarize the findings (Popay et al., 2006). A narrative synthesis was used in this study. Additionally, a thorough assessment of the literature was conducted to investigate generative AI and its impact on employees in the hospitality sector. The literature and data came from a variety of sources, including EBSCO, Google Scholar, Scopus, Web of Science, ScienceDirect books, and research articles.

The present study included various definitions of GAI propounded in different highly reputed journals in the hospitality domain, written and explained in English. The data were collected between January 15th and September 24th, December 2023. To review the database, the researcher identified a set of keywords associated with GAI and their impact on the mental health of hospitality personnel in the hotel industry. The keywords found included "Artificial intelligence," "Generative AI," "Generative AI adoption," "AI technology," "Technological Innovations," "AI ChatGpT," "Hospitality industry," "Mental Health," and "Hospitality employees."

In qualitative systematic reviews, content analysis is a suitable and effective data-analysis technique. According to Jennings (2001), in qualitative content analysis, a researcher is free to examine texts without having to prove or reject a theory. This suggests that researchers are more "open" to a wide range of outcomes. Furthermore, it is recommended that qualitative systematic reviewers aim to produce knowledge and theory change or adapt content analysis techniques to suit data that is, by definition, highly contextualized and arranged (Finfgeld-Connett, 2014). Therefore, in this review chapter, qualitative content analysis was conducted.

12.4 RESULTS

Generative AI has gradually penetrated the hospitality business in many developing and developed nations. The widespread application of AI is a crucial aspect of economic progress. Moreover, Generative AI has been perceived as digital assistants solving the diverse needs of hotel employees and customers. The result of the previous study revealed that both the positive and negative impacts of Generative AI on the mental health of hospitality employees are crucial to deriving the success of the hotel organization. Despite its positive outcomes benefiting employees, the risks posed by the GAI advancements are manifold, creating survival challenges for future human resource development within the hotel sectors.

12.5 DISCUSSION AND CONCLUSION

Generative AI is a novel idea for the hospitality industry and remains in a nascent stage. AI technology impacts the workforce and employment opportunities in hotels, both positively and negatively (Bisoi et al., 2020). Generative AI has enormous potential to transform the hospitality business by enhancing customer satisfaction and operational efficiency, resulting in increased profitability (Buhalis, 2020; Samara et al., 2020). In addition, Generative AI is enhancing guest engagement, loyalty, and satisfaction (Pereira et al., 2022), bringing happiness to the workplace.

The advancement of GAI significantly increases the value for hospitality employees, and also simultaneously brings disruptive changes into the hospitality workforce (Dwivedi et al., 2023a; Sandstrom and Dunn, 2014). Furthermore, overindulgence and over-dependence of employees on GAI lead to the creation of an unfavorable work culture, destroying employees' well-being, and leading to depression and burnout. Furthermore, widespread GAI integration may disrupt the decision-making chain, resulting in value destruction (Grundner and Neuhofer, 2021). In conclusion, generative AI technology is being used by the hospitality sector to enhance marketing, customer service, experience, and employee retention (Citak et al. 2021).

Li et al. (2019) propose that the potential for Generative AI to increase employee turnover may arise from workers' decision to switch employment out of fear of losing their current positions. The increasing frequency and growing prevalence of GAI in the workplace outcome physical health problems such as diabetes, heart and cardiovascular disease, musculoskeletal disorders, and obesity, often in conjunction with concurrent mental health issues (Duncan et al., 2012; Ford and Caspersen, 2012; Waters et al., 2016; Yang et al., 2017). Overall, GAI technology is having a detrimental influence on mental health in the workplace in a variety of ways, including increasing demands, lowering resources, and changing how employees perceive the future, all of which have hidden and direct consequences for businesses and individuals.

The topic of employee mental health induced by Generative AI is becoming more and more significant in the hospitality workplace. Globally, the recent advancement of Generative AI in the hospitality business has aroused immense interest among academics, practitioners, and policymakers to understand the positive and negative impact of GAI on the mental health of employees. By reviewing the literature on mental health and Generative AI, this chapter advances the agenda on workplace mental health of hospitality employees. The chapter examined the current situation of mental health in the workplace using the GAI's latest breakthroughs and talked about the implications for this field. In light of the potential consequences, the study also identifies the main problems and difficulties associated with using Generative AI to the hospitality industry.

12.5.1 KEY ISSUES AND CHALLENGES

Though hospitality business operations have been significantly influenced by the penetration of Generative AI in recent times, its long-term implementation faces significant obstacles. The requirement for cost-effective and time-consuming high-quality data (Paschen et al., 2020; Sivarajah et al., 2017), and the challenge of connecting AI systems with human workers (Thiebes et al., 2021) are essential in the present era. The hospitality industry's historical aversion to adopting new and emerging technology (Chan et al., 2018; Stylos et al., 2021) could also be an impediment. Goel et al. (2022) stress that factors such as psychological, social, financial, technical, and functional act as roadblocks to the adoption of generative artificial intelligence in the hospitality business. Lack of industry knowledge and non-availability of experts are the major barriers to successful implementation of Generative AI. The rapid development of GAI and the increasing customer privacy issues also act as obstacles during the implementation of AI in the hospitality business (Hu and Min, 2023; Prentice et al., 2020).

12.5.2 IMPLICATIONS

This review chapter adds significantly to the body of literature by examining the possible effects on mental health, both positive and negative, as well as the implications for future research and key stakeholders such as hospitality employees and customers. Historically, there exists a lack of research nexus of generative AI on the mental health of hospitality employees. The chapter also explores the linkage of mental health in psychiatry, psychology, occupational stress, management, and the future of work to improve the academic and practical agenda for workplace mental health. Though the beginning of the future hospitality workplaces accelerated by technology-driven innovation such as Generative AI, mental health has been a concerning issue offering both opportunities and threats to the employees.

KEYWORDS

- **challenges**
- **generative artificial intelligence (GAI)**
- **hospitality industry**
- **hotel sector**
- **mental health**
- **skill**
- **technology**

REFERENCES

Aaronson, D., & Phelan, B. (2021). The evolution of technological substitution in low-wage labor markets. *Economic Studies at Brookings*, 2021. https://www.brookings.edu/wp-content/uploads/2020/01/Phelan-Aaronson_Full-Report-Tables.pdf (accessed on 8th July 2025).

American Psychiatric Association. (2000). *Diagnostic and Statistical Manual of Mental Disorders* (4th ed.). Washington, DC: Author.

Antaramian, S. P., Huebner, E. S., Hills, K. J., & Valois, R. F. (2010). A dual-factor model of mental health: Toward a more comprehensive understanding of youth functioning. *American Journal of Orthopsychiatry*, *80*, 462–472. https://doi.org/10.1111/j.1939-0025.2010.01049.x

Baek, S. U., Yoon, J. H., & Won, J. U. (2022). Association between workers' anxiety over technological automation and sleep disturbance: Results from a nationally representative survey. *International Journal of Environmental Research and Public Health*, *19*(16), 10051. https://doi.org/10.3390/ijerph191610051 (accessed on 8th July 2025).

Bankins, S., Formosa, P., Griep, Y., & Richards, D. (2022). AI decision making with dignity? Contrasting workers' justice perceptions of human and AI decision making in a human resource management context. *Information Systems Frontiers*, *24*, 857–875. https://doi.org/10.1007/s10796-021-10223-8 (accessed on 8th July 2025).

Bauer, J. C., John, E., Wood, C. L., Plass, D., & Richardson, D. (2019). Data entry automation improves cost, quality, performance, and job satisfaction in a hospital nursing unit. *JONA: The Journal of Nursing Administration*, *50*(1), 34–39. https://doi.org/10.1097/nna.0000000000000836 (accessed on 8th July 2025).

Bayighomog, S. W., & Arasli, H. (2022). Reviving employees' essence of hospitality through spiritual wellbeing, spiritual leadership, and emotional intelligence. *Tourism Management*, *89*, 104406. https://doi.org/10.1016/j.tourman.2021.104406 (accessed on 8th July 2025).

Black, D. C. (2008). *Working for a Healthier Tomorrow*. London: Cross-Government Health, Work and Well-being Programmed.

Blöcher, K., & Alt, R. (2020). AI and robotics in the European restaurant sector: Assessing potentials for process innovation in a high-contact service industry. *Electronic Markets*, *31*, 529–551. https://doi.org/10.1007/s12525-020-00443-2 (accessed on 8th July 2025).

Bowen, J., & Morosan, C. (2018). Beware hospitality industry: The robots are coming. *Worldwide Hospitality and Tourism Themes*, *10*(6), 726–733. https://doi.org/10.1108/WHATT-07-2018-0045 (accessed on 8th July 2025).

Brougham, D., & Haar, J. (2020). Technological disruption and employment: The influence on job insecurity and turnover intentions: A multi-country study. *Technological Forecasting and Social Change*, *161*, 120276. https://doi.org/10.1016/j.techfore.2020.120276 (accessed on 8th July 2025).

Brynjolfsson, E., & McAfee, A. (2011). *Race Against the Machine: How the Digital Revolution Is Accelerating Innovation, Driving Productivity, And Irreversibly Transforming Employment and The Economy*. Lexington: Digital Frontier Press.

Buhalis, D. (2020). Technology in tourism—from information communication technologies to eTourism and smart tourism towards ambient intelligence tourism: A perspective article. *Tourism Review*, *75*(1), 267–272. https://doi.org/10.1108/TR-06-2019-0258 (accessed on 8th July 2025).

Cain, L. N., Thomas, J. H., & Alonso, M. (2019). From sci-fi to sci-fact: The state of robotics and AI in the hospitality industry. *Journal of Hospitality and Tourism Technology*, *10*(4), 624–650.

Carvalho, I., & Ivanov, S. (2023). ChatGPT for tourism: Applications, benefits, and risks. *Tourism Review*, *7*(2), 290–303. https://doi.org/10.1108/TR-02-2023-0088 (accessed on 8th July 2025).

Chan, E. S., Okumus, F., & Chan, W. (2018). Barriers to environmental technology adoption in hotels. *Journal of Hospitality & Tourism Research, 42*(5), 829–852. https://doi.org/10.1177/1096348015614959 (accessed on 8th July 2025).

Chi, C. G., & Gursoy, D. (2009). Employee satisfaction, customer satisfaction, and financial performance: An empirical examination. *International Journal of Hospitality Management, 28*(2), 245–253. https://doi.org/10.1016/j.ijhm.2008.08.003 (accessed on 8th July 2025).

Citak, J., Owoc, M. L., & Weichbroth, P. (2021). A note on the applications of artificial intelligence in the hospitality industry: Preliminary results of a survey. *Procedia Computer Science, 192*, 4552–4559.

De Witte, H., Pienaar, J., & De Cuyper, N. (2018). Review of 30 years of longitudinal studies on the association between job insecurity and health and well-being: Is there causal evidence? *Australian Psychologist, 51*(1), 18–31. https://doi.org/10.1111/ap.12176 (accessed on 8th July 2025).

Duncan, M. J., Vandelanotte, C., Caperchione, C., et al. (2012). Temporal trends in and relationships between screen time, physical activity, overweight and obesity. *BMC Public Health, 12*, 1060. https://doi.org/10.1186/1471-2458-12-1060 (accessed on 8th July 2025).

Dwivedi, Y. K., Kshetri, N., Hughes, L., et al. (2023). "So what if ChatGPT wrote it?" Multidisciplinary perspectives on opportunities, challenges and implications of generative conversational AI for research, practice and policy. *International Journal of Information Management, 71*, 102642. https://doi.org/10.1016/j.ijinfomgt.2023.102642 (accessed on 8th July 2025).

Finfgeld-Connett, D. (2014). Use of content analysis to conduct knowledge-building–generating qualitative systematic reviews. *Qualitative Research, 14*(3), 341–352.

Ford, E. S., & Caspersen, C. J. (2012). Sedentary behavior and cardiovascular disease: A review of prospective studies. *International Journal of Epidemiology, 41*, 1338–1353.

Frey, C. B., & Osborne, M. A. (2017). The future of employment: How susceptible are jobs to computerization? *Technological Forecasting and Social Change, 114*, 254–280. https://doi.org/10.1016/j.techfore.2016.08.019 (accessed on 8th July 2025).

Gamor, E., Amissah, E. F., & Boakye, K. A. A. (2014). Work–family conflict among hotel employees in Sekondi-Takoradi Metropolis, Ghana. *Tourism Management Perspectives, 12*, 1–8. https://doi.org/10.1016/j.tmp.2014.06.001 (accessed on 8th July 2025).

Goel, P., Kaushik, N., Sivathanu, B., Pillai, R., & Vikas, J. (2022). Consumers' adoption of artificial intelligence and robotics in hospitality and tourism sector: Literature review and future research agenda. *Tourism Review, 77*(4), 1081–1096. https://doi.org/10.1108/TR-03-2021-0138 (accessed on 8th July 2025).

Goforth, A. (2022). Industries hit the hardest by COVID or ever thought about a government job? *BenefitsPRO*. Available at: https://www.benefitspro.com/2021/09/08/industries-hit-the-hardest-by-covid-or-ever-thought-about-a-government-job/?slreturn=20220127223738 (accessed on 8th July 2025).

Harvey, S. B., Deady, M., Wang, M. J., Mykletun, A., Butterworth, P., Christensen, H., & Mitchell, P. B. (2017). Is the prevalence of mental illness increasing in Australia? Evidence from national health surveys and administrative data, 2001–2014. *The Medical Journal of Australia, 206*(11), 490–493. https://doi.org/10.5694/mja16.00295 (accessed on 8th July 2025).

He, H., & Harris, L. (2020). The impact of COVID-19 pandemic on corporate social responsibility and marketing philosophy. *Journal of Business Research, 116*, 176–182. https://doi.org/10.1016/j.jbusres.2020.05.030 (accessed on 8th July 2025).

Henkel, A. P., Čaic, M., Blaurock, M., & Okan, M. (2020). Robotic transformative service research: Deploying social robots for consumer well-being during COVID-19 and beyond. *Journal of Service Management, 31*(6), 1131–1148. https://doi.org/10.1108/JOSM-05-2020-0145 (accessed on 8th July 2025).

Hotelogix. (2018). How artificial intelligence is changing the game for the hotel industry. *Hotel News Resource*. Available at: https://www.hotelnewsresource.com/article98643.html (accessed on 8th July 2025).

Hu, Y., & Min, H. (2023). The dark side of artificial intelligence in service: The "watching-eye" effect and privacy concerns. *International Journal of Hospitality Management, 110*(3). https://doi.org/10.1016/j.ijhm.2023.103437 (accessed on 8th July 2025).

Huang, M. H., & Rust, R. T. (2018). Artificial intelligence in service. *Journal of Service Research, 21*(2), 155–172. https://doi.org/10.1177/1094670517752459 (accessed on 8th July 2025).

Huang, M. H., & Rust, R. T. (2022). A framework for collaborative artificial intelligence in marketing. *Journal of Retailing, 98*(2), 209–223. https://doi.org/10.1016/j.jretai.2021.03.001 (accessed on 8th July 2025).

Ivanov, D. (2020). Predicting the impacts of epidemic outbreaks on global supply chains: A simulation-based analysis on the coronavirus outbreak (COVID-19/SARS-CoV-2) case. *Transportation Research. Part E, Logistics and Transportation Review, 136*, 101922. https://doi.org/10.1016/j.tre.2020.101922 (accessed on 8th July 2025).

Ivanov, S., & Webster, C. (2017). Adoption of robots, artificial intelligence automation by travel, tourism and hospitality companies: A cost-benefit analysis. *International Scientific Conference Contemporary Tourism–Traditions and Innovations*, Sofia University, 19–21 October, 2017.

Ivanov, S., & Webster, C. (2019). Conceptual framework of the use of robots, artificial intelligence and service automation in travel, tourism, and hospitality companies. In S. Ivanov & C. Webster (Eds.), *Robots, Artificial Intelligence, And Service Automation in Travel, Tourism and Hospitality* (pp. 7–37). Emerald Publishing Limited. https://doi.org/10.1108/978-1-78756-687-320191001 (accessed on 8th July 2025).

Jennings, G. (2001). *Tourism Research*. Milton: John Wiley and Sons Australia Ltd.

Kong, H., Yuan, Y., Baruch, Y., Bu, N., Jiang, X., & Wang, K. (2021). Influences of artificial intelligence (AI) awareness on career competency and job burnout. *International Journal of Contemporary Hospitality Management, 33*(2), 717–734. https://doi.org/10.1108/IJCHM-07-2020-0789 (accessed on 8th July 2025).

Koo, B., Curtis, C. R., & Ryan, B. (2020). Examining the impact of artificial intelligence on hotel employees through job insecurity perspectives. *International Journal of Hospitality Management, 95*, 102763.

Lalli, M. (2023). ChatGPT & Co: The revolution of generative AI and big data and the future of the tourism sector. *Data-Driven Destinations*. https://datadrivendestinations.com/2023/01/chat-gpt-co-the-revolution-of-generative-ai-and-big-data-and-the-future-of-the-tourism-sector/ (accessed on 8th July 2025).

Li, J. (J.), Bonn, M. A., & Ye, B. H. (2019). Hotel employee's artificial intelligence and robotics awareness and its impact on turnover intention: The moderating roles of perceived organizational support and competitive psychological climate. *Tourism Management*, *73*, 172–181. https://doi.org/10.1016/j.tourman.2019.02.006 (accessed on 8th July 2025).

Loureiro, S. M. C., Bilro, R. G., & Neto, D. (2022). Working with AI: Can stress bring happiness? *Service Business*, *17*, 233–255. https://doi.org/10.1007/s11628-022-00514-8 (accessed on 8th July 2025).

Luo, Y., & Xu, X. (2021). Comparative study of deep learning models for analyzing online restaurant reviews in the era of the COVID-19 pandemic. *International Journal of Hospitality Management, 94*. https://doi.org/10.1016/j.ijhm.2020.102849 (accessed on 8th July 2025).

Mariani, M. M., Perez-Vega, R., & Wirtz, J. (2022). AI in marketing, consumer research and psychology: A systematic literature review and research agenda. *Psychology & Marketing, 39*(4), 755–776. https://doi.org/10.1002/mar.21619 (accessed on 8th July 2025).

McClure, P. (2018). 'You're fired,' says the robot: The rise of automation in the workplace, technophobes, and fears of unemployment. *Social Science Computer Review, 36*, 139–156.

McMurtrey, M. E., Grover, V., Teng, J. T. C., & Lightner, N. J. (2002). Job satisfaction of information technology workers: The impact of career orientation and task automation in a CASE environment. *Journal of Management Information Systems*, *19*(2), 273–302. https://doi.org/10.1080/07421222.2002.11045719 (accessed on 8th July 2025).

Melián-González, S., Gutiérrez-Taño, D., & Bulchand-Gidumal, J. (2021). Predicting the intentions to use chatbots for travel and tourism. *Current Issues in Tourism, 24*(2), 192–210. https://doi.org/10.1080/13683500.2019.1706457 (accessed on 8th July 2025).

Mich, L., & Garigliano, R. (2023). ChatGPT for e-tourism: A technological perspective. *Information Technology & Tourism, 25*(1), 1–12. https://doi.org/10.1007/s40558-023-00248-x (accessed on 8th July 2025).

Mondal, S., Das, S., & Vrana, V. G. (2023). How to bell the cat? A theoretical review of generative artificial intelligence towards digital disruption in all walks of life. *Technologies, 11*(2), 44. https://doi.org/10.3390/technologies11020044 (accessed on 8th July 2025).

Musk, E. [@elonmusk]. (2014). Tweet. *Elon Musk*. https://twitter.com/elonmusk/status/495759307346952192 (accessed on 8th July 2025).

Mustak, M., Salminen, J., Plé, L., & Wirtz, J. (2021). Artificial intelligence in marketing: Topic modeling, scient metric analysis, and research agenda. *Journal of Business Research, 124*, 389–404. https://doi.org/10.1016/j.jbusres.2020.10.044 (accessed on 8th July 2025).

Nazareno, L., & Schiff, D. S. (2021). The impact of automation and artificial intelligence on worker well-being. *Technology in Society, 67*, 101679. https://doi.org/10.1016/j.techsoc.2021.101679 (accessed on 8th July 2025).

O'Neill, J. W., & Davis, K. (2011). Work stress and well-being in the hotel industry. *International Journal of Hospitality Management, 30*(2), 385–390.

OECD. (2018). *OECD tourism trends and policies 2018*. OECD Publishing. Available at: http://dx.doi.org/10.1787/tour-2018-en (accessed on 8th July 2025).

Orfila-Sintes, F., & Mattsson, J. (2009). Innovation behavior in the hotel industry. *Omega, 37*(2), 380–394.

Ozdemir, O., Dogru, T., Kizildag, M., & Erkmen, E. (2023). A critical reflection on digitalization for the hospitality and tourism industry: Value implications for stakeholders. *International Journal of Contemporary Hospitality Management, 35*(9), 3305–3321. https://doi.org/10.1108/IJCHM-04-2022-0535 (accessed on 8th July 2025).

Paschen, U., Pitt, C., & Kietzmann. (2020). Artificial intelligence: Building blocks and an innovation typology. *Business Horizons, 63*(2), 147–155. https://doi.org/10.1016/j.bushor.2019.10.004 (accessed on 8th July 2025).

Pereira, T., Limberger, P. F., Minasi, S. M., & Buhalis, D. (2022). New insights into consumers' intention to continue using chatbots in the tourism context. *Journal of Quality Assurance in Hospitality & Tourism*, 1–27. https://doi.org/10.1080/1528008X.2022.2136817 (accessed on 8th July 2025).

Popay, J., Roberts, H., Sowden, A., Petticrew, M., Arai, L., Rodgers, M., Britten, N., Roen, K., & Duffy, S. (2006). *Guidance on the conduct of narrative synthesis in systematic reviews. A product from the ESRC Methods Programme*, *1*(1).

Prentice, C., Weaven, S., & Wong, I. A. (2020). Linking AI quality performance and customer engagement: The moderating effect of AI preference. *International Journal of Hospitality Management, 90*(8), 102629. https://doi.org/10.1016/j.ijhm.2020.102629 (accessed on 8th July 2025).

PricewaterhouseCoopers. (2017). *What's Next for the 2017 Data science and analytics job market?* PWC. http://www.pwc.com/us/en/publications/data-science-and-analytics.html (accessed on 8th July 2025).

Productivity Commission. (2020). *Inquiry Report on Mental Health*. https://www.pc.gov.au/inquiries/completed/mental-health/report (accessed on 8th July 2025).

Saah, F. I., Amu, H., & Kissah-Korsah, K. (2021). Prevalence and predictors of work-related depression, anxiety, and stress among waiters: A cross-sectional study in upscale restaurants. *PLOS ONE, 16*(4), e0249597. https://doi.org/10.1371/journal.pone.0249597 (accessed on 8th July 2025).

Samara, D., Magnisalis, I., & Peristeras, V. (2020). Artificial intelligence and big data in tourism: A systematic literature review. *Journal of Hospitality and Tourism Technology, 11*(2), 343–367. https://doi.org/10.1108/JHTT-12-2018-0118 (accessed on 8th July 2025).

Sariisik, M., Sengul, S., Okumus, B., Ceylan, V., Kurnaz, A., & Kapucuoglu, M. I. (2022). In-house responses and anxiety levels of commercial kitchen employees towards new COVID-19 food safety regulations. *Journal of Foodservice Business Research, 26*(1), 1–28. https://doi.org/10.1080/15378020.2022.2027199 (accessed on 8th July 2025).

Schwabe, H., & Castellacci, F. (2020). Automation, workers' skills and job satisfaction. *PLOS ONE, 15*(11), e0242929. https://doi.org/10.1371/journal.pone.0242929 (accessed on 8th July 2025).

Shin, H., Perdue, R. R., & Kang, J. (2019). Front desk technology innovation in hotels: A managerial perspective. *Tourism Management, 74*(3), 310–318. https://doi.org/10.1016/j.tourman.2019.04.004 (accessed on 8th July 2025).

Sivarajah, U., Kamal, M. M., Irani, Z., & Weerakkody, V. (2017). Critical analysis of big data challenges and analytical methods. *Journal of Business Research, 70*, 263–286. https://doi.org/10.1016/j.jbusres.2016.08.001 (accessed on 8th July 2025).

Statista Research Department. (2018). *Global Travel and Tourism Industry: Statistics & Facts*. https://statista.com/topics/962/global-tourism (accessed on 8th July 2025).

Stening, T. (2023). Is "generative" AI the way of the future? Expert explains new models, the need for human involvement. *Techxplore*. Available at: https://techxplore.com/news/2023-04-generative-ai-future-expert-human.html (accessed on 8th July 2025).

Stylos, N., Fotiadis, A. K., Shin, D., & Huan, T. C. (2021). Beyond smart systems adoption: Enabling diffusion and assimilation of smartness in hospitality. *International Journal of Hospitality Management, 98*(2), 103042. https://doi.org/10.1016/j.ijhm.2021.103042 (accessed on 8th July 2025).

Super Friend. (2020). *Spotlight on the Retail Industry Report.* https://SuperFriend.com.au/wp-content/uploads/2020/12/SPF0396_SpotlightReport_Retail_fv-2.pdf (accessed on 8th July 2025).

Super Friend. (2021). *Indicators of a Thriving Workplace: National Report 2021.* https://itw2021.SuperFriend.com.au/#446c7c7674002b8f5 (accessed on 8th July 2025).

Thanem, T., & Elraz, H. (2022). From stress to resistance: Challenging the capitalist underpinnings of mental unhealth in work and organizations. *International Journal of Management Reviews, 24*(5), 1–22. https://doi.org/10.1111/ijmr.12293 (accessed on 8th July 2025).

Thiebes, S., Lins, S., & Sunyaev, A. (2021). Trustworthy artificial intelligence. *Electronic Markets, 31*(2), 447–464. https://doi.org/10.1007/s12525-020-00441-4 (accessed on 8th July 2025).

Tran, L. T. T., Pham, L. M. T., & Le, L. T. (2019). E-satisfaction and continuance intention: The moderator role of online ratings. *International Journal of Hospitality Management, 77*, 311–322. https://doi.org/10.1016/j.ijhm.2018.07.011 (accessed on 8th July 2025).

Tussyadiah, I. (2020). A review of research into automation in tourism: Launching the *Annals of Tourism Research Curated Collection* on artificial intelligence and robotics in tourism. *Annals of Tourism Research, 81*, 5–6.

Varga, S., Mistry, T. G., Ali, F., & Cobanoglu, C. (2021). Employee perceptions of wellness programs in the hospitality industry. *International Journal of Contemporary Hospitality Management, 33*(10), 3331–3354. https://doi.org/10.1108/IJCHM-12-2020-1417 (accessed on 8th July 2025).

Waters, C. N., Ling, E. P., Chu, A. H., Ng, S. H., Chia, A., Lim, Y. W., & Müller-Riemenschneider, F. (2016). Assessing and understanding sedentary behavior in office-based working adults: A mixed-method approach. *BMC Public Health, 16*, 360. https://doi.org/10.1186/s12889-016-3023-z (accessed on 8th July 2025).

Westerhof, G. J., & Keyes, C. L. (2010). Mental illness and mental health: The two continua model across the lifespan. *Journal of Adult Development, 17*(2), 110–119. https://doi.org/10.1007/s10804-009-9082-y (accessed on 8th July 2025).

Whiteford, H. A., Degenhardt, L., Rehm, J., Baxter, A. J., Ferrari, A. J., Erskine, H. E., Charlson, F. J., Norman, R. E., Flaxman, A. D., Johns, N., Burstein, R., Murray, C. J., & Vos, T. (2013). Global burden of disease attributable to mental and substance use disorders: Findings from the Global Burden of Disease Study 2010. *Lancet (London, England), 382*(9904), 1575–1586. https://doi.org/10.1016/S0140-6736(13)61611-6 (accessed on 8th July 2025).

World Health Organization. (2001). *The World Health Report: 2001: Mental Health: New Understanding, New Hope*. World Health Organization.

World Health Organization. (2023). *Coronavirus Disease (COVID-19) Pandemic*. https://www.who.int/europe/emergencies/situations/covid-19 (accessed on 8th July 2025).

World Travel & Tourism Council. (2017). *Travel & Tourism Economic Impact 2017.* https://www.wttc.org/about/media-centre/press-releases/press-releases/2017/tourism-supports-1-in-10-jobs-outpacing-global-economy-for-6th-conservative-year-wttc/ (accessed on 8th July 2025).

Xia, H., Vu, H. Q., Lan, Q., Law, R., & Li, G. (2019). Identifying hotel competitiveness based on hotel feature ratings. *Journal of Hospitality Marketing & Management, 28*(1), 81–100. https://doi.org/10.1080/19368623.2018.1504366 (accessed on 8th July 2025).

Xu, G., Xue, M., & Zhao, J. (2023). The association between artificial intelligence awareness and employee depression: The mediating role of emotional exhaustion and the moderating role of perceived organizational support. *International Journal of Environmental Research and Public Health, 20*(6), 5147. https://doi.org/10.3390/ijerph20065147 (accessed on 8th July 2025).

Yan, J., Kim, S., Zhang, S. X., Foo, M. D., Alvarez-Risco, A., Del-Aguila-Arcentales, S., & Yáñez, J. A. (2021). Hospitality workers' COVID-19 risk perception and depression: A contingent model based on transactional theory of stress model. *International Journal of Hospitality Management, 95*, 102935. https://doi.org/10.1016/j.ijhm.2021.102935 (accessed on 8th July 2025).

Yang, Y., Shin, J. C., Li, D., & An, R. (2017). Sedentary behavior and sleep problems: A systematic review and meta-analysis. *International Journal of Behavioral Medicine, 24*(4), 481–492.

Yun, J., & Park, J. (2022). The effects of chatbot service recovery with emotion words on customer satisfaction, repurchase intention, and positive word-of-mouth. *Frontiers in Psychology, 13*. https://doi.org/10.3389/fpsyg.2022.922503 (accessed on 8th July 2025).

Zhang, S. X., Liu, J., Afshar Jahanshahi, A., Nawaser, K., Yousefi, A., Li, J., & Sun, S. (2020). At the height of the storm: Healthcare staff's health conditions and job satisfaction and their associated predictors during the epidemic peak of COVID-19. *Brain, Behavior, and Immunity, 87*, 144–146. https://doi.org/10.1016/j.bbi.2020.05.010 (accessed on 8th July 2025).

CHAPTER 13

SMART HOTEL: LEADING WITH INNOVATIVE SERVICE EXCELLENCE

SHOHEL MD. NAFI[1] and LITON CHANDRA VOUMIK[2*]

[1] *Department of Tourism and Hospitality Management, Noakhali Science and Technology University, Noakhali, Bangladesh*

[2] *Department of Economics, Noakhali Science and Technology University, Noakhali, Bangladesh; and Central Queensland University, Melbourne, Victoria, Australia*

*Corresponding author

ABSTRACT

The aim of this study is to identify the future trends in the hospitality sector. The future of the hospitality industry relies on the technological adoption of hotels. This study aspires to gain knowledge and understanding of the role of the smart hotel in achieving service excellence. This study is a qualitative study in which data were collected from secondary sources. In this study, different types of smart technologies used in the hotel are described to understand the reformation of this sector. The emerging concept of smart hotels, the importance of smart hotels, Artificial Intelligence (AI) in smart hotels, smart phone usage for smart services, wearable technology, mobile applications in the hotel industry, digital payment and transactions, and data security-related issues are described in this chapter. This study helps to understand the new technologies that are used in the hotel industry to gain guest experience and service excellence. This study will offer useful insights for hotel managers and policymakers to improve their sustainability efforts by identifying the primary drivers of service excellence and their impact on sustainable hotel business practices.

Smart Tourism: Managing Technological Revolution in the Tourism Industry. Santus Kumar Deb (Ed.)

DOI: 10.1201/9781779641090-13

The findings of this study will help to understand how service excellence and sustainable practices interact and affect hotel performance.

13.1 INTRODUCTION

Companies in various service industries are focusing on delivering exceptional customer service to gain their position and build long-lasting client relationships due to the ongoing globalization of services and technological innovation (Gouthier et al., 2012). The hotel business experiences a similar phenomenon, and by proactively adopting such advanced innovations, smart hotels have emerged. When a customer understands that the current service is exceeding their previous expectations, service excellence happens (Horwitz and Neville, 1996). This kind of service excellence creates customer delight. Customer satisfaction was previously ignored by businesses, whereas today's competitive world largely depends on customer satisfaction. Overall, delivering great service that is focused on fulfilling the requirements and surpassing the expectations of each guest is necessary for delivering hotel service excellence. Smart hotel services are also struggling to deliver service excellence to the customers.

Wu and Cheng (2018) describe a smart hotel as a unique type of hotel that operates and uses various innovative technologies for ensuring quality services. Therefore, an inventive hotel that uses cutting-edge technology to achieve operational excellence through service automation and personalized offers is referred to as a smart hotel. Furthermore, novel innovations like robots could manage a variety of activities that humans used to perform, which is why human-free services are frequently associated with smart hotels (Stock and Merkle, 2017; Tussyadiah and Park, 2018). An innovative approach in the hotel can help to achieve better service quality. Service excellence is a key component of hotel design. Using a variety of design decisions, including servicescapes and resource flows, hotels help to satisfy their guests (Bagchi et al., 2021). Hotels may boost revenue flows by strategically designing and managing their operations to increase client satisfaction while also elevating their brand's appeal. The word "service excellence" is derived from the two words "service" and "excellence." Parasuraman et al. (1985) defines services as *"exhibiting certain traits that distinguish them, namely intangibility, inseparability, heterogeneity, and perishability."* Excellence was defined as remarkable performance or meeting or exceeding expectations or standards in a specific field.

Hence, service excellence is attained when knowledge or abilities are put to use in a specialized action or method that goes above and beyond the established norms, expectations, or quality demands of a particular client or business. Service quality and service excellence go hand in hand (Johnston, 2004).

Achieving hotel service excellence is essential to a hotel's accomplishment in the hospitality sector. Delivering outstanding customer service, forecasting and meeting customers' needs, and ensuring they have a memorable experience are all components of service excellence. Numerous studies have emphasized the value of providing excellent customer service. A crucial component of service excellence is service quality. According to Umar et al. (2013), service excellence is built on ongoing value addition for the services provided. Service excellence was described by Cook (2008) as the delivery of great services with the goal of attracting and retaining clients without influencing their view of reality. According to Withiam (2013), prices must be set in such a way as to guarantee customer satisfaction if service excellence is to be achieved. For customers to be happy, the services given must deliver value for the money. After service excellence achievement, another objective is to confirm that delighted customers return and recommend the hotel. Return business entails greater profitability from both new and return business, as well as increased client loyalty. In the hotel industry, the importance of repeat guests is well documented. However, providing excellent customer service is essential to a profitable hotel industry.

In previous studies, only a few research projects focus on the emergence of smart hotels and service excellence in the hospitality sector. Prior studies in the hotel and tourism sectors tended to concentrate on wealthy nations (Fraj et al., 2015; Leonidou et al., 2015). There has been little research on the role of service excellence in promoting sustainable practices, with studies mostly focusing on service quality and its effects on customer satisfaction and loyalty (Kim and Han, 2020; Umar et al., 2013). As a result, there is a research gap in the literature discussing the connection between superior customer service and ethical hotel management. In addition, very little research is conducted about the precise elements that contribute to superior service in tourism. In order to improve service excellence and promote sustainable practices in hotels, factors including employee training and development, leadership and management techniques, and technology adoption may be essential. This study aims to find the smart hotel influence to achieve service excellence.

13.2 THEORETICAL BACKGROUND

The industry must meet and surpass these source markets' diverse service needs and expectations to gain consumer satisfaction, service excellence, and loyalty. Hotel service excellence is the delivery of outstanding guest service that goes above and beyond guest expectations and creates a special experience. Maintaining visitor contentment, enhancing brand reputation, and eventually earning repeat business all depend on excellent hotel service. In the hospitality industry, there is very tough competition to attract guests in the same market (Becerra et al., 2013; Zhang et al., 2020). The speedy growth of nonstandard lodging in several locations with high levels of tourism development has made it extremely difficult for traditional hotels to maintain and improve industry efficiency. In order to advance hotel performance and raise the standard of destination lodging offerings, it is essential to examine the external elements that affect the hotel industry's productivity (Yang et al., 2012). Zhang et al. (2020) examines how service excellence might increase organizational advantage in East African hotels. This study pointed out that management evaluated service standards to satisfy customer needs. This study concluded that customers should be satisfied with the total service delivery process systems, which is known as service excellence.

Ouma (2020) studied hotel service excellence strategy at Nairobi hotels. The study discovered that, after strategic choices, technological advancements have the biggest impact on the level of service quality at star-rated hotels. The following factors, in that order, had a substantial impact on the service excellence of star-rated hotels: technological advancements, strategic choices, service and product innovations, skilled manpower management, and organizational formation. This study suggests that star-ranked hotels should implement the six drivers of service excellence to encourage repeat business, devoted patronage, and brand ambassadors. The study also suggests that businesses should maintain their adaptability to new technological developments to keep up with the rate of technological improvement. In fact, contemporary technology improves the delivery of services and makes it simpler to communicate information.

To gain competitive advantage in the market, businesses are progressively integrating service excellence into their excellent service (Salama et al., 2022). Service excellence, which is regarded as being "simple to do business with," fulfills commitments and shows a high level of satisfaction (Johnston, 2004; Oliver, 1997). However, the main theme of service organizations should not only satisfy the customers but also need to delight them. In this study, the definition of excellent customer service is based on the client's participation, responsiveness, and the service process and delivery (Bettencourt and Gwinner, 1996). Hence, service excellence is attained when knowledge or abilities are put to use in a specialized action or method that goes above and beyond the established norms, expectations, or quality demands of a particular client or business. Service quality and service excellence go hand in hand. Customers greatly value flexibility and customization in the service delivery process (Deb et al., 2022; Lovelock et al., 2005). The physical environment and mood in a service company are referred to as the servicescape, and how they look and feel has an effect on how customers are treated.

A crucial component of service excellence is service quality. According to Umar et al. (2013), service excellence is built on the addition of ongoing value to the services provided. Service excellence was described by Cook (2008) as the delivery of great services with the goal of attracting and retaining clients without influencing their view of reality. According to Withiam (2013), prices must be set in such a way as to guarantee customer satisfaction if service excellence is to be achieved. For customers to be happy, the services given must deliver value for the money. After reaching service excellence, another objective is to ensure that happy clients return and endorse the company. Satisfied customers are a great asset for increasing the business's profitability. It could be attractive to think that the way forward is through moving up—bigger open-space lobbies, more individualized services, and more high-end in-room technology. This is because delivering service excellence is becoming more challenging. The Internet of Things (IoT), facial recognition, and even augmented reality are among the technologies that hotel operators are reportedly trying to adopt in the future of the industry (Berry and Parasuraman, 1992; Yoong, 2022).

According to Kandampully and Suhartanto (2000), customer loyalty and satisfaction are positively correlated with service quality. Furthermore, a study conducted by Paraskevas and Arendell (2007) revealed that providing excellent customer service is crucial for generating positive word-of-mouth recommendations and boosting repeat business. Hotels must make personnel training and development investments if they want to provide exceptional service. A study by Lee et al. (2014) found that improving service quality requires training programs that emphasize interpersonal skills, communication, and problem-solving.

Additionally, exceptional service depends on staff and customer interaction (Carroll and Siguaw, 2003). Finally, in order to sustain excellent customer service, hotels must commit to ongoing development. According to Kim et al. (2015), monitoring customer feedback and making adjustments in response to their recommendations is essential for enhancing service quality and raising customer delight. Providing excellent customer service is essential for success in the hotel business. Hotels that put a high priority on employee development, good communication, and ongoing improvement are more likely to give their visitors an unforgettable experience, boost customer satisfaction and loyalty, and ultimately increase profitability.

When an organization provides better services than its competitor, that gives one kind of competitive advantage in maintaining consumers and safeguarding them from aggressive forces (Carroll and Siguaw, 2003; Deb and Nafi, 2021). Service excellence helps to create a competitive advantage for the customer. To be competitive, sustainable in the industry, and to be successful in the future, organizations must have a thorough analysis of the dynamic and complicated environment (Wijetunge, 2016). With this insight, the organization's primary success factor has been the adoption of the plans to be competitive and fulfill the demand uncertainties. Due to the increased competition, certain hotels, particularly those with five and four stars, are now experiencing liquidity issues, and some have even been placed under receivership (Yoong, 2022; Yunis et al., 2013). Aziz and Wahiddin (2017) studied on the hotel service excellence. They proposed a structural model for understanding the antecedents of service excellence in the hotel (Figure 13.1). In order to comprehend the causes of service excellence in the hotel business, they created a structural model. According to the concept, customer orientation, employee involvement, and service innovation are the three key characteristics that have an impact on service excellence.

As it enables hotels to recognize and cater to their guests' needs, customer orientation is a crucial precondition of service excellence. This includes developing a culture that is focused on the needs of the customer and places a high value on repeat business. Service excellence is influenced by a number of crucial factors, including employee engagement. Employee engagement increases the likelihood that they will be dedicated to providing outstanding service and enhancing the guest experience. In order to provide customers with exceptional service, hotels must be able to anticipate and meet their changing demands. Service innovation is one way to do this. These dimensions can be achieved through innovation, which is provided by the smart hotel.

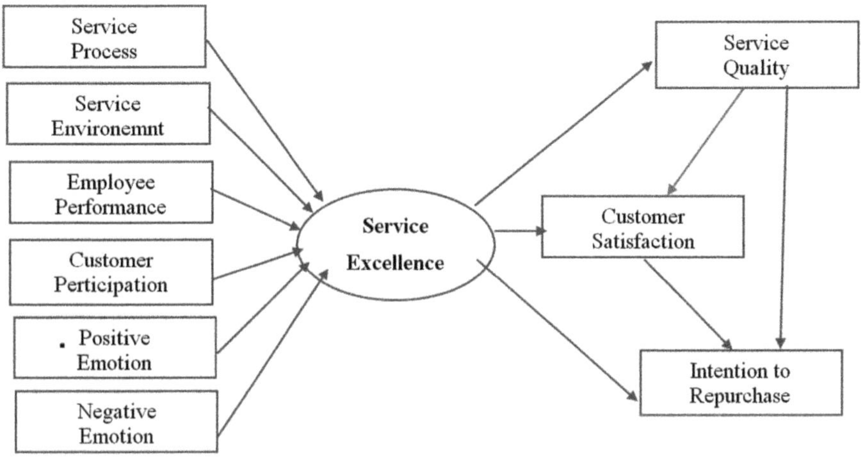

FIGURE 13.1 Structural model of service excellence antecedents.

Source: Reprinted from Aziz and Wahiddin (2017). https://creativecommons.org/licenses/by/4.0/

13.3 METHODOLOGY

The aspiration of this study is to gain a better understanding of the role of the smart hotel in achieving service excellence. This study is a qualitative study where data were collected from secondary sources. Data were gathered from secondary sources, including journal articles, relevant literature, websites, and various published works (Deb et al., 2023), to provide a comprehensive picture of smart hotels and service excellence for the tourism and hospitality industries. Secondary data can be defined as information that was obtained by a third party without the author's involvement or as information that, with certain restrictions, can be utilized to address new problems (Boslaugh, 2007; Clark, 2013; Deb et al., 2024; Vartanian, 2010).

13.4 EMERGENCE OF SMART HOTEL

The advent of smart hotels, which seamlessly integrate technology with visitor experiences, represents a revolutionary change in the hospitality sector. In recent years, smart technology has completely changed the human living standard. Therefore, it seems sensible that intellectual advances such as self-monitoring analysis and reporting technology have now permeated the hotel sector. Particularly, hotels want to match—and ideally surpass—the degree of luxury and convenience that their visitors may be accustomed to from home. The hotel sector has seen a transformation due to the introduction of novel technologies in recent times (Cobos et al., 2016; Ivanov and Webster, 2019).

In a similar vein, Law et al. (2020) claimed that the hotel industry's technological advancements offer limitless opportunities for hotel practitioners to operate intelligent systems that enhance effective administration. The development of a smart hotel has been aided by the widespread acceptance of disruptive technologies, even though hotel chains have varied degrees of technological adoption. These forward-thinking businesses use automation, AI, and the IoT to improve customer satisfaction and operational efficiency. Through smartphone apps for check-in, room customization, and even virtual concierge services, guests have individualized experiences (Chan et al., 2014). Voice-activated assistants and automated climate management are just two examples of the incredibly convenient features available in smart rooms. Furthermore, hotels can save operating costs, optimize services, and predict visitor preferences thanks to data analytics. In addition to efficiency, the smart hotel revolution promises modern travelers a new standard of customized luxury. Kim and Han (2020) provide examples of smart hotels as "Yotel Singapore Orchard Road and Flyzoo Hotel in China." In addition to offering smart TVs, technowalls with mood lighting that can be adjusted, smartbeds, and service robots (Yoshi and Yolanda), Yotel also took first place in the Robotics for Hospitality and Leisure category of the 2019 Technology Excellence Award. At the Flyzoo Hotel in Hangzhou, China, visitors can use face recognition to access their rooms, while food and amenities are delivered by service robots. Interactive screens and guided exercises provide guests with enjoyable digital experiences. Service excellence can be attained with the use of these contemporary technologies.

13.5 IMPORTANCE OF SERVICE EXCELLENCE

The process of providing service is made up of several intangible activities that happen during interactions between customers and service providers, as well as any physical resources, items, or systems of the service provider that are offered as a remedy for problems with customers (Grönroos, 2000). The word "service excellence" refers to a company's ability to consistently fulfill and, on certain occasions, surpass the expectations of its customers. Service excellence expresses to customers that they are respected, understood, and that their needs will be well addressed. Service excellence is the capacity to deliver outstanding service for the purpose of attracting as well as keeping clients; customers need more options and will not be influenced based on their understanding of reality itself (Umar et al., 2013). To enhance the customers' expectations and after that fulfilling their needs are not the main meaning of service excellence. It basically defines that we provide more of what we have promised to our guests or customers. When a business generates services rather than products, service quality is extremely important for the business's continued existence and development. The vast majority of businesses supply satisfactory customer service, but customers return to those that go beyond what is necessary to deliver service excellence. That is why, for better repeat business and to keep potential and existing customers, service excellence is vital for any business. Tieng (2022) identified processes of providing service excellence, shown in Figure 13.2. Five stages have been identified by Tieng (2022) of good service that can be evaluated by the customer experience.

This process starts with *convenience*. The first step is to take action in order to take into account the demands of the customer, such as when and how they would want to contact executives. In every communication channel—such as social media, phone calls, and conversations—hotel employees should be present. Use a customer service model that optimizes agent availability as much as possible to ensure that a client can always get in touch with you and speak with an agent. The second process is *agility*; do not keep the clients waiting. Because patience is not a virtue in the customer service sector, customers value rapid replies. Another crucial factor is how quickly you answer questions from customers. Even if some problems are harder to address than others, it's best to tackle them as soon as possible. To get the ideal operating speed, any bottlenecks such as an excess of manual tasks that slow down your agents or a shortage of resources to manage the volume of calls should be addressed.

The third stage is to *do the right thing*. Customer service staff has to be given the freedom to do whatever it takes to reach a final resolution, even if it involves making some compromises or going above and beyond in order to genuinely reach a level of service excellence. The fourth stage is *taking the initiative.* It shows that you prioritize the matter at hand and act with urgency and empathy when you call a customer when they have a concern before they have an opportunity to contact you. An excellent service organization may be distinguished from others by its level of diligence. The final stage is to *evaluate and improve*, gather and analyze the operational data, and make a report. In this stage, the hotel takes corrective actions and improves the efficiency of the hotel operation. However, in the smart hotel environment, technology is involved in all five stages of service excellence. Technological innovation upgrades the communication process, service delivery process, feedback process, and evaluation process, making them easier and enhancing customer satisfaction. Thus, globally, hotels adopt new technologies for gaining more customer satisfaction and making operation processes more efficient.

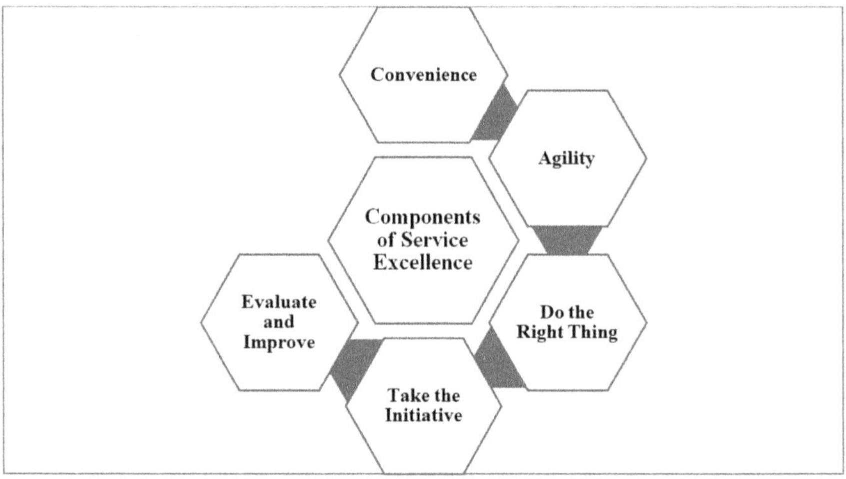

FIGURE 13.2 Process of providing service excellence.

Source: Adapted from Tieng (2022).

Figure 13.3 shows the service excellence model developed by Johnston in 2007. Johnston (2007) defined that service excellence is dependent on four characteristics: promised service delivery, providing a branded service, exceeding the service user's expectations, and dealing with issues and inquiries. These four characteristics must be carefully reviewed with regard to maintaining the promises of service to customers. If these four attributes fulfill the service excellence's requirement, then the service excellence will increase the revenue of the business and competitive advantage. It will also increase the satisfaction of employees and customers and help in repeat business and keep existing and potential customers.

FIGURE 13.3 Model for service excellence.

Source: Reprinted from Johnston (2007). https://creativecommons.org/licenses/by/4.0/

13.6 ARTIFICIAL INTELLIGENCE (AI) AND SMART HOTEL

Almost every industry is changing as a result of AI, which gives computers the ability to decide for themselves and create more efficient processes (Leung, 2019). AI will inevitably be used to show how quickly the hospitality industry is growing and how it helps corporate operations by providing services and creating value. AI may be used to understand guest demands better and address issues impacting the hospitality industry. It could provide unique, customized experiences and even encourage possibilities to spend more time. Additionally, AI technologies may ensure hotel managers are more efficient, innovative, and sustainable. The hotel industry currently uses AI extensively, from personalized recommendations to automated room booking processes. AI-powered smart tourism is altering the way we plan, book, and enjoy our holidays in hotels. While there are many benefits to this new technology, there are also certain issues that need to be resolved for travelers to get the most out of their vacations. Here are some of the most popular AI technologies being used to enhance the tourist experience are:

1. **Virtual Tour Guides:** Visitors are becoming more and more fascinated by AI-powered virtual tour guides. These internet tour guides can provide advice and details regarding the hotel facilities, local customs, and tourist attractions. Some of the most well-known virtual tour guides are Google Maps, TripAdvisor, and Yelp.

2. **Chatbots:** Computer programs called chatbots use natural language processing to simulate human conversation. They may be used to solve problems, offer advice, book reservations, and place orders for tickets. Several travel companies are increasingly utilizing chatbots to give their customers a more customized experience.
3. **Personalized Recommendations:** Tourists are given customized recommendations using AI-powered systems. These algorithms are able to examine a traveler's prior preferences and behavior.
4. **Augmented Reality (AR):** It is being used to enhance the visitor experience through immersive experiences. AR may be applied to create interactive experiences and to provide information about a particular area. AR helps to remove the service intangibility of the hotel.

13.7 SMART PHONES AND SMART SERVICES IN HOTELS

The hotel industry has witnessed a notable surge in mobile commerce, commonly known as m-commerce, and smartphones have emerged as a new kind of mobile technology that helps businesses engage with their customers (Kamboj and Gupta, 2020; Kleijnen et al., 2007; Morosan and DeFranco, 2016; Sarmah et al., 2017). The increasing use of smart phones and smart phone-based services helps to provide more sophisticated services to its users. Nowadays, hotels use smartphone apps to let their customers communicate with service personnel and create new services based on their requirements and preferences. Sarmah et al. (2017) stated that smart phones help customers to co-create services at the hotel. A tourist is traveling to a destination or hotel to meet his or her personal needs and wants. Morosan and DeFranco (2016) stated that tourist involvement in the service development and service co-creation process may help to satisfy the tourists and affect further adoption intention. These types of interaction are only possible through the smart phone apps which are gaining popularity with different hotels like Hilton and Marriott. Sarmah et al. (2017) mentioned that *"co-creation can occur with a multitude of tourist and context-specific variables based on tourist innovativeness, nature of interaction and involvement, and it can be influenced by mobile technology and m-commerce; this study presents a comprehensive view of co-creative service innovation in Indian luxury hotels."* In this process, the customers or tourists must be innovative to interact with the smart device and the hotel employees. Nonetheless, a small number of smartphone applications are created specifically for the travel and hospitality industry (Dickinson et al., 2014; Kamboj and Gupta, 2020). Thus, the hospitality industry must focus on developing smart phone apps and influence the tourist to actively use these apps for better service co-creation. Furthermore, these types of innovation are one of the actors of smart hotels.

13.8 WEARABLE TECHNOLOGIES IN HOTELS

The hospitality industry is one of the pioneer industries for adopting wearable technology for delivering services (Ahram and Falcão, 2020; Atembe and Abdalla, 2015). An increasing number of holidaymakers are selecting hotels based on unique experiences. One of the finest ways to leave an impression is to include the newest smart technologies for the comfort and convenience of visitors. Wearable technologies will be the next revolutionary thing in the hospitality industry, as research indicates that 53% of millennials and 54% of early adopters are very much excited about the new wearable technologies (Elphick, 2023). Smart glass technology provides hotels with the much-needed "wow" factor. The selection of wearable technologies must consider six factors, which are given in Figure 13.4. Thus, selecting a wearable technology should consider the usability, unique design, comfort, performance, conformity, and sustainability.

The usage of wearables in the hospitality sector is growing as a way to enhance overall customer satisfaction and optimize operations (Formolo, 2024). The hospitality sector adopts these technologies as a way of enhancing and achieving service excellence. Wearables, including wristbands and smartwatches, can be used as room keys in the hotel industry, displacing traditional keycards and enabling guests to enter their rooms. Additionally, wearables may be utilized to give visitors in-room amenities like temperature control and room service ordering. These types of technologies help to revolutionize the way guests interact with the hotel.

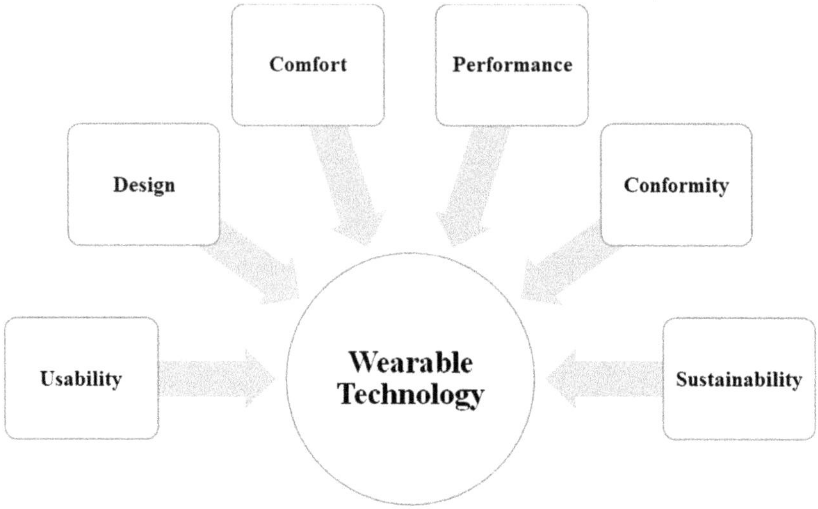

FIGURE 13.4 Consideration of selecting wearable technologies.

Source: Author

13.8.1 STARWOOD: PIONEER OF ADOPTING WEARABLE TECHNOLOGIES

Starwood developed the first-ever guest-facing application for Google Glass. Through the app, visitors may reserve a room at any Starwood hotel and discover other locations. When visitors arrive, it detects them and directs them to explore and share pictures of the hotel's features, rooms, and surrounding area. They can also access the most recent account information, which includes starpoint balances and forthcoming stay specials. Additionally, Starwood and iWatch have teamed up to enable watch users to make mobile payments as well as open their hotel entrance by simply waving the watch in front of the door.

Another wearable technology adopted by Starwood Hotel is Google Glass. Starwood Hotels' pilot program primarily encourages guests to exchange physical keys for virtual keys to make it easier to use their smartphones. Starwood Hotels has deployed the Starwood Preferred Guest (SPG) app for Google Glass (Nelson and Stolterman, 2014). The Starwood Hotels SPG app allows guests to get directions to the hotel, explore and share hotel images with friends. Additionally, guests can access booking information, star balance, and upcoming events (Moscaritolo, 2014).

13.9 MOBILE APPLICATION IN THE HOTEL INDUSTRY

Mobile applications have become indispensable tools in the hospitality sector, offering numerous advantages to both customers and hoteliers. Mobile applications help to increase productivity and ease the way to service delivery processes. Previous studies have shown various mobile technologies used in reservation, room customization, in-room entertainment, billing, payment, automated check-in/check-out, marketing, promotion, and other relevant activities (Mo Kwon et al., 2013; Huang et al., 2019; Tomislav and Stifanich, 2020). Here is an explanation of how hotel mobile applications are used:

1. **Booking and Reservations:** Travelers may easily browse and reserve hotel rooms using mobile apps. Real-time booking for both guests and hotels allows them to see available rooms, check their rates, and make bookings.
2. **Customization of the Hotel Room:** Prior to their arrival, guests are frequently able to define their hotel room preferences using mobile apps, including the type of bed, floor, and room view. The guest experience is improved by the customization.
3. **Billing and Payment:** Using the mobile application, visitors may check their charges and pay their bills, speeding up check-out and lowering the need to stop by the front desk.
4. **Contactless Check-In and Check-Out:** Contactless check-in and check-out processes are enabled through mobile apps. Guests can conduct these chores via the app, eliminating the need for actual interaction with hotel staff and shortening wait times.
5. **Keyless Entry:** Some hotels provide keyless entry through their mobile apps. Guests are given a digital room key on their cellphones, which allows them to enter their rooms without the need for a traditional key card.
6. **Feedback and Reviews:** Many hotel apps allow guests to submit comments and post reviews. This feedback is useful for enhancing services and responding quickly to guest issues.
7. **Loyalty Program:** Many hotel brands provide loyalty programs through their smartphones, enabling customers to accrue points, have access to exclusive discounts, and receive tailored offers.
8. **Safety and Security:** Hotels can employ mobile apps to share crucial information and safety recommendations with visitors during emergencies or unplanned incidents.

9. **Data analytics:** Apps gather information on visitor preferences and behavior that hotels can utilize for tailored advertising, bettering services, and operational decision-making.
10. **Marketing and Promotions:** Hotels are able to use mobile applications to directly advertise their services, deals, and loyalty programs to guests, improving the likelihood of upselling and repeat reservations.

13.10 DIGITAL PAYMENT AND TRANSACTION

The incorporation of digital payment solutions into the smart tourism ecosystem is significantly transforming the manner in which passengers conduct transactions and handle their expenditures while traveling. Digital payments comprise a diverse array of electronic and non-cash techniques, which include mobile wallets, contactless cards, online payment platforms, and peer-to-peer (P2P) payment applications. In the realm of smart tourism, the utilization of payment technology is fundamentally transforming the manner in which travelers engage in activities such as reserving accommodations, acquiring tickets, dining, and engaging in retail transactions during their journeys (Adamba, 2024; Arafa et al., 2022; Kim et al., 2022; Zohry and Alsaied, 2021).

The key features and applications of digital payments and transactions in smart tourism are:

1. **Cashless Convenience:** Digital payment systems offer a convenient and safe alternative to real cash, enhancing the ease of transactions for visitors throughout their travels. This practice contributes to the promotion of safety by mitigating the potential for theft and obviating the necessity of carrying substantial amounts of cash.
2. **Contactless and Mobile Payment:** Mobile payment applications and digital wallets, such as Apple Pay, Google Pay, and other payment services, enable travelers to securely save their payment details on their smartphones and conveniently make contactless transactions via Near Field Communication (NFC) at various establishments, encompassing hotels, restaurants, and retail outlets. This innovative approach allows tourists to effortlessly make transactions by merely tapping or waving their cards or smartphones near a payment terminal.

3. **Online Booking:** The utilization of digital payment networks enables travelers to make online reservations for lodgings, flights, tours, and activities. These platforms frequently include secure payment methods in order to safeguard sensitive financial information.
4. **QR Code and P2P Payments:** QR code-based payment systems, like Alipay and WeChat Pay are commonly utilized by tourists for the purpose of purchasing products and services. This is especially prominent in regions where these platforms are widely adopted. P2P payment applications, such as PayPal, Venmo, Bkash, Nagad, and other Apps developed by individual countries, offer the convenience for visitors to divide expenses, distribute costs, and transfer funds to individuals or relatives instantly, thereby enhancing the feasibility of group travel.
5. **Local Payment Integration:** The integration of local digital payment solutions into the infrastructure of smart tourism destinations is witnessing a notable surge. These developments are primarily driven by the desire to provide access to the diverse preferences of both international and domestic travelers.

13.11 DATA SECURITY AND PRIVACY

The security and privacy of data are important in the context of smart tourism, given its heavy reliance on data gathering, processing, and sharing. The growth of smart tourism involves strict adherence to established standards and procedures in order to safeguard the personal data of visitors and maintain the integrity of tourism-related systems (Magalhães et al., 2017; Moon et al., 2022; Nain, 2018). In the context of a smart hotel, the following is an overview of the most important aspects of data security and privacy:

1. **Secure Data Storage:** The secure storage of sensitive data, such as personal information, payment details, and trip plans, is crucial for the effective functioning of smart tourism systems. The implementation of secure encryption and access restrictions is important in order to effectively mitigate the risk of unwanted access.
2. **Data Encryption:** It is essential to employ industry-standard practices for encrypting data transmitted among visitors, service

providers, and destinations in order to effectively protect it from interception and theft.

3. **Compliance with Regulations:** In order to adhere to regulatory requirements, smart tourism initiatives are obligated to comply with appropriate data protection legislation, including the General Data Protection Regulation (GDPR) in Europe, the California Consumer Privacy Act (CCPA) in the United States, and applicable municipal privacy laws in other geographical areas. This entails the acquisition of appropriate consent for the gathering and processing of data.
4. **Transparent Privacy Policies:** Tourists must be provided with transparent privacy rules that offer easily comprehensible and brief clarifications about the utilization, dissemination, and retention of their personal data. Explicit and readily revocable consent should be obtained for the purpose of data collection.
5. **Data Minimization:** The principle of data minimization advocates for the collection of just the essential data that is specifically required for certain reasons. The act of minimizing data mitigates the potential risks that arise from the storage of excessive information.
6. **User Access Control:** It is imperative that tourists are granted authority over their personal data. People should be able to access, modify, and remove the personal information that businesses connected to the tourism industry collect.
7. **Secure Payment Processing:** The safe processing of payments necessitates the implementation of measures that adhere to the Payment Card Industry Data Security Standard (PCI DSS) and employ tokenization techniques to substitute sensitive card information with distinct tokens.

13.12 CONCLUSION

The study gives a thorough description of the idea of a smart hotel and excellent customer service. This study describes how smart technology changes hotel operations. In the current year, it has been identified that service quality is not enough to ensure guest satisfaction. However, in this aspect, service excellence will be the most prominent thing among hotel customers. Now, the business environment is very competitive.

Technological innovation and the appearance of social media make it more complex. These help customers be more empowered to make decisions about the service. Thus, this study will help to achieve service excellence in the hotel sector all over the world. This study is important because it would clarify the current level of service excellence in the hotel industry and point up development opportunities. The findings of this study can be used by hotel managers and policymakers to create plans for promoting sustainable business practices and enhancing service excellence in the hotel industry. It will make a significant addition to the field of hospitality research, and the hotel sector and elsewhere may benefit practically.

The effects of researching service excellence contribution to a sustainable hotel industry are extensive. First, this study helps to understand the new technologies that are used in the hotel industry to gain guest experience and service excellence. Second, this study will help to create sustainable tourism practices, which are essential for the long-term survival of the sector. This study will offer useful insights for hotel managers and policymakers to improve their sustainability efforts by identifying the primary drivers of service excellence and their impact on sustainable hotel business practices. Third, this study will help to add new knowledge on superior customer service and environmentally friendly corporate practices. The study's findings will offer empirical evidence for future researchers and will add value to the present of service quality in promoting sustainable behaviors. The findings of this study will help to better understand how service excellence and sustainable practices interact and affect hotel performance.

KEYWORDS

- **artificial intelligence (AI)**
- **guest**
- **hospitality industry**
- **policymakers**
- **service excellence**
- **smart hotel**
- **smart phone**
- **wearable technology**

REFERENCES

Adamba, F. L. (2024). Effect of digital banking technology on loan uptake in hotels industry in Kenya. *African Journal of Commercial Studies, 4*(2), 166–177.

Ahram, T., & Falcão, C. (2020). *Advances in Human Factors in Wearable Technologies and Game Design*. Springer International Publishing.

Arafa, H. F., El-latif, A., & Fathey, M. (2020). Challenges and opportunities of the transmission to use cashless payment methods in Egyptian hotels and travel agencies. *International Journal of Heritage, Tourism and Hospitality, 16*(2), 88–100.

Atembe, R., & Abdalla, F. (2015). The use of smart technology in tourism: Evidence from wearable devices. In *ISCONTOUR 2015–Tourism Research Perspectives: Proceedings of the International Student Conference in Tourism Research* (p. 23).

Aziz, Y. A., & Wahiddin, K. (2017). Conceptualizing the service excellence and its antecedents: The development of the structural equation model. *Journal of Tourism, Hospitality & Culinary Arts, 9*(2), 47–58.

Bagchi, S., He, J., Chen, C. Y. F., Zhang, M., & Bhati, A. (2021). Creating excellent guest experiences: Servicescape and processes. In *Service Excellence in Tourism and Hospitality: Insights from Asia* (pp. 27–39).

Becerra, M., Santaló, J., & Silva, R. (2013). Being better vs. being different: Differentiation, competition, and pricing strategies in the Spanish hotel industry. *Tourism Management, 34*, 71–79.

Berry, L. L., & Parasuraman, A. (1992). Prescriptions for a service quality resolution in America. *Organizational Dynamics, 20*(4), 5–15.

Bettencourt, L. A., & Gwinner, K. (1996). Customization of the service experience: The role of the frontline employee. *International Journal of Service Industry Management, 7*(2), 3–20.

Boslaugh, S. (2007). *Secondary Data Sources for Public Health: A Practical Guide*. Cambridge University Press.

Carroll, B., & Siguaw, J. A. (2003). The evolution of hotel front office operating models. *International Journal of Hospitality Management, 22*(2), 103–121.

Chan, E. S. W., Hon, A. H. Y., Chan, W., & Okumus, F. (2014). What drives employees' intentions to implement green practices in hotels? The role of knowledge, awareness, concern and ecological behavior. *International Journal of Hospitality Management, 40*(Supplement C), 20–28. https://doi.org/10.1016/j.ijhm.2014.03.001 (accessed on 8th July 2025).

Clark, G. (n.d.). Secondary data. In *Methods in Human Geography* (pp. 57–73).

Cobos, L. M., Mejia, C., Ozturk, A. B., & Wang, Y. (2016). A technology adoption and implementation process in an independent hotel chain. *International Journal of Hospitality Management, 57*, 93–105.

Cook, S. (2008). *Customer Care Excellence: How To Create an Effective Customer Focus* (5th ed.). Kogan Page Limited.

Deb, S. K., & Nafi, S. M. (2021). Assessing critical success factors of ICT based learning: An analysis. *The Cost and Management, 49*(04).

Deb, S. K., Kuri, B. C., & Nafi, S. M. (2024). Application of knowledge management in tourism and hospitality industry: A sustainable approach. In *Knowledge Management and*

Knowledge Sharing: Business Strategies and An Emerging Theoretical Field (pp. 99–116). Springer Nature Switzerland.

Deb, S. K., Nafi, S. M., & Valeri, M. (2022). Promoting tourism business through digital marketing in the new normal era: A sustainable approach. *European Journal of Innovation Management, 27*(3), 775–799. https://doi.org/10.1108/EJIM-04-2022-0218 (accessed on 8th July 2025).

Deb, S. K., Nafi, S. M., Mallik, N., & Valeri, M. (2023). Mediating effect of emotional intelligence on the relationship between employee job satisfaction and firm performance of small business. *European Business Review, 35*(5), 624–651. https://doi.org/10.1108/EBR-12-2022-0249 (accessed on 8th July 2025).

Dickinson, J. E., Ghali, K., Cherrett, T., Speed, C., Davies, N., & Norgate, S. (2014). Tourism and the smartphone app: Capabilities, emerging practice and scope in the travel domain. *Current Issues in Tourism, 17*(1), 84–101.

Elphick, D. (2023). 3 examples of what wearable technology might mean to a small hotel marketer. https://www.littlehotelier.com/blog/running-your-property/3-examples-wearable-technology-might-mean-small-hotel-marketer/ (accessed on 8th July 2025).

Formolo, D. (2024). The benefits and challenges of wearables in hospitality. https://www.viqal.com/blog/the-benefits-and-challenges-of-wearables-in-hospitality#:~:text=In%20hospitality%2C%20wearables%20such%20as,service%20or%20controlling%20room%20temperature. (accessed on 8th July 2025).

Fraj, E., Matute, J., & Melero, I. (2015). Environmental strategies and organizational competitiveness in the hotel industry: The role of learning and innovation as determinants of environmental success. *Tourism Management, 46*, 30–42. http://dx.doi.org/10.1016/j.tourman.2014.05.009 (accessed on 8th July 2025).

Gouthier, M., Giese, A., & Bartl, C. (2012). Service excellence models: A critical discussion and comparison. *Managing Service Quality: An International Journal, 22*(5), 447–464. https://doi.org/10.1108/09604521211281378 (accessed on 8th July 2025).

Horwitz, F. M., & Neville, M. A. (1996). Organization design for service excellence: A review of the literature. *Human Resource Management, 35*(4), 471–492.

Huang, Y. C., Chang, L. L., Yu, C. P., & Chen, J. (2019). Examining an extended technology acceptance model with experience construct on hotel consumers' adoption of mobile applications. *Journal of Hospitality Marketing & Management, 28*(8), 957–980.

Ivanov, S., & Webster, C. (2019). Conceptual framework of the use of robots, artificial intelligence and service automation in travel, tourism, and hospitality companies. In *Robots, Artificial Intelligence, and Service Automation in Travel, Tourism and Hospitality* (pp. 7–37). Emerald Publishing Limited.

Johnston, R. (2004). Towards a better understanding of service excellence. *Managing Service Quality: An International Journal, 14*(2/3), 129–133.

Johnston, R. (2007). Insights into service excellence. In Gouthier, M. H. J., Coenen, C., Schulze, H. S., & Wegmann, C. (Eds.), *Service Excellence Als Impulsgeber* (pp. 17–35). GablerVerlag.

Kamboj, S., & Gupta, S. (2020). Use of smart phone apps in co-creative hotel service innovation: Evidence from India. *Current Issues in Tourism, 23*(3), 323–344.

Kandampully, J., & Suhartanto, D. (2000). Customer loyalty in the hotel industry: The role of customer satisfaction and image. *International Journal of Contemporary Hospitality Management, 12*(6), 346–351.

Kim, J. J., & Han, H. (2020). Hotel of the future: Exploring the attributes of a smart hotel adopting a mixed-methods approach. *Journal of Travel & Tourism Marketing, 37*, 804–822.

Kim, J. J., Radic, A., Chua, B. L., Koo, B., & Han, H. (2022). Digital currency and payment innovation in the hospitality and tourism industry. *International Journal of Hospitality Management, 107*, 103314.

Kim, W. G., Lee, C., & Lee, K. (2015). Hotel guests' preferences for room attributes: Importance–performance analysis and implications for service quality improvement. *Journal of Quality Assurance in Hospitality & Tourism, 16*(3), 222–241.

Kleijnen, M., De Ruyter, K., & Wetzels, M. (2007). An assessment of value creation in mobile service delivery and the moderating role of time consciousness. *Journal of Retailing, 83*, 33–46.

Law, R., Sun, S., & Chan, I. C. C. (2020). Hotel technology: A perspective article. *Tourism Review, 75*(1), 286–289.

Lee, J., Kao, H. A., & Yang, S. (2014). Service innovation and smart analytics for Industry 4.0 and big data environment. *Procedia CIRP, 16*, 3–8.

Leonidou, L. C., Leonidou, C. N., Fotiadis, T. A., & Aykol, B. (2015). Dynamic capabilities driving an eco-based advantage and performance in global hotel chains: The moderating effect of international strategy. *Tourism Management, 50*, 268–280. https://doi.org/10.1016/j.tourman.2015.03.005 (accessed on 8th July 2025).

Leung, R. (2019). Smart hospitality: Taiwan hotel stakeholder perspectives. *Tourism Review, 74*(1), 50–62.

Lovelock, C. H., Keh, H. T., Lu, X., & Wirtz, J. (2005). *Services Marketing in Asia: Managing People, Technology, and Strategy* (2nd ed.). Prentice Hall.

Magalhães, M. J., de Magalhães, S. T., Revett, K., & Jahankhani, H. (2017). A review on privacy issues in hotels: A contribution to the definition of information security policies and marketing strategies. In *International Conference on Global Security, Safety, and Sustainability* (pp. 205–217). Springer International Publishing.

Mo Kwon, J., Bae, J. I., & Blum, S. C. (2013). Mobile applications in the hospitality industry. *Journal of Hospitality and Tourism Technology, 4*(1), 81–92.

Moon, H., Yu, J., Chua, B. L., & Han, H. (2022). Hotel privacy management and guest trust building: A relational signaling perspective. *International Journal of Hospitality Management, 102*, 103171.

Morosan, C., & DeFranco, A. (2016). Modeling guests' intentions to use mobile apps in hotels: The roles of personalization, privacy, and involvement. *International Journal of Contemporary Hospitality Management, 28*(9), 1968–1991.

Moscaritolo, L. M. (2009). Interventional strategies to decrease nursing student anxiety in the clinical learning environment. *Journal of Nursing Education, 48*(1), 17–23.

Nain, A. (n.d.). A study on major challenges faced by hotel industry globally. *International Journal of Creative Research Thoughts, 6*(1), 561–571.

Nelson, H. G., & Stolterman, E. (n.d.). *The Design Way: Intentional Change in an Unpredictable World*. MIT Press.

Oliver, R. L. (1997). *Satisfaction: A Behavioral Perspective on the Consumer*. Irwin/McGraw-Hill.

Ouma, B. R. (2023). *Competitive Strategies and Performance of Star-Rated Hotels in Nairobi City County, Kenya* (Doctoral dissertation, Kenyatta University).

Paraskevas, A., & Arendell, B. (2007). The role of service excellence in tourism development: Towards a new service quality framework. *Tourism and Hospitality Research, 7*(3), 205–222.
Parasuraman, A., Zeithaml, V. A., & Berry, L. L. (1985). A conceptual model of service quality and its implications for future research. *Journal of Marketing, 49*(4), 41–50.
Salama, W., Nor El Deen, M., Albakhit, A., & Zaki, K. (2022). Understanding the connection between sustainable human resource management and the hotel business outcomes: Evidence from the green-certified hotels of Egypt. *Sustainability, 14*, 5647. https://doi.org/10.3390/su14095647 (accessed on 8th July 2025).
Sarmah, B., Kamboj, S., & Rahman, Z. (2017). Co-creation in hotel service innovation using smart phone apps: An empirical study. *International Journal of Contemporary Hospitality Management, 29*(10), 2647–2667.
Stock, R. M., & Merkle, M. (2017). Service robot acceptance model: User acceptance of humanoid robots during service encounters. In *Proceedings of the 2017 IEEE International Conference on Pervasive Computing and Communications Workshops (PerCom Workshops), Kona, HI, USA, 13–17 March* (pp. 339–344).
Tieng, S. (2022). *Service Excellence in Tourism and Hospitality*. Society Publishing.
Tomislav, C. A. R., & Stifanich, L. P. (2020). Mobile technology: A tool to increase hotel sales and revenue. *Economics Business and Organization Research*, 454–465.
Tussyadiah, I. P., & Park, S. (2018). Consumer evaluation of hotel service robots. In *Information and Communication Technologies in Tourism* (pp. 308–320). Springer.
Umar, M. A., Kasim, R., & Martin, D. (2013). A review of service excellence models. In *2nd International Conference on Technology Management, Business and Entrepreneurship.*
Vartanian, T. P. (2010). *Secondary Data Analysis*. Oxford University Press.
Wijetunge, W. A. D. S. (2016). Service quality, competitive advantage and business performance in service providing SMEs in Sri Lanka. *International Journal of Scientific and Research Publications, 6*(7), 720–728.
Withiam, G. (2013). Building service excellence for customer satisfaction. *Cornell Hospitality Proceedings, 5*(12).
Wu, H. C., & Cheng, C. C. (2018). Relationships between technology attachment, experiential relationship quality, experiential risk and experiential sharing intentions in a smart hotel. *Journal of Hospitality and Tourism Management, 37*, 42–58.
Yang, Y., Wong, K. K. F., & Wang, T. (2012). How do hotels choose their location? Evidence from hotels in Beijing. *International Journal of Hospitality Management, 31*(3), 675–685.
Yoong, K. (2022). What is service excellence? The service excellence definition for 2022. Available at: https://hospitalityinsights.ehl.edu/what-is-service-excellence#:~:text=By%20definition%2C%20service%20excellence%20refers,occasionally%20even%20exceed%20customers'%20expectations (accessed on 8th July 2025).
Yunis, M., Jung, J., & Chen, S. (2013). TQM, strategy, and performance: A firm-level analysis. *International Journal of Quality & Reliability Management.*
Zhang, D., Tu, J., Zhou, L., & Yu, Z. (2020). Higher tourism specialization, better hotel industry efficiency. *International Journal of Hospitality Management, 87*, 102509.
Zohry, M. A. E. F., & Alsaied, M. (2021). The impact of bitcoin electronic trust factors on hotel transactions as a mechanism for digital transformation in Egypt. *Journal of Association of Arab Universities for Tourism and Hospitality, 20*(4), 505–532.

CHAPTER 14

VALUE CREATION THROUGH SMART TOURISM SERVICE MANAGEMENT

TANVIR AHMED and SANJOY KUMAR ROY*

Department of Tourism and Hospitality Management, Noakhali Science and Technology University, Noakhali, Bangladesh

*Corresponding author

ABSTRACT

This book chapter delves into the value creation paradigm in the tourism sector, focusing on intelligent tourism service management. It explores the dynamic interaction between stakeholders, technology, and customer engagement in shaping the tourism experience, embracing the values of service, value, and innovation. The chapter explains the basic ideas and elements of smart tourism service management, emphasizing the critical function of service delivery and the integration of "smartness" into tourism exchange. It highlights the complex process of designing, delivering, and maintaining value for various stakeholders within the tourism ecosystem, using insights from big data analysis, technological innovation, and stakeholder engagement. It emphasizes the importance of active communication, collaboration, and feedback mechanisms for successful value creation. The chapter also delves into value co-creation, highlighting channels and measurements that enable significant exchanges and participation between travelers and tourism stakeholders. It emphasizes the importance of using cutting-edge technology and data analytics to boost value creation and encourage stakeholder participation in astute tourism service management. The book chapter advocates for an integrated approach to service management in the travel and tourism sector,

Smart Tourism: Managing Technological Revolution in the Tourism Industry. Santus Kumar Deb (Ed.)

DOI: 10.1201/9781779641090-14

promoting a customer-centric mindset and cutting-edge technologies for long-term tourism sustainability.

14.1 INTRODUCTION

In the competitive landscape, many companies aim to surpass their competitors by consistently providing exceptional value propositions. It is important to understand that the idea of value is relative and depends on the situation. It's important to consider that the effectiveness of a solution can vary depending on the context (Cronin Jr et al., 2000). What may work well in one situation may not necessarily be as valuable in another (Albrecht et al., 2005). For example, in a business-to-consumer (B2C) context, McDonald's can be a great option for hosting an 8-year-old's birthday party. McDonald's restaurants are a popular choice for various family occasions due to their affordability, convenience, and welcoming atmosphere for families with children. Nevertheless, the McDonald's option may not be the most suitable choice when considering where to celebrate your girlfriend's 27th birthday. Considering the circumstances, elements like the overall atmosphere, high-quality food, and a romantic setting are likely to be more significant, resulting in a wider range of options beyond just fast-food places.

14.2 UNDERSTANDING VALUE

The term "value" has been employed throughout various contexts. Value can be defined from various perspectives, including those of customers, employees, stakeholders, or shareholders. By established norms, we define value using economic criteria (Baker et al., 2002). When a company efficiently produces quality goods that rivals cannot match or find too expensive to copy, it thereby creates superior value (Corvellec and Hultman, 2014). It is the discrepancy between the cost of producing a product and the price at which customers are inclined to purchase it (Mukerjee, 1964).

Value and benefits are occasionally applied interchangeably, and several concepts appear to be overlapping and ambiguous, including worth (Zwikael and Smyrk, 2012), success (Angus et al., 2005), value creation (Laursen and Svejvig, 2016), benefits management (Ward and Daniel, 2006), and benefits realization management (Breese, 2012; Love et al., 2014).

From the perspective of business, it can be seen as the customer's perception of the balance between the advantages gained from a product or service and the efforts made to obtain those advantages. This perception can be encapsulated in the form of an equation:

$$\text{Value} = \text{Benefits Received} / \text{Sacrifices Made}$$

Customers encounter various forms of sacrifices during their consumption experiences, including financial sacrifice, search costs, and psychic costs. **Monetary sacrifice** refers to the additional cost of goods or services beyond the listed price, including supplementary expenses such as surcharges, interest rates, and warranty (Michael Minkov et al., 2019). **Search costs** involve thorough pre-purchase research and comparison, which can be time-consuming and have financial consequences, especially in business-to-business settings (Susan et al., 2004). For example, employees researching and evaluating suitable venues during a corporate retreat incur tangible costs. **Psychic costs** involve anxiety and annoyance associated with the purchasing process, such as brainstorming gift ideas, navigating crowded areas, interacting with uninformed personnel, and enduring unfavorable weather conditions (Weitzl et al., 2016). Perceived risks, such as difficulties with performance, finances, social interactions, and psychological health, also impact psychic costs. For example, a visitor feeling anxious about safety standards and potential physical dangers may cancel the purchase or choose a less risky alternative.

Comprehending and dealing with these aspects of sacrifice in the tourist sector are essential for providers to customize their services, simplify procedures, and efficiently handle consumer expectations, thereby improving the perceived worth in the eyes of the clientele (Hsiao and Chen, 2012). Customers often feel uncomfortable with higher levels of risk, leading them to seek ways to reduce it. Suppliers can help by implementing solutions such as performance guarantees and transparent pricing structures (Vilar-Zanon et al., 2017).

14.3 HISTORY OF VALUE-BASED MANAGEMENT

Value-based management in smart tourism service management has been shaped by important academic papers that have improved our grasp of value creation in the tourism sector. Although the relevant literature on value-based management can be traced back to the early 1980s as benefits or worth management, the focus on value in tourism experiences started to attract academic interest in the late 20th century, highlighted by Crompton's influential publication in "Tourism Experience: A New Research Paradigm" (Otto and Ritchie, 1996) (Figure 14.1).

The evolution of value-based management in smart tourism service management has been shaped by a diverse array of academic contributions spanning several decades. From the early conceptualizations of value in tourism experiences to the integration of digital technologies and sustainability principles, these seminal papers have collectively advanced our understanding of how value is created, delivered, and perceived within the tourism industry. By building upon the theoretical foundations laid by these pioneering works, future research endeavors can continue to push the boundaries of value-based management in the context of smart tourism service management.

14.4 THE CORE PRINCIPLES OF VALUE CREATION

The Service-Dominant Logic (SDL), a concept developed by Grönroos (2008) and Vargo and Lusch (2004), suggests that value is not inherent in a product or service but is co-created through its utilization by various stakeholders (Vargo and Lusch, 2017). Consumers develop an understanding of value throughout their entire interaction, rather than immediately

Author and Year	Title	Key Insights
JL Crompton (1979)	Motivations for pleasure vacation	This this paper explored the intricate aspects of tourism experiences and paved the way for future dialogues on the importance of perceived value in shaping tourist behaviors and preferences.
Parasuraman, Zeithaml, and Berry's (1988)	SERVQUAL: A Multiple-Item Scale for Measuring Consumer Perceptions of Service Quality	This study presented the SERVQUAL model, highlighting the important connection between service quality, customer satisfaction, and perceived value. This paper emphasized the significance of meeting or surpassing customer expectations in service delivery, offering a theoretical framework to comprehend how tourists perceive and assess value.
Pine and Gilmore (1998)	Welcome to the experience economy	Our understanding of how value is generated in the tourism sector changed dramatically with this paper. Value comes from more than just consuming goods and services. It also stems from the memorable and transformative experiences provided by businesses. It is crucial for tourism service providers to concentrate on creating immersive and engaging experiences that deeply connect with tourists.

Author and Year	Title	Key Insights
Vargo and Lusch (2004)	Evolving to a New Dominant Logic for Marketing	A fundamental shift from the conventional goods-dominant logic of marketing was made with the introduction of the Service-Dominant Logic (SDL) by this groundbreaking work. SDL supplied a theoretical framework for comprehending value as a cooperative and dynamic process by highlighting the co-creation of value through interactions between service providers and customers. This paradigm change reinforced the significance of involving tourists as active contributors to value creation, which had far-reaching effects on the tourism sector.
Prahalad and Ramaswamy (2004)	Co-creation experiences: The next practice in value creation	This paper emphasized the significance of immersive and personalized experiences in promoting visitor satisfaction and loyalty by highlighting the role of tourists as co-creators of value with the extended use of SDL principles to tourism-related services.. This change in viewpoint paved the way for the tourism sector to adopt a more experiential and participatory approach to value creation.
Boes, K., Buhalis, D., & Inversini, A. (2016)	Smart tourism destinations: ecosystems for tourism destination competitiveness	This paper presented the idea of how "smart tourism destinations" can provide highly customized experiences that increase visitor satisfaction and loyalty by utilizing data analytics, mobile applications, and social media platforms.
Zhang, Li, & Chen (2019)	Enhancing Tourist Value Co-Creation Through Augmented Reality Experiences: A Framework for Smart Tourism Destinations	The paper introduces a new framework for enhancing tourist value co-creation in smart tourism destinations through augmented reality (AR) experiences. It highlights the role of AR in bridging the physical and digital worlds, enhancing visitor engagement and collaboration. The paper uses case studies and empirical analyses to demonstrate how AR can create personalized experiences, driving satisfaction and loyalty in smart tourism destinations.
Vikas Gupta, Manohar Sajnani, Saurabh Kumar Dixit (2020)	Impact of augmented and virtual reality on the value co-creation of visitor's tourism experience	paper explores the impact of augmented and virtual reality (AR) and VR technologies on tourism value co-creation. They argue that these technologies enhance visitor engagement, satisfaction, and loyalty, redefining the boundaries of value creation and collaboration in the digital age.
Dimitrios Buhalis, Peter O' Connor, Rosanna Leung (2023)	Smart hospitality: from smart cities and smart tourism towards agile business ecosystems in networked destinations	This paper introduced the concept of smart tourism destinations, where technology is leveraged to create seamless and personalized experiences for visitors. By harnessing the power of data analytics, mobile applications, and social media platforms, smart tourism destinations can deliver highly tailored experiences that enhance tourist satisfaction and loyalty.

FIGURE 14.1 Evolution of value concept in tourism management.

Source: Authors.

recognizing its precise worth (Warhurst et al., 2005). Morris Holbrook, a pioneer in value establishment research, is credited with formulating the fundamental elements of consumer behavior and expenditure (Mieli, 2017) (Figure 14.2).

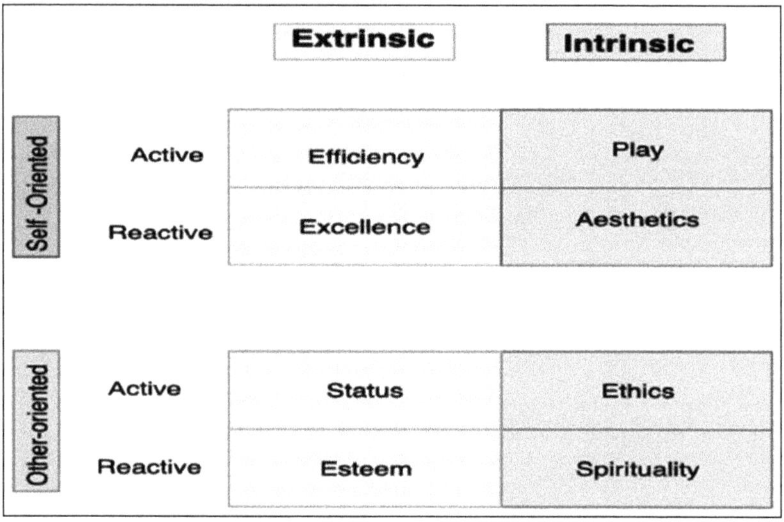

FIGURE 14.2 Typology of consumer value.

Value can be categorized into two main classifications: intrinsic and external. External value is primarily derived from the practical utility of a commodity or service, while intrinsic value is associated with feelings of satisfaction and personal experience stemming from consumption (Sánchez-Fernández et al., 2009). Holbrook's concept of value can be further categorized based on self-focus and focus on others (Heinonen, 2006). Self-centered value is established through personal experiences and revolves around how consumption directly impacts the individual encounter such as (Holbrook, 1999):

1. **Efficiency (Convenience):** Efficiency is a self-oriented, active value often measured by output to input ratio. It is often linked to convenience, such as toothpaste designed for oral hygiene and keys' ability to easily access for unlocking. However, consumers often choose products with additional psychological time-related value (Willskytt, 2019), like a high-quality fridge for longer preservation.

2. **Excellence (Quality):** It refers to an individual's acknowledgement of an object's characteristics, practicality, or functionality. When a product, service, or device is able to fulfil its intended objective while preserving the expected performance, it is generally considered to have significant worth (Parasuraman and Grewal, 2000). For example, a genuine leather shoe is usually favored over an artificial one due to its durability, usability, and coziness, which are the characteristics that an individual would instinctively value and link with superiority, thereby resulting in exceptional worth.
3. **Status (Success, Image, Impression):** This type of value is extrinsic, active, and other-oriented. In some cases, consumers select specific goods or activities in order to present a particular impression of themselves to others, displaying their accomplishments and social standing. This is commonly observed in the acquisition of fashion items created by renowned designers and high-end accessories (Michael Minkov et al., 2019).
4. **Esteem (Reputation, Possessions):** The worth of esteem lies in its external nature, its capacity for response, and its emphasis on other individuals. The esteem value, however, is not connected to a superficial or shallow representation of success like "status" (Shohel Md and Tanvir, 2019). Rather, it takes into account the high value that society places on owning or consuming goods. It combines a person's sense of value and materialism to create a positive perception of oneself (Chen and Barnes, 2007). As an instance, having a distinct piece of art or antique item evokes a deeper sense of satisfaction and elevates an individual's status within a select community.
5. **Play (Fun):** Play shows intrinsic, self-oriented, and active characteristics. This word is linked to the idea of wanting to have fun or feel personally satisfied when using a product, service, or interaction. Furthermore, this type of value is also known as situational because it changes depending on the person and the circumstances of the task (Nafi and Ahmed, 2018). For example, most people think of skydiving as a fun activity, but soldiers who use parachute strategies to attack enemies see it as a duty to their country.
6. **Aesthetics (Beauty):** Aesthetic value is self-centered, reactive, and intrinsic, influenced by beauty in products or experiences. People appreciate fine arts like music, painting, dance, and

poetry (Eckman and Wagner, 1994). However, when valued for its intended use, value can shift from aesthetic to practical, such as a sophisticated supercar's efficiency value for traveling long distance in a short time (Cuesta-Valiño et al., 2023).

7. **Ethics (Virtue):** Ethics is a changing, inherent moral principle that focuses on others. It involves following moral and social norms and considering others' well-being. This value classification can be described by virtue, fairness, and morality (Barry et al., 1994). Ethics emphasizes moral behavior regardless of personal gain. This could include volunteering or donating blood to save lives.
8. **Spirituality (Faith and Sacredness):** This value is produced by an individual's religious beliefs or dedication, allowing them to experience a sense of connection with a divine or higher power (Sánchez-Fernández et al., 2009). Fundamentally, the valuation of spiritual worth is identified by an individual's desire to form a link with a superior entity, thereby rendering it an adaptable and outwardly directed value.

14.5 TOOLS OF SMART TOURISM VALUE CREATION

The value chain, a framework by Michael Porter, explains how companies create value through primary and secondary activities (Simatupang et al., 2017). Key components include inbound logistics, operations, marketing, sales, and service, with support activities like procurement, technology development, human resource management, and firm infrastructure (Porter, 2001). Companies must effectively manage the value chain, improving communication and collaboration between activities (Kaplinsky and Morris, 2000). This helps them establish and maintain a competitive advantage by understanding the potential and performance of resources and capabilities. Value-chain analysis helps managers evaluate factors contributing to value creation, enabling them to develop strategies and address challenges in designing, manufacturing, transporting, selling, and servicing products (Normann and Ramirez, 1993) (Figure 14.3).

Value chain analysis is a valuable tool for smart tourism service management, helping managers understand the internal cost structure of tourist companies (Kaplinsky, 2004). It assesses operations at every level, from pre-trip planning to customer service, allowing for improvement. Hotel chains can use this to evaluate their operations, including booking management, room preparation, guest services, and post-stay feedback, enhancing customer satisfaction and loyalty (Normann and Ramirez, 1993) (Figure 14.4).

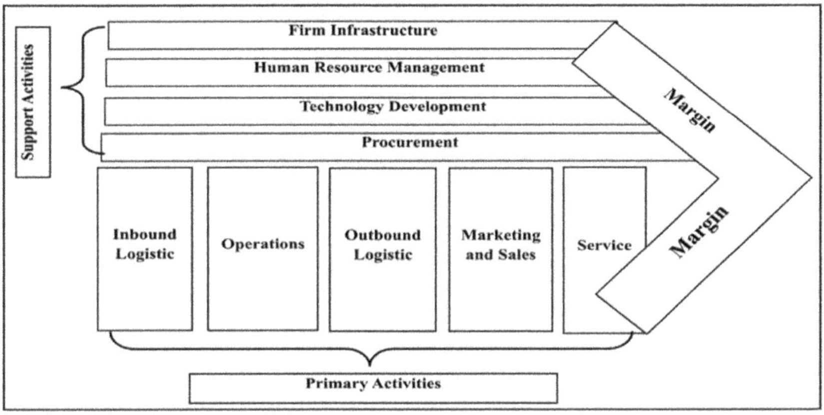

FIGURE 14.3 Michael Porter's value chain model.

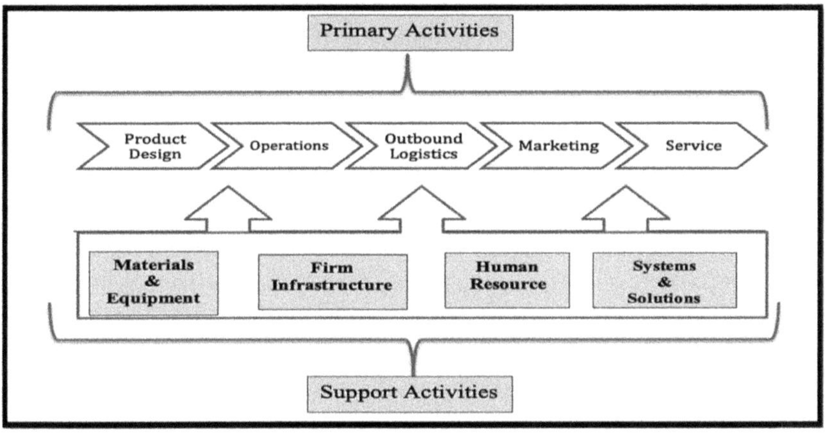

FIGURE 14.4 Visualizing the value chain.

Smart tourism's value chain analysis encompasses the entire tourism ecosystem, including transportation, attractions, lodging, and visitor services (Lien et al., 2015). DMOs can assess a destination's value chain, enabling stakeholders to collaborate and innovate for enhanced competitiveness and visitor experience (Flint et al., 2002).

14.5.1 SOURCES OF CUSTOMER VALUE

Customers form a strong connection with a company when their personal beliefs align with the organization's values, which are influenced by cultural norms (Cronin Jr et al., 2000). These core beliefs, such as environmental consciousness, honesty, and family-centeredness, cultivate loyalty and support (Umair et al., 2018). Michael Treacey and Fred Wiersema identified three fundamental types of value propositions: product leadership, customer intimacy, and operational excellence (Treacy and Wiersema, 1993). These strategies can be applied to the tourism industry, particularly in smart tourism service management.

Operational excellence is demonstrated by certain companies in the tourism industry. Budget hotel chains like Motel 6 and ibis, as well as tour operators like Contiki Tours, are great examples of this. They have successfully managed to provide standardized accommodations and affordable package deals without compromising on service quality (Oakland, 2014). Brands like Disney Parks and Resorts and luxury hotel chains such as Four Seasons and Ritz-Carlton are prime examples of **Product leadership** in the smart tourism industry. These brands are constantly pushing the boundaries of innovation in attractions, entertainment experiences, and service (Cooper, 1998). They are renowned for their ability to deliver exceptional and personalized experiences that go above and beyond customer expectations. **Customer intimacy** is exemplified by boutique hotels such as Ace Hotel and Kimpton Hotels. These establishments go above and beyond to create personalized and one-of-a-kind experiences that cater to the specific preferences of each guest. Similarly, online travel agencies like Airbnb utilize customer data to provide personalized accommodation options. By understanding and embracing these value disciplines, tourism businesses can strategically position themselves to cater to the diverse needs and preferences of travelers. This approach fosters long-term competitiveness and value creation within the industry (Treacy and Wiersema, 1993).

14.5.2 CREATING VALUE THROUGH SERVICE MANAGEMENT

Tourism, a multidimensional industry, requires unique management approaches compared to traditional business models (Buonincontri and Micera, 2016). Its service-oriented enterprises, including transportation and accommodation, require unique approaches (Buhalis and Amaranggana, 2013). Advancements in technology and market dependence have significantly changed service provision, focusing on guest satisfaction and enhancing the overall experience for tourists.

Luxury hospitality companies face competition from other companies due to the value they provide, not just physical facilities (Tankovic and Benazic, 2018). Customers derive value from products through innovation, added benefits, bundling, strong branding, and synergies (Edvardsson et al., 2011). In the luxury hotel industry, customers seek solutions to their travel needs and desires, such as personalized concierge services, exclusive access to amenities, and unique opportunities like private excursions or cultural immersion programs (Manuel et al., 2017). To maximize customer value, companies must find the right balance between benefits and sacrifices. Products or services are classified into various levels within a customer value hierarchy (Sanjoy Kumar and Tanvir, 2023), such as the core product, enabling product, and augmented product (Shukla and Mishra, 2023). The core product includes extravagant accommodations, exquisite dining choices, and flawless service, while the enabling product includes all necessary physical facilities and services (Huang et al., 2017). The augmented product includes extra elements that elevate the overall customer experience.

In the tourism industry, service management pertains to the methodical and planned synchronization of tasks, procedures, and assets with the aim of providing tourists with excellent and fulfilling services (Testa and Sipe, 2012). Planning, arranging, and maximizing many aspects of the tourism experience is necessary to either meet or surpass travelers' expectations. Tourism service management includes all services offered to visitors, including transportation, attractions, entertainment, and customer support, and goes beyond the conventional facets of hospitality and lodging. The hospitality industry is a service-oriented industry that relies on the delivery of quality service to customers (Mok et al., 2013). Service management is the process that ensures that service quality is maintained and improved.

Especially in the hotel sector, service management plays a crucial role in maintaining and enhancing the quality of services provided. The hotel sector would not be able to offer the kind of excellent, timely, and efficient service that consumers have come to expect without efficient service management (Hong and Zailani, 2011).

There are several sectors where we need to provide emphasis in ensuring better provision of services. Figure 14.5 will give an overview of the supply chain of the tourism industry.

In each of these stages from Figure 14.5, there are various opportunities for the service suppliers to enhance their offerings in a way so that it becomes one of their guests' attracting tools for potential customers. Nonetheless, the components of service management in the tourism industry need to be identified to understand the points where improvement can be brought with the assistance of smart technology.

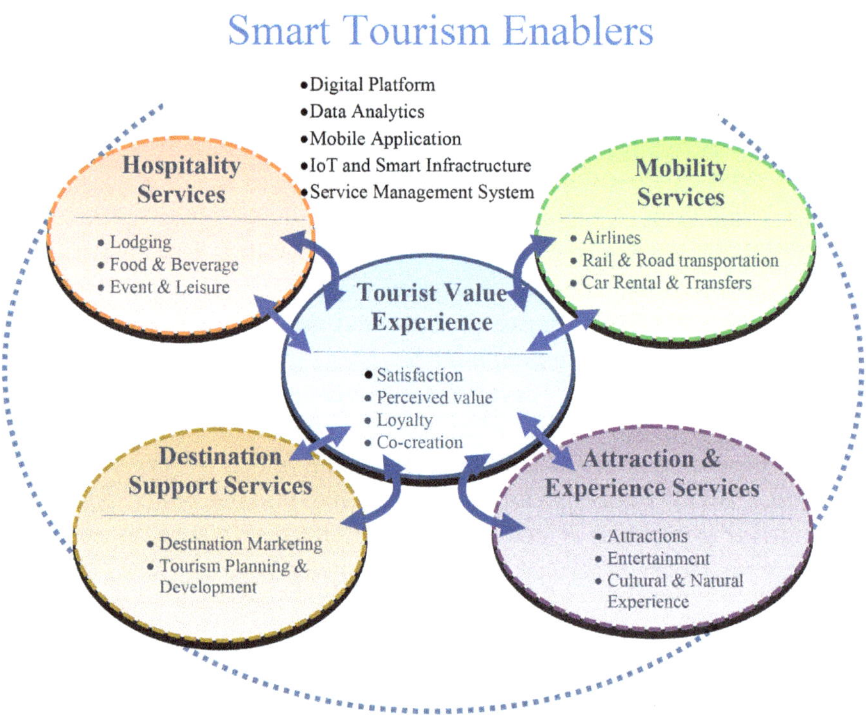

FIGURE 14.5 Conceptual overview of tourism service interfaces in value creation

Source: Author's Own work

In order to ensure client pleasure, loyalty, and overall success in a variety of industries, including tourism, value creation through service management is essential. If a service can be immediately linked to financial outcomes for company outcomes, then quantifying its financial value would be easy. When the business results are not directly correlated with a monetary worth, it is more difficult to assess a service's value (Arslan and Kaymaz, 2020).

A customer's assessment of a service's value is influenced by a number of additional elements in addition to its capacity to assist in achieving company goals. These are the inclinations and viewpoints of the client (Figure 14.6).

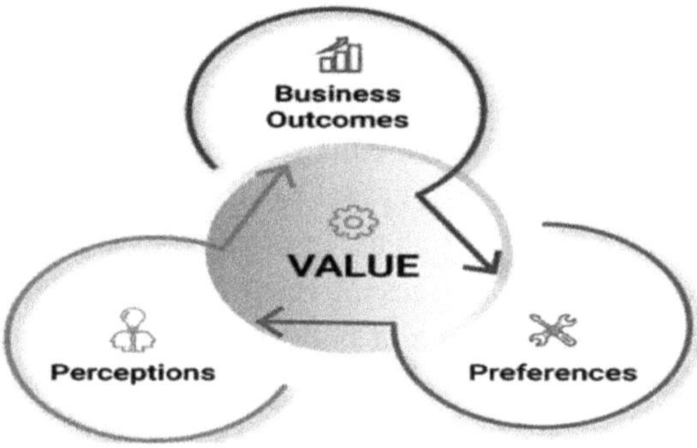

FIGURE 14.6 Relationship among business outcomes, perceptions, preferences, and the value of the services.

Source: Adapted from Gillingham (2023).

Here's how efficient service management generates value:

1. **Customer-First Strategy:** Comprehending Client Needs: Comprehensive market research and an awareness of client expectations are key components of service management. Value is created when services are tailored to the desires of the consumer. This increases satisfaction. For example, an upscale resort conducts market research to understand the preferences of its target audience. By offering personalized vacation packages, spa treatments, and dining options, the resort caters to individual customer needs, creating a unique and satisfying experience.
2. **High-Quality Provision of Services:** Consistency and Reliability: Organizations can guarantee the consistent and dependable provision of services by implementing an efficient service management system. Reliable services increase consumer happiness and perceived value, and consistency fosters confidence. A multinational hotel chain keeps uniform service standards at all of its locations. Travelers visiting hotels

in Tokyo or New York may anticipate the same caliber of service. This consistency raises the brand's perceived value and fosters trust.

3. **Individualization and Customization:** Customizing and personalizing services according to the interests of each individual consumer is made possible by service management. Customers will feel that their particular needs are being met, which will increase the perceived value. Customers can alter their trip plans with a tour operator in accordance with their own interests and preferences. Every tourist has a more fulfilling and joyful experience thanks to this customized approach.
4. **Effective Functions:** Process Optimization: In order to increase productivity, service management entails streamlining internal procedures. Organizations can create value by offering better services at lower costs overall through the implementation of streamlined operations. An airline streamlines the boarding and check-in procedures to cut down on wait times and boost overall effectiveness. This raises client satisfaction levels and enables the airline to provide cheap pricing without sacrificing the caliber of its services.
5. **Proactively Solving Issues:** Effective Issue Handling: Proactive issue identification and resolution techniques are part of service management. Rapid and effective problem solving improves the client experience and highlights the importance of the service provider's dedication to client happiness. A car rental company proactively addresses potential issues by implementing a real-time tracking system for its vehicles. This system allows the company to identify and resolve issues such as maintenance needs before they impact the customer's experience.
6. **Integration of Technology:** Enhanced Convenience: Including digital communication, smartphone apps, and online reservation platforms in service delivery can improve clients' overall convenience. The incorporation of technology enhances value by improving accessibility and user-friendliness of services. A travel agency employs a user-friendly mobile app that allows customers to book flights, accommodations, and activities seamlessly. The app's integration with navigation tools enhances convenience for travelers, adding value to the services offered.
7. **Ethical and Responsible Behavior:** Socially conscious clients are drawn to service management that integrates sustainable

and ethical behavior. By promoting social and environmental well-being, this adds value and is consistent with the values of contemporary customers. Eco-friendly programs, including trash reduction, wildlife conservation, and ethical tourism, are implemented in national parks. The park's dedication to sustainability is valued by visitors, adding to the experience's total worth.

8. **Optimal Brand Image:** Reputation management: A positive brand image is a result of effective service management. Because clients are more likely to select service providers with a solid track record, a great reputation enhances the value of the services provided. A cruise line that demonstrates a strong dedication to environmental responsibility, safety, and customer pleasure enhances its reputation. This reputation keeps current clients loyal and draws in new ones.
9. **Constant Enhancement:** Adaptation to Changing Needs: Service management entails ongoing observation and adjustment in response to shifting consumer demands and industry developments. For customers, this adaptability and dedication to development result in continuous value. The tourism board of a city aggressively solicits visitor input via social media and questionnaires. The board may continuously enhance the overall tourism experience by refining its infrastructure, services, and advertising methods through analysis of this input.
10. **Utilization of Customer Feedback:** Learning from Customer Input: By using Service Management to collect and evaluate customer feedback, businesses may pinpoint areas that need improvement. Making use of this feedback shows that you are dedicated to satisfying client needs and adds value by making timely service improvements. A chain of restaurants frequently monitors online evaluations and comments left by patrons. The chain adds new menu items, improves service procedures, and makes sure that patrons have a better time dining by implementing these suggestions.

The Grönroos-Voima value model separates the process into three value spheres: a joint sphere, a customer sphere closed to the provider, and a provider sphere closed to the customer (Grönroos and Voima, 2013). Figure 14.7 shows an illustration of the model. The roles and objectives of the provider and the client vary within each of the value spheres. It is important to note that the roles of the characters might shift, for instance, when the client provides feedback and other useful information on how the business's procedures should be improved (Figure 14.7).

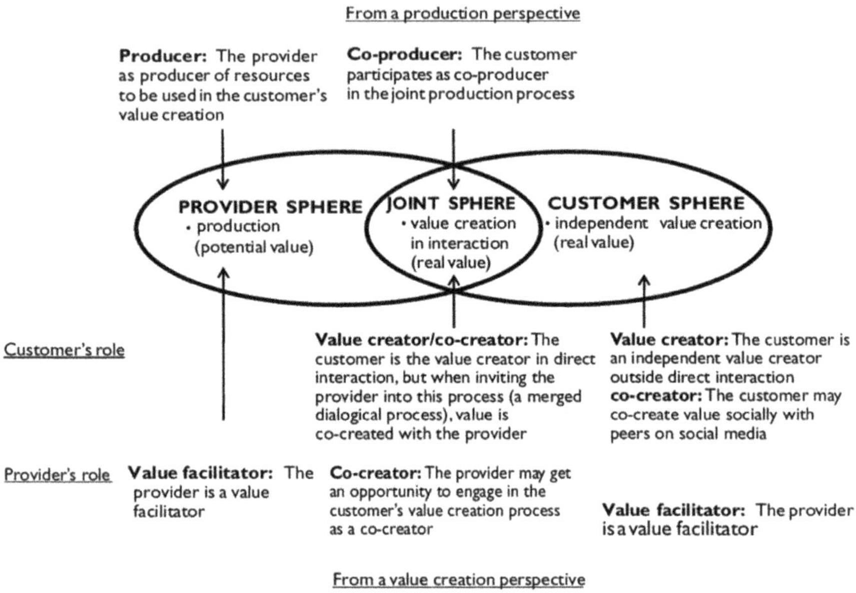

FIGURE 14.7 The Grönroos-Voima value model.

Source: Reprinted with permission from Grönroos and Voima (2013). Copyright © 2012, Academy of Marketing Science

To summarize, Service Management generates value by matching services to customer requirements, guaranteeing dependability and quality, streamlining processes, making efficient use of technology, and upholding a customer-focused mindset. Establishing a competitive edge and fostering enduring relationships with clients are possible for firms that continuously provide extraordinary experiences.

14.5.3 ENHANCING VALUE THROUGH SMART TOURISM SERVICES

Technology breakthroughs have altered the way locations interact with visitors and add value through smart tourism. The capabilities, which range from immersive technologies to data analytics, create an ecosystem that is integrated and improves the traveler experience in general. The value generation tools for smart tourism are an advanced integration of data and technology. A dynamic ecosystem is supported by the interplay of mobile applications, cloud computing, IoT devices, data analytics, and immersive technologies. The potential for developing unique, creative, and engaging visitor experiences is becoming more and more evident as locations continue to use these tools, which will influence how the tourism industry develops in the future.

14.5.3.1 ECOSYSTEM OF SMART TRAVEL

The ecosystem of smart travel encompasses a dynamic synergy of interconnected technologies such as IoT devices, data analytics, cloud computing, and mobile applications. These elements work in harmony to enhance tourist experiences, providing real-time data insights and personalized services. From smart sensors collecting data (Gössling, 2021) to advanced analytics shaping decision-making (Xiang and Fesenmaier, 2017), this ecosystem revolutionizes the travel industry. As cloud computing facilitates seamless data accessibility (Gretzel et al., 2015), mobile applications and immersive technologies like AR and VR enrich engagements (Egger and Neuburger, 2020; Flavián et al., 2021; Nhan et al., 2022; Shin and Jeong, 2021). The smart travel ecosystem exemplifies a technological convergence fostering a new era in travel and tourism.

The ecosystem of smart tourism stands for an all-encompassing strategy to use technology to the advantage of all parties involved. Incorporating diverse tools and technologies not only improves the effectiveness of tourism services but also makes travel for tourists more sustainable, entertaining, and individualized. The idea behind a smart tourism ecosystem is the smooth integration of different technologies to improve the whole vacation experience. It goes beyond discrete elements, promoting a network that is integrated and meets the demands of travelers, companies, and destination management. A smart tourism ecosystem is developed with the purpose of ensuring tailored, context-aware, and real-time monitoring of the sustainable tourism experience. The relationships between micro and macro level stakeholders for creating tourism experiences through shared information and services, technology, human organizations, and resource exchange based on pre-, delivery-, and post-delivery experiences have been referred to as a smart ecosystem. One study had noted that the goal of smart tourism systems is to function as dynamic networks (Boes et al., 2016). A smart ecosystem's foundation must be built on smart technologies' capacity to create value (Bhuiyan et al., 2022).

Travelers can readily and actively contribute data via their activities, engagement, queries, and content with the aid of personal technology, and this highlights the most important components in addition to the government, media, and locals in terms of developing a smart ecosystem. It is defined that the smart tourism ecosystem is a platform for developing, controlling, and providing tourism services through technological innovation, resulting in the exchange of knowledge and the generation of value (Gretzel et al., 2015). The main components of an ecosystem are service providers and recipients, platforms, technologies, support services, NGOs, and businesses from different industries. Based on tangibility/intangibility (tools, software, infrastructure), human resources (skill, knowledge, and commitment), and relational resources (network among the resources), the resource is also categorized among actors (Del Chiappa and Baggio, 2015). From Figure 14.8, the key elements such as IoT devices, mobile applications, data analytics, cloud computing, and user interfaces are illustrated in the next section in detail.

The field of data analytics is at the forefront of changing the tourist industry. The industry is moving toward a more data-driven and visitor-centric future thanks in large part to the insights gained from data analytics. These insights range from streamlining corporate operations to offering tailored experiences and supporting sustainable practices. Data analytics will play an even more significant role in the tourism industry as technology develops, influencing how businesses and destinations respond to the demands and expectations of contemporary tourists. Figure 14.9 represents several areas where the uses of data analytics are on the rise, especially for the hospitality sector. The next section will provide specific practical implications of data analytics in the tourism and hospitality industry to enhance customer satisfaction.

14.5.3.2 REVENUE MANAGEMENT (RM)

RM is the process of utilizing analytics and customer performance data to assist individuals in the hotel sector in forecasting the behavior of their clientele. Analytics encompasses machine learning methods including

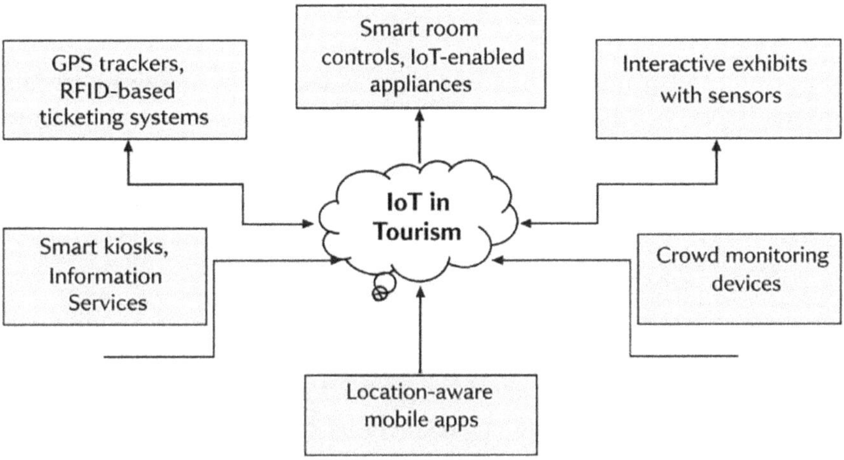

FIGURE 14.8 Dimensions of Internet of Things (IoT) in tourism services.

Source: Adapted by Authors (2024).

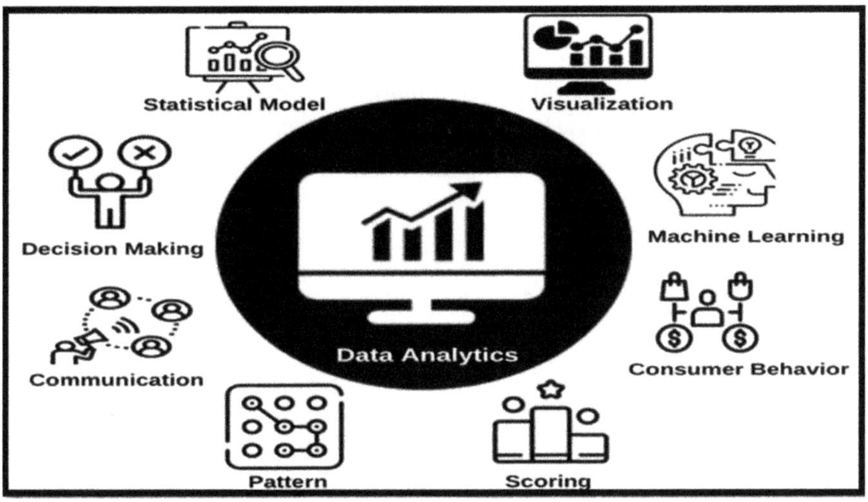

FIGURE 14.9 Data analytics in the hospitality industry.

Source: Latentview (2024). https://www.latentview.com/blog/how-data-analytics-is-transforming-the-hospitality-industry/

classifier models, clustering algorithms for customer segmentation, and demand forecasting. The distribution and price of the product can be optimized with the use of this data.

14.5.3.3 DEMAND FORECASTING

Forecasting and examining trends in traveler demand by utilizing past data, seasonality, events, and outside variables. Makes it possible for companies to prepare for peak seasons, manage resources efficiently, and optimize pricing tactics.

14.5.3.4 PERSONALIZATION OF SERVICES

Services like making recommendations, trip plans, and targeted marketing based on client data to provide individualized experiences. This increases client happiness, loyalty, and engagement by offering services that are in line with personal preferences.

14.5.3.5 CUSTOMER SEGMENTATION

Dividing up travelers into groups according to their interests, habits, and demographics to use them in targeted marketing campaigns is a significant addition of data analytics. It also helps companies to maximize relevance and effect by customizing marketing messages, promotions, and services for consumer categories.

14.5.3.6 PREDICTIVE MAINTENANCE

It is common nowadays to use data analytics to anticipate and stop maintenance problems and equipment failures in hotels, attractions, and transportation. It prolongs the life of assets, decreases downtime, and increases safety by proactively attending to maintenance needs.

14.5.3.7 SMART DESTINATIONS AND VISITOR MANAGEMENT

Data is used in putting into practice smart destination projects to track visitor density, evaluate visitor flows, and enhance the traveler experience. This increases guest expectations by enhancing safety, optimizing traffic management, and enabling better event and attraction planning.

14.5.3.8 REVIEWS AND SENTIMENT ANALYSIS

Examining comments, attitudes on social media, and internet reviews to determine what customers are satisfied with and what needs to be improved is supported by data analytics. It gives information about the advantages and disadvantages of tourism services, assisting companies in improving their reputation and overall quality.

14.5.3.9 SUSTAINABILITY AND ENVIRONMENTAL IMPACT

Employing data analytics to track and control how tourism affects the environment, including trash production, energy use, and carbon footprint encourages the use of eco-friendly initiatives and the identification of areas for improvement to support sustainable tourism practices.

14.5.3.10 FRAUD DETECTION AND SECURITY

Using analytics to find trends and abnormalities suggestive of fraud, strengthening security protocols in areas such as reservations and payments ensures a safe and reliable tourism environment by shielding travelers and companies from dishonest practices.

14.5.3.11 DYNAMIC PRICING STRATEGIES

The use of data analytics to instantly modify prices according to variables including supply, demand, and market dynamics which confirms competitiveness in a changing market, optimizes occupancy rates, and maximizes revenue.

To explore the uses of cloud computing in the travel industry, Lemax is one of the recognized brands in dealing with services of tour operators. Since its founding in 1990 in Croatia, Lemax has been successfully assisting clients in creating and implementing digital business solutions. The area of expertise of this entity has been offering software solutions for the tourism sector and giving their clients the chance to enhance their communication and customer loyalty programs for more than 20 years (Figure 14.10).

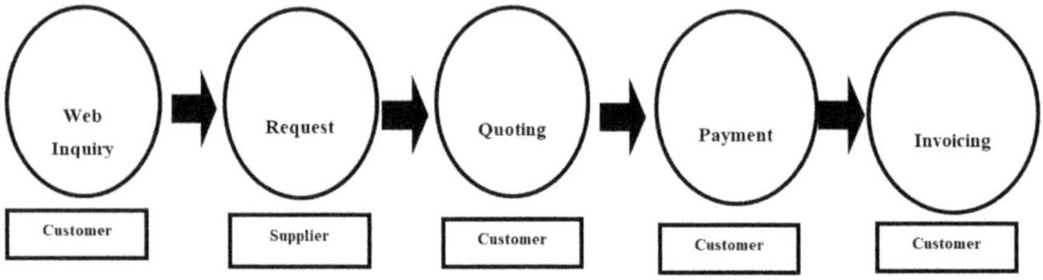

FIGURE 14.10 Automated service ordering with cloud support.

Source: Adapted from Lemax (2024).

New developments, such as edge computing, move processing power closer to the data source, improving productivity and cutting down on latency in travel industry processes. The tourism industry could benefit from increased security and transparency in areas like identity verification and payment processing through the combination of blockchain technology with cloud computing.

14.6 MOBILE APPLICATIONS

Travelers can use mobile-friendly smart tourism applications as individualized tour guides. These apps include functions including real-time communication, virtual tours, reserving accommodations, and navigation. According to Wang et al. (2016), the incorporation of mobile applications improves visitor engagement and streamlines the trip experience. Mobile applications have revolutionized the way tourists plan, experience, and share their travels in the digital era, making them essential to the tourism sector (Nhan et al., 2022). This chapter examines the various uses of mobile applications in tourism, including destination marketing, tailored experiences, navigation, and trip planning. There are numerous benefits of mobile applications in the travel industry where it enhances the experiences of tourists as well as services suppliers. The followings are some of the instances of such:

1. **Travel Arrangements and Scheduling:** By enabling users to arrange and oversee their trip schedules on a single, centralized platform, mobile applications make itinerary planning easier. The ability to add flights, lodging, and activities allows users to

construct an itinerary conveniently. Mobile apps help to streamline the booking procedure. By booking travel-related activities, hotels, flights, and other services directly through the app, users may streamline the entire trip planning process.

2. **Location-Based Services and Navigation:** Interactive maps are available for users to better grasp their environment. Talking about the ways that augmented reality (AR) features improve navigation. By pointing their device at the real world, users can get information about sites of interest, making for an engaging and educational experience.
3. **Tailored Experiences:** Using user information and preferences, mobile apps make tailored suggestions for activities, eateries, and attractions. Through consideration of unique interests, this customization improves the user experience. The built-in cultural awareness and language translation tools in travel apps assist in supplying cultural context and bridging language difficulties; these features help visitors from abroad.
4. **Instantaneous Communication and Evaluations:** Real-time contact between visitors and service providers can be facilitated using mobile apps that include messaging functionality. This improves customer service and expedites inquiries. Decision-making is influenced by reviews and ratings since they offer information on the caliber of hotels, activities, and services.
5. **Tourism Promotion and Cross-Cultural Understanding:** Travel agencies use mobile applications to highlight the distinctive experiences, attractions, and cultures of a place. This draws in potential tourists and enhances destination promotion. Applications that help bridge cultural gaps by bringing visitors and locals together. This involvement promotes more in-depth knowledge of regional cultures while supporting eco-friendly and community-based tourism projects. Figure 14.11 presented below shows the different uses of mobile applications in travel arrangements.

As technology advances, the ongoing evolution of mobile applications in tourism promises to create a more connected, informed, and enjoyable travel environment. Since they offer a comprehensive toolkit for seamless and enriched travel experiences, mobile applications have become indispensable companions for modern travelers.

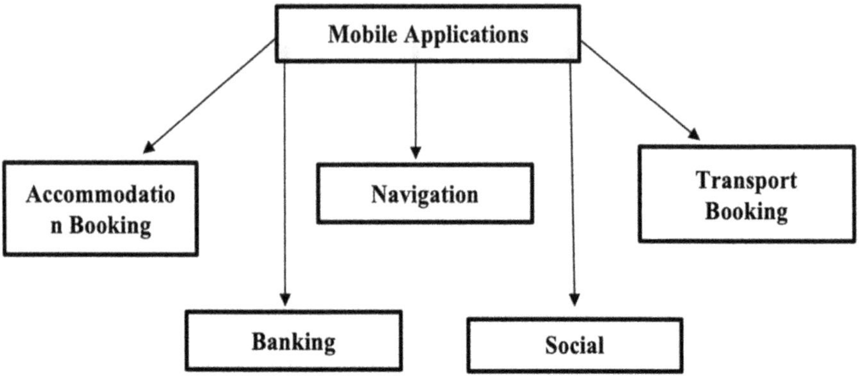

FIGURE 14.11 Multi-dimensional application of mobile apps in tourism.

Source: Authors.

14.7 AUGMENTED REALITY (AR) AND VIRTUAL REALITY (VR)

Immersion experiences provided by AR and VR technologies change how travelers view and engage with travel locations. According to Pencarelli (2020), virtual tours, interactive guides, and 3D visualizations all improve user engagement. These technologies offer engaging and novel experiences, which create a distinctive channel for value development. Travelers' perceptions and interactions with places have changed dramatically because of the introduction of AR and VR into the tourist sector. By superimposing digital information on top of the physical world, augmented reality improves it for visitors by providing them with real-time information on tourist attractions, historical context, and interactive guides. Through the direct provision of contextual information inside the user's actual surroundings, this technology enhances the trip experience. In contrast, virtual reality submerges users in fully digital worlds, enabling them to practically inspect trip destinations, lodging options, and attractions prior to making travel-related decisions. VR in tourism makes virtual tours possible, giving visitors a sense of presence and empowering them to make better decisions. Both augmented reality and VR add to an increased degree of involvement, personalization, and anticipation, altering the parameters of conventional travel and creating novel opportunities for immersive and unforgettable experiences. The following are some of the most common uses of augmented reality in today's world of tourism which includes:

1. **Interactive Navigation:** By giving travelers access to real-time information about their surroundings, augmented reality applications improve their navigational experiences. For example, AR is used by apps like Google Maps to superimpose directions over the real world, providing users with simulated street names and arrows.
2. **Historical Insights:** By utilizing augmented reality, museums and historical places may provide tourists with fully immersive historical experiences. For instance, the "Smartify" app leverages augmented reality to provide comprehensive details about artworks in museums, resulting in an engaging and instructive experience.
3. **Language Translation:** By using language translation features, AR makes cross-cultural communication easier. Using augmented reality, apps such as "Google Translate" can translate text quickly, enabling passengers to view menus, signage, and other written content in their local tongue.
4. **Interactive Guides:** AR makes standard guidebooks interactive through the creation of interactive guides. Applications such as "AR City" overlay real-time information on restaurants, sights, and landmarks to give users interactive, context-aware guides.

While virtual reality has been playing a significant role in bringing unthinkable experiences to potential travelers to visit new destinations, the specific facilities which can be produced with the help of virtual reality are discussed below:

1. **Virtual Reality (VR) Tours of Destinations:** Before deciding where to go, prospective visitors can take virtual tours. With virtual reality experiences that highlight Scotland's breathtaking landscapes, "VisitScotland" invites users to explore and make travel plans.
2. **Hotel Preview:** VR is used by hotels to provide virtual tours of their rooms. For example, Marriott Hotels offers a service called "VRoom Service," which lets visitors take 360° virtual tours of potential destinations before making a reservation. This increases expectations and transparency.
3. **Immersive Travel Experiences:** VR allows users to travel to different destinations through immersive experiences. Qantas Airways launched the "Qantas VR" app, which offers users virtual

flights to multiple Australian destinations, encouraging curiosity and anticipation.

4. **Cultural Immersion:** VR enables users to fully immerse themselves in cultural experiences. Virtual travel experiences such as "Discovery VR" allow users to witness festivals, events, and customs from around the globe, promoting a deeper understanding of global cultures.

Because AR and VR technologies offer personalized, immersive, and interactive experiences, they have completely transformed the tourism sector. These technologies, which can be used to digitally tour locations, lead visitors through historical places, or improve hotel previews, have become indispensable resources for businesses and travelers alike, influencing the way that travel is planned, experienced, and shared in the future.

14.8 IMPLICATIONS OF STUDY

Service management in the tourism industry involves identifying and enhancing components such as customer focus, quality control, crisis management, technology integration, sustainability practices, cross-sector collaboration, and feedback mechanisms. These components aim to make visitors' stays enjoyable and unforgettable, promoting the prosperity and sustainability of a location. Customer focus is crucial in understanding and customizing services to meet tourists' requirements, preferences, and happiness. Quality control ensures excellence and consistency in service delivery, while crisis management prepares for unforeseen circumstances like crises or natural disasters. Technology integration can be achieved through online booking systems, mobile apps, and digital communication platforms. Sustainability practices involve incorporating socially and ecologically conscious methods into service management, minimizing environmental impact and supporting local populations. Cross-sector collaboration is essential in ensuring a smooth and delightful experience for tourists.

Feedback mechanisms enable continuous improvement in service quality in response to changing demands and expectations of travelers. The goal of service management in the tourism industry is to make visitors' stays enjoyable and memorable, recognizing the interdependence of different service providers and maximizing their combined efforts to improve travelers' pleasure and loyalty.

Tourism as a Service (TAS) combines services across various tools and services, such as information, route planning, and tickets for public transportation and tourist attractions. A study on e-mobile ticketing reveals that excessive tourism in many cities leads to new challenges, such as mobility issues and overuse of public transportation or road networks. A study on mobile ticketing solutions in tourism validates the proposed solution's usability testing validation, demonstrating its potential to relieve clients of the burden of making difficult decisions about purchases and making services more accessible.

According to Gretzel et al. (2015), a smart tourism ecosystem is a platform for developing, controlling, and providing tourism services through technological progress that promotes value generation and information sharing. Through contact and reciprocity with tourists through tourism activities, businesses can co-create value and network relationship value, creating a cycle of high-quality tourism. Tourists' value creation process can be ensured through the formation of a smart tourism ecosystem, in which they can have technology-mediated tourism experiences of personalization, context awareness, and real-time monitoring before, during, and after a trip (Bhuiyan et al., 2022). Furthermore, by concentrating on technology-denominated services, sustainability can be produced through the value co-creation process (Chuang, 2023). The economy, sociocultural, and environment are the three facets of sustainability benefits that are included in the value co-creation behaviors of tourists. Thus, a healthy smart tourism ecosystem is necessary to value co-created behaviors (Buhalis and Foerste, 2015), and businesses and travelers both contribute equally to the development of sustainable value (Bhuiyan et al., 2022). Figure 14.12 shows how STS fits into the broader conceptual domains of the relationship among service dominant logic, smart tourism ecosystems, and tourist value co-creation behaviors. The service principle's emphasis on the smart tourist ecosystem can be seen via the lens of S-D logic. Six components of tourism services make up the ecosystem: purchase, transit, lodging, food, and attraction. By working together and sharing resources, tourism service providers can improve travelers' values-based co-creation behaviors, which will lead to sustainability.

Ultimately, the development of the tourism economy and society is influenced by smart tourism technology, which is the primary factor influencing tourists' experiences, satisfaction, and intentions to return (Jeong and Shin, 2020). Travelers can influence other travelers and future travelers through the real-time tourism services provided by STS platforms and their receptiveness to the same, particularly through social connections made online. In addition to enabling activity among tourists, tourism practitioners can offer spaces where tourists can share views, preferences, and experiences with others.

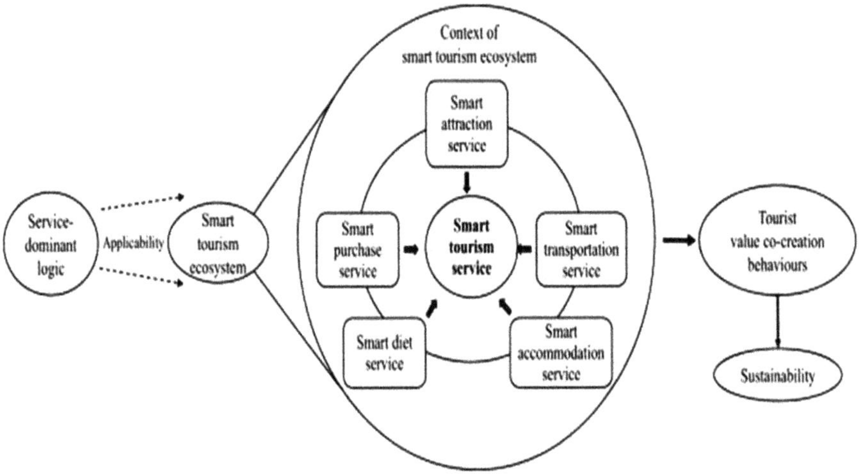

FIGURE 14.12 The relationship among S-D logic, smart tourism ecosystem, and tourist value co-creation behaviors.

Source: Reprinted from Chuang (2023). https://creativecommons.org/licenses/by/4.0/

KEYWORDS

- **augmented reality**
- **co-creation**
- **service management**
- **service-dominant logic**
- **smart tourism**
- **stakeholders**
- **tourism ecosystem**
- **value creation**
- **virtual reality**

REFERENCES

Albrecht, C. C., Dean, D. L., & Hansen, J. V. (2005). Marketplace and technology standards for B2B e-commerce: Progress, challenges, and the state of the art. *Information & Management, 42*(6), 865–875.

Angus, G. Y., Flett, P. D., & Bowers, J. A. (2005). Developing a value-centered proposal for assessing project success. *International Journal of Project Management, 23*(6), 428–436.

Arslan, E., & Kaymaz, I. (2020). Visitor perception of recreational ecosystem services and their role in landscape management of Gölcük Nature Park, Turkey. *International Journal of Sustainable Development & World Ecology, 27*(3), 202–213.

Baker, J., Parasuraman, A., Grewal, D., & Voss, G. B. (2002). The influence of multiple store environment cues on perceived merchandise value and patronage intentions. *Journal of Marketing, 66*(2), 120–141.

Barry, J. B., William, R. D., & Mitch, G. (1994). Work and/or fun: Measuring hedonic and utilitarian shopping value. *Journal of Consumer Research, 4*, 644.

Bhuiyan, K. H., Jahan, I., Zayed, N. M., Islam, K. M. A., Suyaiya, S., Tkachenko, O., & Nitsenko, V. (2022). Smart tourism ecosystem: A new dimension toward sustainable value co-creation. *Sustainability, 14*(22), 150–153.

Boes, K., Buhalis, D., & Inversini, A. (2016). Smart tourism destinations: Ecosystems for tourism destination competitiveness. *International Journal of Tourism Cities, 2*(2), 108–124.

Breese, R. (2012). Benefits realization management: Panacea or false dawn? *International Journal of Project Management, 30*(3), 341–351.

Buhalis, D., & Amaranggana, A. (2014). Smart tourism destinations. In *Information and Communication Technologies in Tourism 2014: Proceedings of the International Conference in Dublin, Ireland* (pp. 553–564). Springer International Publishing.

Buhalis, D., & Foerste, M. (2015). SoCoMo marketing for travel and tourism: Empowering co-creation of value. *Journal of Destination Marketing & Management, 4*(3), 151–161.

Buonincontri, P., & Micera, R. (2016). The experience co-creation in smart tourism destinations: A multiple case analysis of European destinations. *Information Technology & Tourism, 16*, 285–315.

Chen, Y. H., & Barnes, S. (2007). Initial trust and online buyer behavior. *Industrial Management & Data Systems, 107*(1), 21–36.

Chuang, C. M. (2023). The conceptualization of smart tourism service platforms on tourist value co-creation behaviors: An integrative perspective of smart tourism services. *Humanities and Social Sciences Communications, 10*(1), 1–16.

Cooper, R. G. (1998). *Product Leadership: Creating and Launching Superior New Products* (pp. 189–351). Perseus Books.

Corvellec, H., & Hultman, J. (2014). Managing the politics of value propositions. *Marketing Theory, 14*(4), 355–375.

Cronin Jr, J. J., Brady, M. K., & Hult, G. T. M. (2000). Assessing the effects of quality, value, and customer satisfaction on consumer behavioral intentions in service environments. *Journal of Retailing, 76*(2), 193–218.

Cuesta-Valiño, P., Kazakov, S., Gutiérrez-Rodríguez, P., & Rua, O. L. (2023). The effects of the aesthetics and composition of hotels' digital photo images on online booking decisions. *Humanities and Social Sciences Communications, 10*(1), 1–11.

Del Chiappa, G., & Baggio, R. (2015). Knowledge transfer in smart tourism destinations: Analyzing the effects of a network structure. *Journal of Destination Marketing & Management, 4*(3), 145–150.
Eckman, M. (1994). Judging the attractiveness of product design: The effect of visual attributes and consumer characteristics. *Advances in Consumer Research, 21*(1).
Edvardsson, B., Tronvoll, B., & Gruber, T. (2011). Expanding understanding of service exchange and value co-creation: A social construction approach. *Journal of the Academy of Marketing Science, 39*(2), 327–339.
Egger, R., & Neuburger, L. (2020). Augmented, virtual, and mixed reality in tourism. In *Handbook of e-Tourism* (pp. 1–25). Springer.
Flavián, C., Ibáñez-Sánchez, S., & Orús, C. (2021). Integrating virtual reality devices into the body: Effects of technological embodiment on customer engagement and behavioral intentions toward the destination. In *Future of Tourism Marketing* (pp. 79–94). Routledge.
Flint, D. J., Woodruff, R. B., & Gardial, S. F. (2002). Exploring the phenomenon of customers' desired value change in a business-to-business context. *Journal of Marketing, 66*(4), 102–117.
Gössling, S. (2021). Tourism, technology and ICT: A critical review of affordances and concessions. *Journal of Sustainable Tourism, 29*(5), 733–750.
Gretzel, U., Sigala, M., Xiang, Z., & Koo, C. (2015). Smart tourism: Foundations and developments. *Electronic Markets, 25*, 179–188.
Grönroos, C., & Voima, P. (2013). Critical service logic: Making sense of value creation and co-creation. *Journal of the Academy of Marketing Science, 41*, 133–150.
Heinonen, K. (2006). The role of customer participation in creating e-service value. In Maula, M., Hannula, M., Seppä, M., & Tommila, J. (Eds.), *FeBR 2006* (Tampere).
Holbrook, M. B. (1996). Customer value—A framework for analysis and research. *Advances in Consumer Research, 23*(1), 138–142.
Hong, T. K., & Zailani, S. (2011). Service supply chain practices from the perspective of Malaysian tourism industry. In *2011 IEEE International Conference on Industrial Engineering and Engineering Management* (pp. 539–543).
Hsiao, L., & Chen, Y. J. (2012). Returns policy and quality risk in e-business. *Production & Operations Management, 21*(3), 489–503.
Huang, D., Li, Z., Mou, J., & Liu, X. (2017). Effects of flow on young Chinese consumers' purchase intention: A study of e-servicescape in hotel booking context. *Information Technology & Tourism, 17*(2), 203.
Jeong, M., & Shin, H. H. (2020). Tourists' experiences with smart tourism technology at smart destinations and their behavior intentions. *Journal of Travel Research, 59*(8), 1464–1477.
Kaplinsky, R. (2004). Spreading the gains from globalization: What can be learned from value-chain analysis? *Problems of Economic Transition, 47*(2), 74–115.
Laursen, M., & Svejvig, P. (2016). Taking stock of project value creation: A structured literature review with future directions for research and practice. *International Journal of Project Management, 34*(4), 736–747.
Lien, C. H., Wen, M. J., Huang, L. C., & Wu, K. L. (2015). Online hotel booking: The effects of brand image, price, trust and value on purchase intentions. *Asia Pacific Management Review, 20*(4), 210–218.

Love, P. E., Matthews, J., Simpson, I., Hill, A., & Olatunji, O. A. (2014). A benefits realization management building information modeling framework for asset owners. *Automation in Construction, 37*, 1–10.

Manuel, R. D., Rosa, R. D., Ana Cristina, R. V., & Crina Isabel, R. V. (2017). A model of market positioning of destinations based on online customer reviews of lodgings. *Sustainability, 10*(1), 78.

Mieli, M. (2018). The value of travel guidebooks in the digital age. Abstract from *the 27th Nordic Symposium on Tourism and Hospitality: Tourism Implications and Dilemmas*, Alta, Norway.

Minkov, M., Dutt, P., Schachner, M., Jandosova, J., Khassenbekov, Y., Morales, O., & Blagoev, V. (2019). What would people do with their money if they were rich? A search for Hofstede dimensions across 52 countries. *Cross Cultural & Strategic Management, 26*(1), 93–117.

Mok, C., Sparks, B., & Kadampully, J. (2001). *Service Quality Management in Hospitality, Tourism, and Leisure* (pp. 339). Routledge.

Mukerjee, R. (1964). *The Dimensions of Values: A Unified Theory* (1st ed.). Allen & Unwin.

Nafi, S., & Ahmed, T. (2018). Travel and leisure activities, motivation and behavior of young Bangladeshi tourists. *European Scientific Journal, 14*(5), 341–358.

Nhan, V. K., Dung, H. T., & Vu, N. T. (2022). A conceptual model for studying the immersive mobile augmented reality application-enhanced experience. *Helion, 8*(8).

Normann, R., & Ramirez, R. (1993). From value chain to value constellation: Designing interactive strategy. *Harvard Business Review, 71*(4), 65–77.

Oakland, J. S. (2014). *Total Quality Management and Operational Excellence: Text with Cases*. Routledge.

Otto, J. E., & Ritchie, J. B. (1996). The service experience in tourism. *Tourism Management, 17*(3), 165–174.

Parasuraman, A., & Grewal, D. (2000). The impact of technology on the quality–value–loyalty chain: A research agenda. *Journal of the Academy of Marketing Science, 28*(1), 168–174.

Pencarelli, T. (2020). The digital revolution in the travel and tourism industry. *Information Technology & Tourism, 22*(3), 455–476.

Porter, M. E. (2001). The value chain and competitive advantage. *Understanding Business Processes, 2*, 50–66.

Sánchez-Fernández, R., Iniesta-Bonillo, M. Á., & Holbrook, M. B. (2009). The conceptualization and measurement of consumer value in services. *International Journal of Market Research, 51*(1), 1–17.

Sanjoy Kumar, A., & Tanvir, A. (2023). The impact of social media on tourists' decision-making process: An empirical study based on Bangladesh. *Journal of Social Sciences and Management Studies, 3*(1), 56–71.

Shin, H. H., & Jeong, M. (2021). Travelers' motivations to adopt augmented reality (AR) applications in a tourism destination. *Journal of Hospitality and Tourism Technology, 12*(2), 389–405.

Shohel Md, N., & Tanvir, A. (2019). The ethical standpoint of social influencers on hotel e-services cape: A theoretical perspective on the existing literature. *African Journal of Hospitality, Tourism and Leisure, 8*(1), Article a17.

Shukla, A., & Mishra, A. (2023). Effects of visual information and argument concreteness on purchase intention of consumers towards online hotel booking. *International Journal of Consumer Studies, 27*(5), 639–649.
Simatupang, T. M., Piboonrungroj, P., & Williams, S. J. (2017). The emergence of value chain thinking. *International Journal of Value Chain Management, 8*(1), 40–57.
Susan, P., Lisa, L. G., Arturo, E., & Andrew, G. (2004). Chapter 7: The political ecology of consumption: Beyond greed and guilt. In *Handbook of Sustainable Consumption* (pp. [specific pages not provided]). Rutgers University Press.
Tankovic, A. C., & Benazic, D. (2018). The perception of e-servicescape and its influence on perceived e-shopping value and customer loyalty. *Online Information Review, 42*(7), 1124–1145.
Testa, M. R., & Sipe, L. (2012). Service-leadership competencies for hospitality and tourism management. *International Journal of Hospitality Management, 31*(3), 648–658.
Treacy, M., & Wiersema, F. (1993). Customer intimacy and other value disciplines. *Harvard Business Review, 71*(1), 84–93.
Umair, A., Peng, H., Muhammad Kaleem, K., Chen, Y., & Zubair, A. (2018). Factors affecting online impulse buying: Evidence from Chinese social commerce environment. *Sustainability, 10*(2), 352. https://doi.org/10.3390/su10020352 (accessed on 8th July 2025).
Vargo, S. L., & Lusch, R. F. (2017). Service-dominant logic 2025. *International Journal of Research in Marketing, 34*(1), 46–67.
Vilar-Zanon, J. L., Vilar, E., & Heras, A. (2017). Online product returns risk assessment and management. *TOP, 25*(3), 445–466.
Wang, X., Li, X. R., Zhen, F., & Zhang, J. (2016). How smart is your tourist attraction? Measuring tourist preferences of smart tourism attractions via a FCEM-AHP and IPA approach. *Tourism Management, 54*, 309–320.
Ward, J., & Daniel, E. (2006). *Benefits Management: Delivering Value from IS & IT Investments* (Vol. 30). John Wiley & Sons.
Warhurst, C., Nickson, D., & Dutton, E. (2005). The importance of attitude and appearance in the service encounter in retail and hospitality. *Managing Service Quality: An International Journal, 15*(2), 195–208.
Weitzl, W., Wolfsteiner, E., Einwiller, S., & Wagner, U. (2016). When credibility truly matters online: Investigating the role of source credibility for the impact of customer reviews. *Advances in Consumer Research, 44*, 353–358.
Willskytt, S., & Tillman, A. M. (2019). Resource efficiency of consumables–Life cycle assessment of incontinence products. *Resources, Conservation and Recycling, 144*, 13–23.
Xiang, Z., & Fesenmaier, D. R. (2017). *Analytics in Smart Tourism Design: Concepts and Methods*. Springer International Publishing.
Zwikael, O., & Smyrk, J. (2012). A general framework for gauging the performance of initiatives to enhance organizational value. *British Journal of Management, 23*, S6–S22.

CHAPTER 15

EMPOWERING SUSTAINABLE ECONOMIC GROWTH: SMART TOURIST E-SAFETY TECHNOLOGIES AND THEIR ROLE IN PROPELLING THE TOURISM INDUSTRY

CHHANDA BISWAS[1*], TOHOMINA KHATUN[2], SHADIA SHARMIN[1], and MD. MANOWER HOSSAIN[1]

[1]*Department of Tourism and Hospitality Management, Gopalganj Science & Technology University, Gopalganj, Bangladesh*

[2]*Department of Anthropology, Comilla University, Cumilla, Bangladesh*

*Corresponding author

ABSTRACT

This chapter aims to explore the pivotal role of smart tourist e-safety technologies in the sustainable economic growth of the tourism industry in Bangladesh. Utilizing natural resources and smart technological development, as well as managing proper investment, can contribute to sustainable economic development so that it can meet the needs of present and future potentials. In the pursuit of this study, the author will employ a qualitative research methodology, chosen for its suitability in providing a comprehensive and nuanced understanding of the subject matter. This approach involves utilizing semi structured interviews to collect data from both tourists and various stakeholders in the tourism industry. The author will engage with a minimum of 20 interviewees to elicit their perspectives and insights on the matter at hand. However, the study on how smart tourist

Smart Tourism: Managing Technological Revolution in the Tourism Industry. Santus Kumar Deb (Ed.)

DOI: 10.1201/9781779641090-15

e-safety contributes to sustainable economic development in the context of Bangladesh has not been explored earlier, and thus, it can add theoretical and conceptual value to the literature. Therefore, future research endeavors may benefit from adopting a quantitative methodology, which would allow for the collection of a more extensive sample size.

15.1 INTRODUCTION

The entirety of the tourism system has undergone a transformative shift because of technology, influencing every aspect from planning a trip to on-site experiences (Buhalis et al., 2019). Safety and security, coupled with digitalization, authentic experiences, and the sharing economy, stand out as prevalent global trends in the tourism industry (Mouratidis et al., 2021). E-safety involves safeguarding individuals and data, prudent and responsible usage of technology, and extending not only internet utilization but also interactions in the electronic domain, addressing risks and threats associated with technology and the internet (Lombok et al., 2023).

In a study by Popescu (2011), tourism plays a significant role in the economic value of many countries, forming a substantial portion of the national GDP. Amid escalating global threats, tourists are now increasingly giving precedence to choosing secure destinations, highlighting the crucial requirement for safety education among foreign visitors. The advent of smart tourist e-safety technologies, incorporating advancements like IoT (Internet of Things) devices, mobile applications, data analytics, and AI-driven safety solutions, has surfaced to elevate safety measures and advocate for sustainability in the tourism industry (Buhalis et al., 2019).

Smart tourist e-safety encompasses protecting tourists from cyber threats, ensuring the security of their digital transactions, real-time monitoring, and providing a secure environment for tourists. This assurance of safety features can significantly increase tourists' confidence in choosing a destination, resulting in a rise in visitor numbers. In a study by Glensor and Peak (2019), ensuring the safety of tourists and maintaining control over crime is recognized as the foremost requirement for a flourishing tourism industry. A secure environment draws in additional tourists, fostering economic growth, while supporting the sustainable economic growth of the tourism industry. Tourists with a sense of security are more likely to participate in various activities, support local businesses, and elevate their expenditure, thereby directly contributing to the economic growth within the destination.

Previously, numerous studies on smart mobility (Khamis, 2021), tourism, smart specialization, sustainable development (Romão, 2021), safe and seamless travel relating to improved experience (Popescu, 2011), technological contributions to the sustainable future of tourism (Loureiro and Nascimento, 2021), the role of information and communication technology (ICT) for sustainable tourism development (STD) (Ganguly, 2018), the link between digitalization and safety for the tourism industry (Marcelo, 2023), and the role of technology in securing and improving safety in the tourism destinations in Malaysia (Norizawati and Masron, 2014) have been conducted. However, there is insufficient research on fostering sustainable economic growth in the tourism industry by integrating smart tourist e-safety technology, leveraging advanced technology to enhance the safety and security of tourists. Therefore, this study aims to delve into the intricate dynamics and consequential impacts of these cutting-edge technologies in advancing the tourism sector.

15.2 LITERATURE REVIEW

15.2.1 SMART TOURISM TECHNOLOGY

According to the European Commission (2019), smart tourism refers to a destination that utilizes ICT tools to enhance accessibility to tourism and hospitality products, services, spaces, and experiences. Digital technology contributes to the development of tourism in a unique way, fostering various entrepreneurial ventures. The involvement of youth and local communities in the tourism industry is steadily rising, primarily as a result of the impact of digital technology, strengthening a city's intelligence and fostering increased engagement with visitors. Smart tourism technology represents a revolutionary influence in the tourism industry, leveraging cutting-edge digital solutions to increase the overall travel experience for tourists while concurrently streamlining operations for industry stakeholders. In a study by Zhang et al. (2022), it is stated that the emerging trend of the adoption and implementation of smart technologies in tourism destinations and visitor attractions is to improve tourists' experiences and raise their satisfaction level. Within the tourism industry, technological innovation serves not only as a substantial catalyst for economic growth but also as a means to meet the increasing demand for unique tourism experiences (Yovcheva et al., 2013).

Being designated as a smart tourism city entails delivering innovative tourism and hospitality information, products, services, experiences, and spaces tailored to consumer needs through the application of ICT-based solutions and digital tools (Baggio et al., 2020). Key components of smart tourism technology include the Internet of Things (IoT), artificial intelligence, mobile applications, augmented reality, and big data analytics. According to Farid et al. (2023), smart tourism relies on the IoT, enabling smooth connectivity between devices and systems. IoT applications include automated services in hotels, smart transportation systems, and infrastructure enhancements that increase the overall efficiency of tourist destinations. In their study, Lee et al. (2018) developed a model that suggests that tourist happiness is assessed by the perceived experiences of tourists, and these are formed by two key motivational values: Smart Tourism Technology (STT) and destination value. The satisfaction of tourists is closely interlinked to their perceived value of smart technologies, resulting in a positive effect on word-of-mouth recommendations, revisit intention, and willingness to pay a premium price (Zhang et al., 2022).

15.2.2 TOURIST E-SAFETY

E-safety, also referred to as "internet safety," "online safety," or "web safety," involves the secure and responsible utilization of technology, encompassing not just internet usage but also diverse modes of communication through electronic media like text messages, gaming devices, and email (Glasgow et al., 2023). Tourist e-safety, or electronic safety for tourists, is known as the measures and precautions implemented to ensure the well-being of tourists in the digital realm. This encompasses protecting tourists from online threats, potential risks linked to the use of electronic devices, cybercrimes, vulnerability of personal information, and online services throughout their journeys.

In the age of social media, products, and services are effortlessly marketed globally using a few clicks online. In this digital landscape, tourist decision-making is heavily advocated by trust and familiarity, and the conveniences offered in online transactions come with an implicit expectation that service providers will give paramount importance to the security of user data (Farid et al., 2023). In their study, Ramanathan et al. (2022) analyzed that tourist e-safety is formed between tourists and the tourism industry in the course of online purchases. Simultaneously, the perceived quality of online information and the security of online payments contribute to fostering trust in the service company for subsequent purchases. Technological advancement is pivotal in enhancing tourist e-safety, using effective tools such as secure payment gateways, encryption methods, mobile device security, and virtual private networks (VPN). Implementing these technologies is essential to ensure the safeguarding of tourist data.

The use of public Wi-Fi poses a notable risk owing to its potential lack of security, leaving it open to attacks. Exploiting this vulnerability, hackers can discreetly gain access to and pilfer personal information or implant malicious software on tourists' devices (Yovcheva et al., 2013). The impact of social media on tourist e-safety is multifaceted, involving various aspects of digital safety during travel which are sharing too much information, geotagging risks, privacy concerns on disclosing personal information (current location, accommodation details, daily activities), cyber security threats, influence on travel decisions (misleading reviews or endorsements), and so on (European Commission, 2019).

15.2.3 COMPONENTS OF TOURIST E-SAFETY

Tourist e-safety, which involves ensuring the secure and responsible application of technology in travel, comprises several components that collaboratively work towards protecting tourists in the digital environment.

A breached network can grant hackers the ability to intercept, read, and alter the internet traffic within it. This exploitation enables them to engage in several malicious activities, ranging from password theft to installing malware on the devices of unsuspecting victims (Perwej et al., 2021). Secure Wi-Fi connections play a crucial role in thwarting unauthorized access, safeguarding sensitive data, and fostering responsible online practices for a secure digital travel experience.

Incorporating a VPN is a key tourist e-safety element, offering an encrypted and safe online connection that shields travelers from potential cyber threats and guarantees data confidentiality during journeys, particularly when connecting to public Wi-Fi (Venkateswaran, 2021).

Securing online transactions is a cornerstone of tourist e-safety, ensuring the safeguarding of financial information and personal data during digital transactions, creating a reliable and safe online atmosphere for travelers, as well as utilizing secure payment gateways.

Mobile security devices encompass features such as anti-malware measures, leveraging biometrics, password protection, secure browsing, and device tracking to guarantee a secure digital experience for travelers on the go.

Awareness of scams encourages travelers to recognize and avoid fraudulent schemes such as ATM skimmers, radio frequency identification information theft, and juice jacking (avoiding plugging into public charging stations).

Social media privacy involves the implementation of privacy settings to protect personal information and safeguard it from unauthorized access (Deb and Mallik, 2023).

Integral to tourist e-safety, data backup enables the protection of important information, photos, and documents, as well as facilitates swift recovery in situations like device loss, theft, or technical issues, thereby enhancing overall digital security during travel (Choi et al., 2022).

An emergency communication plan provides a structured strategy for swift communication with local authorities, embassies, and emergency services so that tourists have the means to reach out to family and friends in case of an emergency.

In the realm of tourist e-safety, secure passwords stand as a frontline defense against unauthorized access, protect personal information, as well as fortify the resilience of tourists' digital accounts, reducing potential cyber threats.

Location services in the context of tourist e-safety enable effective management and user awareness to balance advantages and risks associated with location services during travel.

Cyber security education empowers travelers with the knowledge to recognize and reduce online risks and safeguard a safe digital experience throughout their journey.

Adhering to local laws and regulations is a fundamental component of tourist e-safety that ensures digital practices comply with legal standards, as well as fostering a safe online environment and bolstering cybersecurity within a specific destination (Figure 15.1).

15.2.4 FACTORS INFLUENCING SMART TOURISM TECHNOLOGIES (STT) ON TOURIST E-SAFETY

The varied and interconnected factors that influence the impact of STT on tourist e-safety underscore the complex interplay between technological advancements, tourism practices, and the ever-evolving landscape of digital threat. In subsections, it delves into key elements shaping the contribution of STT to the enhancement of tourist e-safety.

15.2.4.1 TECHNOLOGICAL INFRASTRUCTURE

1. **Connectivity and Network Security:** The efficacy of STT in guaranteeing e-safety significantly relies on the strength of the technological infrastructure, with a particular emphasis on seamless connectivity across destinations, high-speed internet access, and secure networks. Reliable connectivity forms the foundation for the seamless operation of these technologies, facilitating real-time monitoring of digital activities, communication among various systems, and incorporation of vital safety measures (Huang et al., 2017). In tourist e-safety, a strong digital infrastructure enables quick detection of potential cybersecurity risks and vulnerability management, facilitating prompt responses to minimize threats. It guarantees that tourists can access and leverage smart technologies without disruptions, enhancing their travel experience with a positive and secure environment. Moreover, the secure networks embedded in the technological infrastructure are instrumental in protecting sensitive tourist data, preventing unauthorized entry, and maintaining the integrity of online transactions.

15.2.4.2 DATA ENCRYPTION AND PRIVACY MEASURES

1. **Secure Transactions:** Online transactions are essential to the functionality of smart tourism technologies, especially in the booking of flights, accommodations, and other activities. Employing robust data encryption and privacy protocols guarantees the security of sensitive information, covering financial details and personal data, and shielding it from unauthorized access and potential cybersecurity risks (Farid et al., 2023).

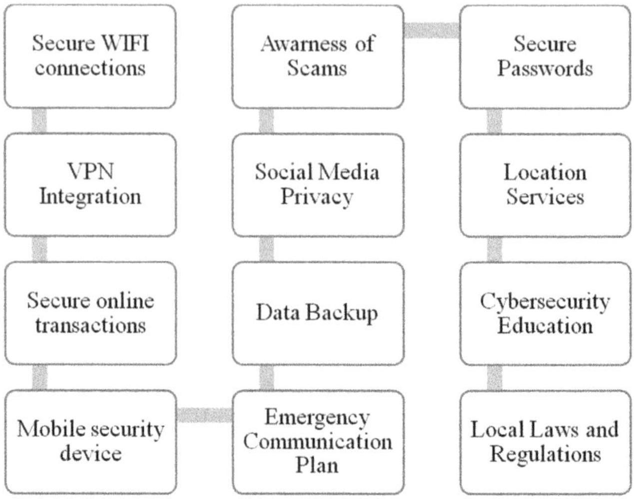

FIGURE 15.1 Components of tourist e-Safety.

Source: Adapted from Lawton (2019).

15.2.4.3 ARTIFICIAL INTELLIGENCE (AI) AND PREDICTIVE ANALYTICS

1. **Risk Assessment and Prevention:** The amalgamation of AI and predictive analytics facilitates the proactive detection of potential security threats. Smart technologies can analyze patterns, detect anomalies (unusual spikes in data traffic), and predictive threat modeling (if a pattern emerges of increased phishing attacks in well-frequented tourist destinations during specific seasons, the system can predict the probability of encountering analogous threats in the future), fraud prevention (AI blocks suspicious transactions and financial institutions), dynamic risk assessment (based on real-time data feeds), and enabling timely interventions to enhance the safety for tourists.

15.2.4.4 INTERNET OF THINGS (IOT) INTEGRATION

1. **Smart Devices and Wearables:** IoT devices, such as smart wearable devices, actively monitor tourists' activities in real-time. In

emergencies, these devices provide instant alerts, allowing swift response and assistance. The integration of location tracking enhances overall security, enabling authorities to pinpoint tourists' whereabouts. Moreover, IoT devices facilitate seamless communication that ensures efficient coordination between tourists and relevant authorities, thereby contributing to the overall safety and quality of the travel experience.

15.2.4.5 GEOSPATIAL TECHNOLOGIES

1. **Location-based Services:** Geospatial technologies, like GPS and location-based services, improve tourist safety through accurate navigation, emergency location tracking, and geofencing capabilities. These technologies play a crucial role in establishing a comprehensive safety framework for travelers exploring unfamiliar destinations (Otebolaku and Lee, 2018).

15.2.4.6 CYBER SECURITY PROTOCOLS

1. **Firewalls and Intrusion Detection Systems:** Effective cyber security protocols, encompassing firewalls and intrusion detection systems, are crucial for safeguarding digital platforms associated with tourism. These measures act as a protective barrier against cyber threats, ensuring the integrity and confidentiality of tourist information, vulnerability management, and detection and response.

15.2.4.7 COLLABORATION AND INFORMATION SHARING

1. **Public-Private Partnerships:** A comprehensive approach to tourist e-safety is nurtured through the collaborative efforts of government agencies, private enterprises, and technology providers. The exchange of information regarding emerging risks, best practices, and technological advancements enhances the collective endeavor to defend against cyber threats.

15.2.4.8 USER EDUCATION AND AWARENESS

1. **Training Programs:** Educating both tourists and industry stakeholders about the potential threats and best practices for e-safety is essential. Training programs and awareness campaigns enable individuals to make informed decisions and adopt safe digital behaviors while traveling.

15.2.4.9 REGULATORY FRAMEWORKS

1. **Compliance and Standards:** Transparent regulatory frameworks and industry standards offer guidance for the responsible development and implementation of smart tourism technologies. Compliance with established norms guarantees that e-safety measures are aligned with ethical and legal considerations (McClelland et al., 2023).

15.2.4.10 CRISIS MANAGEMENT AND RESPONSE SYSTEMS

1. **Emergency Communication Channels:** Smart technologies are instrumental in effective crisis management by facilitating prompt communication among tourists, emergency services, and relevant authorities. This enhances overall tourist safety during unforeseen situations. A clear understanding and proper optimization of these influencing factors can significantly contribute to the success of the efficient integration of smart tourism technologies, prioritizing and enhancing e-safety in an evolving digital travel landscape (Figure 15.2).

15.2.5 SUSTAINABLE ECONOMIC GROWTH

Safeguarding the e-safety of tourists through smart technologies can play a significant role in promoting sustainable economic growth in various ways:

- Utilizing advanced encryption, biometric authentication, and secure payment gateways are examples of how smart technologies can boost

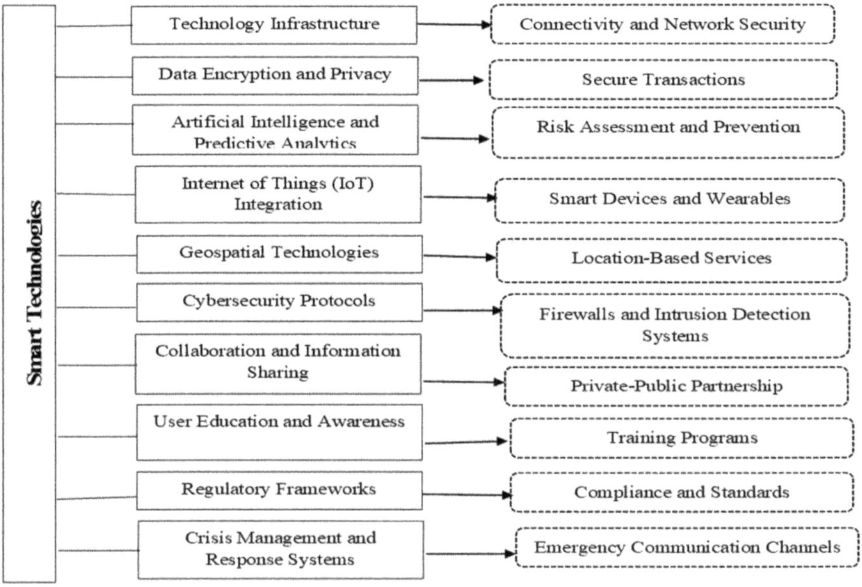

FIGURE 15.2 Factors affecting how smart technologies improve tourist e-Safety.

Source: Adapted from Huang et al. (2017).

the security of online transactions and information (Hamidi, 2019). These measures foster trust and promote confidence in engaging in electronic transactions.

- Enforcing stringent data protection measures ensures the privacy and safety of tourists' data. This practice is crucial in cultivating trust and loyalty, as tourists are more likely to be inclined to engage in online activities when assured that their data is well-protected (Agarwal et al., 2021).
- Smart technologies empower real-time monitoring and communication systems, facilitating swift responses during emergencies or crises. This capability is essential for ensuring tourist safety and mitigating potential risks, ultimately ensuring a positive and secure travel experience.
- Leveraging smart technologies for destination management is demonstrated by employing the application of data analytics and smart city technologies. These tools empower authorities to optimize traffic flow, regulate crowd density, and enhance overall infrastructure,

ultimately elevating the tourist experience and fostering sustainable development at the destination (Ali et al., 2023).

- Utilizing smart technologies to encourage responsible tourism practices is demonstrated through smart applications guiding tourists to eco-friendly attractions, promoting waste reduction, and endorsing sustainable transportation options (Archi et al., 2023). These practices collectively contribute to the destination's overall sustainability.
- Smart technologies drive targeted digital marketing strategies and personalized experiences by utilizing data analytics to understand tourists' preferences and behaviors (Zhang et al., 2022). This tailored approach improves engagement and satisfaction, fostering positive contributions to the economic growth of the destinations through repeat visits and positive word-of-mouth.
- The incorporation of smart technologies in tourism businesses, encompassing efficient booking systems, inventory management, and automated customer service, provides instrumental advantages. This efficiency not only translates to cost savings but also improves services, thereby contributing to the business's economic sustainability.
- Smart technologies play a role in mitigating tourism's environmental impact, especially through the implementation of smart transportation systems. These systems optimize routes, minimize fuel consumption, and alleviate traffic congestion, associating with the principles of sustainable tourism and influencing the industry's environmental footprint positively (Archi et al., 2023).

However, ensuring tourist e-safety with smart technologies not only shields tourists and their personal information but also cultivates a positive and sustainable tourism ecosystem. As tourists feel secure and experience enhanced services, they become more inclined to contribute to the economic growth of destinations, fostering a beneficial relationship between tourism, technology, and sustainability.

15.2.6 OVERVIEW OF INTERRELATIONSHIPS BETWEEN SMART TECHNOLOGY, TOURIST E-SAFETY, SUSTAINABILITY, AND ECONOMIC GROWTH

The complex and interconnected relationship among smart technologies, tourist e-safety, sustainability, and economic growth is notable. The integration of smart technologies in the tourism sector can exert profound effects on each of these components, resulting in a synergistic and positive impact on the overall tourism ecosystem. It discusses an overview of the interrelations among these elements in subsections.

15.2.6.1 SMART TECHNOLOGIES AND TOURIST E-SAFETY

Smart technologies are crucial in safeguarding tourists in the digital realm by employing advanced encryption, biometric authentication, and a secure payment gateway for protecting personal information during online transactions. Furthermore, real-time monitoring and communication systems enhance emergency response capabilities, ensuring the physical safety of tourists (Lawton, 2019). When tourists perceive this heightened security, they become inclined to participate in online activities, fostering the expansion of the digital tourism economy.

15.2.6.2 TOURIST E-SAFETY AND SUSTAINABILITY

Ensuring the e-safety of tourists, both in the physical and digital aspects, is fundamental to the tourism industry's sustainability. Establishing trust through secure digital transactions and safeguarding sensitive information cultivates a positive perception of the destination. A favorable perception results in encouraging responsible tourism practices, as tourists are more inclined to show respect for local environments and cultures when they feel secure and well-informed. The incorporation of sustainable tourism practices contributes to the preservation of natural and cultural resources in the long run (Huang et al., 2017).

15.2.6.3 SUSTAINABILITY AND ECONOMIC GROWTH

Sustainable tourism practices and economic growth are intricately linked to each other. Destinations that prioritize environmental conservation, community engagement, and cultural preservation tend to appeal to more tourists over the long term. Sustainable tourism not only safeguards the natural and cultural assets of a destination but also enhances the well-being of local communities. As the destination flourishes and retains its attractiveness, it sustains economic growth through heightened tourism revenue, employment generation, and the growth of local businesses (León-Gómez et al., 2021).

15.2.6.4 SMART TECHNOLOGIES AND SUSTAINABILITY

Leveraging smart technologies can be crucial in advancing sustainable practices in the tourism industry. For instance, utilizing data analytics can help optimize transportation routes, mitigating carbon emissions. The implementation of smart city technologies can enhance waste management and resource allocation, thereby contributing to environmental sustainability (Sustacha et al., 2023). Smart technologies can play a role in fostering a more sustainable tourism experience by providing tourists with information on eco-friendly activities and encouraging responsible behavior.

15.2.6.5 SMART TECHNOLOGIES, TOURIST E-SAFETY, SUSTAINABILITY, AND ECONOMIC GROWTH

The incorporation of smart technologies elevates tourist e-safety, which results in fostering trust and encouraging responsible tourism practices. These practices play a crucial role in the sustainability of the destination, ensuring the conservation of its natural and cultural resources. Destinations adopting sustainable practices become more appealing to tourists seeking experiences grounded in responsibility and significance (Solis-Radilla et al., 2019). The favorable loop perpetuates as increased tourist engagement fuels the destination's economic growth.

In conclusion, the interplay among smart technologies, tourist e-safety, sustainability, and economic growth is a dynamic and symbiotically reinforcing relationship. The successful integration of smart technologies not only increases the safety and security of tourists but also plays an instrumental role in the enduring sustainability and economic viability of tourism destinations. This comprehensive approach is crucial to building a robust and thriving tourism industry in the era of the digital age (Figure 15.3).

15.3 METHODOLOGY

In this chapter, the qualitative research method is used to acquire in-depth information about smart tourist e-safety technologies and their role in driving the tourism industry in the context of empowering sustainable economic growth. Hence, qualitative data were gathered from multiple sources, employing in-depth interviews and document analysis. Applicable reports, policy-making strategy records, and sustainable economic growth guidelines were used to gain insights into the existing approaches and strategies connected with smart tourism technology, tourist e-safety, components of tourist e-safety, factors influencing smart tourism technologies on tourist e-safety, and effects of smart technology for tourist e-safety on sustainable economic growth. Therefore, to conduct in-depth interviews, semi-structured interviews were undertaken with tourism industry stakeholders, including tourists, destination managers, academic experts, tourism agencies, as well as the local community.

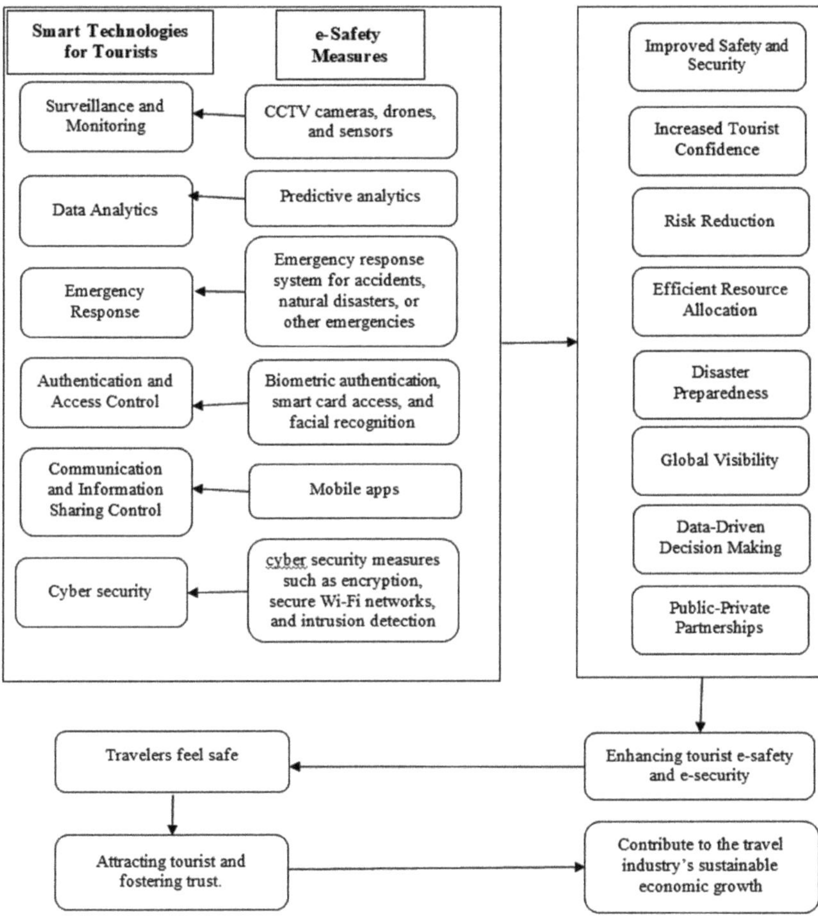

FIGURE 15.3 Integrated model of smart tourist e-safety technologies and tourism industry's sustainable economic growth.

Source: Developed by Authors.

A convenient sampling method was used for collecting data. Therefore, respondents forwarded interview questions to individuals within their network who were engaged in the tourism industry. A total of 15 respondents shared their perspectives. They were asked a total of 10 questions aligned with the research topic. These interviewing questions were adapted from (Huang et al., 2017; Sustacha et al., 2023). First, they were briefed about smart tourism and tourist e-safety technologies, as well as sustainable economic growth concepts, although the majority of the respondents were acquainted with these concepts. Subsequently, they were asked, "Can you please introduce yourself and your role in the tourism industry or related field?" After assuring respondents of the confidentiality of their personal information, they started to ask questions. The next section presents the study's findings based on the questions posed to the respondents.

15.4 FINDINGS OF THE STUDY

Respondents were asked 10 questions on smart tourism technologies, smart e-safety measures, and sustainable economic growth to understand their perspectives.

In your opinion, what role do you believe smart tourism technologies play in the modern tourism industry?

Respondents from various sectors in the tourism industry unanimously acknowledged the crucial role played by smart tourism technologies in elevating the overall tourism experience. Most respondents mentioned effortless bookings, tailored recommendations, real-time information access, and enhanced connectivity that emerge a more enjoyable and technologically advanced travel landscape.

Can you provide examples of smart technologies that have been implemented successfully in the tourism sector?

According to most of the respondents, mobile applications such as real-time information, personalized itineraries, IoT applications in transportation, and navigation assistance are the smart technologies that have been implemented successfully in the tourism sector.

What measures are in place to ensure the secure handling of tourist data and online transactions?
The majority of the respondents mention encryption methods, adherence to data protection laws and regulations, and secure payment gateways are in place to strengthen tourist data security.

With the increasing reliance on digital platforms, what emerging threats to tourist e-safety do you foresee, and how are these being addressed?
According to most of the respondents, the increasing dependence on digital platforms presents emerging threats to tourist e-safety including data breaches, phishing attacks, and ransomware. These are being addressed by raising awareness about the risks associated with digital platforms, AI-driven threat detection, undertaking advanced cyber security measures, and so on.

In your experience, what are the most common challenges or concerns related to ensuring e-safety for tourists?
From most of the respondents' experience, prevalent challenges related to ensuring e-safety for tourists include the risk of unauthorized access to personal information, Wi-Fi vulnerabilities, and potential exposure to cyber threats during online transactions.

From your perspective, how can tourism stakeholders contribute to sustainable practices that promote economic growth while ensuring tourist e-safety through smart technologies?
From the perspective of the majority of the respondents, tourism stakeholders can contribute to sustainable practices by prioritizing cybersecurity infrastructure, implementing secure and user-friendly digital technologies, and participating in collaborative efforts to address emerging e-safety issues. They think that emphasizing eco-friendly and socially responsible tourism practices coupled with technological advancements will create a comprehensive approach to sustainability and e-safety in the tourism sector.

How do you involve local communities in tourism-related decisions to ensure that economic benefits are distributed equitably?
According to most of the respondents, they involve local communities in tourism-related decisions by engaging them in open communication channels, conducting community consultations, as well as incorporating their input in decision-making processes to ensure that economic benefits are distributed equitably.

How do you see the integration of smart technologies, tourist e-safety measures, and sustainability practices working together to create a positive impact on tourism?
The majority of the respondents answered that the integration of smart technologies not only enhances tourist e-safety through efficient and secure digital experiences, instilling trust among tourists, but also ensures responsible and eco-friendly tourism when combined by contributing to a comprehensive strategy that sustains economic growth while conserving cultural and natural resources for a travel experience that is both sustainable and enjoyable.

Are there challenges or conflicts that arise when trying to balance these three aspects?
As per the majority of respondents, balancing smart technologies, tourist e-safety measures, and sustainability practices raises concerns about data privacy. Ensuring e-safety might require constant adaptation to cyber threats. Moreover, aligning with sustainable practices requires striking a delicate equilibrium to avoid environmental impacts, which constitutes a challenge in attaining harmonious synergy among these vital components.

Can you provide specific examples or instances where the integration of smart e-safety measures has had a direct or indirect impact on the economic sustainability of a destination?
Most of the respondents answered that the integration of smart e-safety measures has notably influenced economic sustainability. For instance, they mentioned that secure mobile payment systems increased tourist spending, and the incorporation of smart surveillance and emergency response systems enhanced destination safety, cultivating a positive reputation, and attracting more tourists, resulting in contributing to economic sustainability.

15.5 CONCLUSION

The integration of smart tourist e-safety technologies for tourists serves as an instrumental force in propelling the tourism industry toward a path of sustainable economic growth. The findings of this chapter suggest that these cutting-edge technologies not only enhance the safety and security of tourists but also contribute a substantial impact on the comprehensive development and resilience of the tourism sector. A key discovery from this study is the transformative impact of smart e-safety technologies on the safety dynamics of tourist destinations. Ranging from advanced surveillance systems and real-time communication platforms to emergency response mechanisms, these technologies function as a formidable defense, guaranteeing the well-being of tourists and protecting against potential threats. This heightened security awareness significantly contributes to nurturing confidence among travelers, encouraging them to discover new destinations and immerse themselves in a variety of cultural experiences. Improved safety measures attract a greater number of travelers, leading to augmented revenue for local communities and businesses. Additionally, the implementation of these technologies stimulates innovation and employment generation in the technology sector, bolstering economic growth.

The smooth integration of technology into the tourism framework fosters a competitive advantage for destinations, positioning them as desirable and progressive places to visit. In the process of preparing this chapter on the topic, specific limitations were encountered. Initially, stakeholders within the tourism sector were hesitant to provide genuine information. Additionally, time constraints restricted the depth of the study. Much research has been conducted on the impact of new technologies on the evolution of the greener aviation industry and the emerging new urban air mobility (UAM) (Farid et al., 2023), the role of perceived smart tourism technology experience for tourist satisfaction, happiness, and revisit intention (Yovcheva et al., 2013), investigating the impact of smart tourism technologies on tourists' experiences (Zhang et al., 2022), smart wellness technology for tourism destination based-on evolving tourist expectation model (Ramanathan et al., 2022); fewer researchers have taken smart tourist e-safety technologies and sustainable economic growth into consideration. This chapter aims to enhance the understanding of individuals in the tourism industry, academia, and policymakers regarding the incorporation of smart tourist e-safety technologies to attain sustainable economic growth.

15.6 IMPLICATIONS OF THE STUDY

15.6.1 THEORETICAL IMPLICATIONS

The book chapter plays an instrumental role in the establishment of a smart tourist e-safety framework and its role in attaining sustainable economic growth. At the core of the theoretical implications is the amalgamation of safety theories with technological advancements. While conventional safety theories predominantly focused on physical measures and personnel, the incorporation of smart e-safety technologies represents a revolutionary paradigm shift. The theoretical framework has evolved to include components such as data analytics, real-time communication, emergency response mechanisms, and artificial intelligence, broadening the understanding of how technology can be utilized to enhance safety in intricate tourism environments (Ordóñez-Martínez et al., 2024).

Moreover, this chapter aids in establishing a theoretical framework for the implementation of e-safety technologies in the realm of tourism. This framework includes theoretical concepts associated with surveillance and monitoring, data analytics, emergency response, authentication and access control, communication and information sharing, and cybersecurity. This book chapter delves into the dimensions of smart tourism and factors that are influencing smart tourism technologies on tourist e-safety. The deployment of smart technologies has the potential to enhance waste management and resource allocation, resulting in a substantial contribution to environmental sustainability (Sustacha et al., 2023). Additionally, this model is poised to offer valuable insights into the creation and fair allocation of value across the enhancement of tourist e-safety and e-security, providing tourists a sense of safety, appealing to an influx of tourists, and fostering confidence regarding their overall trip, as well as contributing to the tourism industry's sustainable economic growth.

15.6.2 PRACTICAL IMPLICATIONS

One of the key practical implications revolves around its direct impact on the safety of tourists. The implementation of smart e-safety technologies, including advanced surveillance systems and real-time communication platforms, substantially enhances the safety structure of tourist destinations (Zhang et al., 2022). For practical implications, this implies that destinations must invest in incorporating these technologies to create a secure environment for tourists. Essential to the process is the collaboration among local authorities, policymakers, and businesses, working together to develop and execute safety protocols reinforced by these technologies, ultimately guaranteeing the well-being of tourists and enhancing a positive image of the destination (Baggio et al., 2020).

Travel agencies, tour operators, accommodation providers, and other service providers can elevate their competitive edge by integrating these technologies into their operational frameworks. For example, hotels can choose to invest in secure and contactless check-in systems (Huang et al., 2017), and travel agencies can use real-time communication platforms to keep tourists updated about safety measures and changing circumstances. The practical inference is that businesses need to innovate and adjust to meet dynamic safety expectations, augmenting their array of services and overall competitiveness.

Furthermore, there are practical considerations concerning data privacy and cybersecurity. As smart technologies entail the collection and application of extensive data, there is a need for sturdy regulations and implications to protect the privacy of tourists (Sustacha et al., 2023). This encompasses the development and enforcement of data protection policies, guaranteeing that the incorporation of these technologies aligns with ethical standards and legal requisites.

KEYWORDS

- **qualitative method**
- **real-time monitoring**
- **smart technologies**
- **stakeholders**
- **sustainable economic development**
- **tourism industry**
- **tourist e-safety**

REFERENCES

Agarwal, S., Page, S. J., & Mawby, R. (2021). Tourist security, terrorism risk management and tourist safety. *Annals of Tourism Research, 89*, 103207.

Ali, S. A., Elsaid, S. A., Ateya, A. A., ElAffendi, M., & El-Latif, A. A. A. (2023). Enabling technologies for next-generation smart cities: A comprehensive review and research directions. *Future Internet, 15*(12), 1–43.

Archi, Y. El, Benbba, B., & Kabil, M. (2023). Digital technologies for sustainable tourism destinations: State of the art and research agenda. *Administrative Sciences, 13*(4), 1–22.

Baggio, R., Micera, R., & Del Chiappa, G. (2020). Smart tourism destinations: A critical reflection. *Journal of Hospitality and Tourism Technology, 11*(3), 407–423.

Buhalis, D., Harwood, T., Bogicevic, V., Viglia, G., Beldona, S., & Hofacker, C. (2019). Technological disruptions in services: Lessons from tourism and hospitality. *Journal of Service Management, 30*(4), 484–506.

Choi, Y., Hickerson, B., Lee, J., Lee, H., & Choe, Y. (2022). Digital tourism and wellbeing: Conceptual framework to examine technology effects of online travel media. *International Journal of Environmental Research and Public Health, 19*(9), 1–13.

Deb, S. K., & Mallik, N. (2023). Effects of social media in tourism marketing: Outlook on user generated content. *Journal of Digital Marketing and Communication, 3*(2), 49–65.

European Commission. (2019). *European Capital of Smart Tourism.* Available at: https://ec.europa.eu/growth/sectors/tourism/offer/smart-tourism_en (accessed on 8th July 2025).

Farid, S., Boudia, M. A., & Mwangi, G. (2023). Revolutionizing tourism: Harnessing the power of IoT in smart destinations. *Journal of Digital Marketing and Communication, 3*(2), 91–99.

Ganguly, P. (2018). Role of information and communication technologies (ICTs) to sustainable tourism development in Santiniketan, West Bengal–India. *International Journal of Management Studies, 5*(Special Issue 2), 54.

Glasgow, G., Chinwe, B., & Ile, P. N. M. (2023). Digital life and business transformation of gasoline firms in Rivers State of Nigeria. *European Journal of Finance and Management Sciences, 7*(3), 31–47.

Glensor, R. W., & Peak, K. J. (2019). *Crimes Against Tourists*. US Department of Justice, Office of Community Oriented Policing Services.

Hamidi, H. (2019). An approach to develop the smart health using Internet of Things and authentication based on biometric technology. *Future Generation Computer Systems, 91*, 434–449.

Huang, C. D., Goo, J., Nam, K., & Yoo, C. W. (2017). Smart tourism technologies in travel planning: The role of exploration and exploitation. *Information and Management, 54*(6), 757–770.

Khamis, A. (2021). *Smart Mobility: Exploring Foundational Technologies and Wider Impacts*. Apress.

Lawton, G. R. (2019). Tourist safety and security: Practical measures for destinations. *Tourism Management, 18*(6).

Lee, H., Lee, J., Chung, N., & Koo, C. (2018). Tourists' happiness: Are there smart tourism technology effects? *Asia Pacific Journal of Tourism Research, 23*(5), 486–501.

León-Gómez, A., Ruiz-Palomo, D., Fernández-Gámez, M. A., & García-Revilla, M. R. (2021). Sustainable tourism development and economic growth: Bibliometric review and analysis. *Sustainability, 14*(2), 1–19.

Lombok, P. S., Barat, N. T., Hamzanwadi, U., Barat, N. T., Info, A., & Tourism, M. (2023). Managing tourism in Rinjani circle region: How tourism business owners understand digital literacy. *International Journal on Social Science, Economics and Art, 13*(2), 101–111.

Loureiro, S. M. C., & Nascimento, J. (2021). Shaping a view on the influence of technologies on sustainable tourism. *Sustainability (Switzerland), 13*(22), 1–18.

Marcelo, J. S. (2023). Digitalized tourism practices of selected entrepreneurial hotels in the Philippines towards a safe travel experience. *Journal of Social Entrepreneurship Theory and Practice, 2*(1), 1–12.

McClelland, M., Grobbelaar, S., & Sacks. (2023). Exploring innovation system dynamics: Event history analysis of the evolution of the South African additive manufacturing industry. *Rapid Prototyping Journal, 29*(10), 2109–2133.

Mouratidis, K., Peters, S., & van Wee, B. (2021). Transportation technologies, sharing economy, and tele activities: Implications for built environment and travel. *Transportation Research Part D: Transport and Environment, 92*(1), 102716.

Norizawati, M. A., & Masron, T. (2014). Issues of safety and security: New challenging to Malaysia tourism industry. *SHS Web of Conferences, 12*, 01083.

Ordóñez-Martínez, D., Seguí-Pons, J. M., & Ruiz-Pérez, M. (2024). Conceptual framework and prospective analysis of EU tourism data spaces. *Sustainability (Switzerland), 16*(1), 1–33.

Otebolaku, A., & Lee, G. M. (2018). A framework for exploiting Internet of Things for context-aware trust-based personalized services. *Mobile Information Systems, 23*(3), 1–25.

Perwej, D. Y., Qamar Abbas, S., Pratap Dixit, J., Akhtar, D. N., & Kumar Jaiswal, A. (2021). A systematic literature review on the cyber security. *International Journal of Scientific Research and Management, 9*(12), 669–710.

Popescu, L. (2011). Safety and security in tourism. Case study: Romania. *Forum Geographic, 10*(2), 322–328.

Ramanathan, U., Williams, N. L., Zhang, M., Sa-Nguanjin, P., Garza-Reyes, J. A., & Borges, L. A. (2022). A new perspective of e-trust in the era of social media: Insights from customer satisfaction data. *IEEE Transactions on Engineering Management, 69*(4), 1417–1431.

Romão, J. (2021). Tourism, smart specialization and sustainable development. In *Tourism, Smart Specialization and Sustainable Development* (pp. 672–679).

Solis-Radilla, M. M., Hernández-Lobato, L., Callarisa Fiol, L. J., & Pastor Durán, H. T. (2019). The importance of sustainability in the loyalty to a tourist destination through the management of expectations and experiences. *Sustainability, 11*(4), 1–23.

Sustacha, I., Baños-Pino, J. F., & Del Valle, E. (2023). The role of technology in enhancing the tourism experience in smart destinations: A meta-analysis. *Journal of Destination Marketing and Management, 30*(3), 1–12.

Venkateswaran, R. (2021). Motivations for VPNs. *Technology Analysis and Strategic Management, 3*(1), 11–15.

Yovcheva, Z., Buhalis, D., & Gatzidis, C. (2013). Information and communication technologies in tourism 2013. *Information and Communication Technologies in Tourism, 2*(3), 24–35.

Zhang, Y., Sotiriadis, M., & Shen, S. (2022). Investigating the impact of smart tourism technologies on tourists' experiences. *Sustainability (Switzerland), 14*(5), 1–21.

CHAPTER 16

THE FUTURE OF BLOCKCHAIN IN RESTAURANT INDUSTRY: OPINION PERSPECTIVES

SARA ALAVIYEH[1] and HIRAN ROY[2*]

[1]*International School of Hospitality, Sports, and Tourism Management, Silberman College of Business, Fairleigh Dickinson University, Vancouver, Canada*

[2]*School of Management, Canadian University Dubai, Dubai, UAE*

*Corresponding author

ABSTRACT

Blockchain technology, with its core attributes of transparency, traceability, decentralization, and immutability, has significantly attracted scholarly and industry attention across various sectors. However, its potential applications in the restaurant sector remain insufficiently examined. This chapter provides a critical examination of blockchain's future role in transforming restaurant operations through enhanced supply chain management, food safety, authentication, waste reduction, and customer engagement mechanisms. The chapter presents an opinion-based perspective, drawing on existing literature and recent advancements to demonstrate how blockchain can address persistent challenges, from food traceability and quality assurance to payment security and sustainability practices. Furthermore, the study emphasizes how blockchain could foster consumer trust by enabling transparency across the entire food supply chain, which would be in line with current calls for environmentally conscious and ethical consumption. While the opportunities are promising, the chapter also outlines significant

Smart Tourism: Managing Technological Revolution in the Tourism Industry. Santus Kumar Deb (Ed.)

DOI: 10.1201/9781779641090-16

barriers including technical limitations, regulatory ambiguities, organizational resistance, and the requirement for industry-wide collaboration. These difficulties highlight the need for context-specific, strategic methods when incorporating blockchain technology into restaurant ecosystems. Finally, by clarifying blockchain's potential role in fostering more transparent, effective, and sustainable restaurant operations, this chapter adds to the growing conversation on digital innovation in the hospitality industry.

16.1 INTRODUCTION

The hospitality and tourism industries are inherently information-based. The main reason for this is that people want more information sabout the places they want to eat or visit. Nowadays, consumers are not just passive recipients of services; they are more knowledgeable, experienced, focused on their goals, and independent (Rejeb and Rejeb, 2019). The wide applications of blockchain technology in various industries arise from its ability to ensure immutability, transparency, programmability, decentralization, and distributed trust (Treiblmaier, 2019). Extant literature has shown that large organizations such as tour operators, airlines, and other services in the hospitality and tourism industry are using this technology (Leng et al., 2020); however, a few studies have been conducted in the restaurants context (Jang et al., 2024). All hospitality and tourism intermediaries must adjust to these rapid changes even more quickly, which is problematic since most stakeholders are unaware of the benefits and applications of blockchain technology (Melkic and Cvlek, 2020). Blockchain offers practical ways and solutions to transfer authority among economic participants and establish fairer competition between major and minor players in the hospitality and tourism industry (Tham and Sigala, 2020).

Although many studies have explored the advantages and the challenges of using blockchain in their business, for instance, Tham and Sigala (2020) and Thompson and Rust (2023) argued that consumer trust enhancing through the traceability of the food supply chain can be beneficial through the implementation of blockchain technology. Shirely (2022) further stressed that food safety hazards can be detected through this technology implementation in businesses. On the other hand, Jang et al. (2023) stated the challenges of implementation of blockchain technology, especially the concern of sharing the information among the business partners.

However, in the context of the food sector, especially restaurants, can significantly benefit from the implementation of blockchain technology.

This technology enhances transparency and visibility across the whole supply chain by enabling data traceability from start to finish in the food retail network (Mona Hussain, 2023). Blockchain can monitor and record information about food shipments, such as the amount and type of food, and keep relevant information about the restaurant owner or food provider (Filimonau and Naumova, 2020). Therefore, this chapter aims to present in detail the use of blockchain technology and how it would help the restaurant businesses.

16.2 BLOCKCHAIN TECHNOLOGY

16.2.1 DEFINITIONS AND CHARACTERISTICS

According to Treiblmaier (2019), a blockchain is a digital, decentralized, and distributed ledger that keeps and adds records of events and values in a way that cannot be changed. The main new thing about blockchain technology is that it provides distributed trust. This means that: (i) transactions do not need a trusted intermediary to go through; (ii) trade costs are lower; and (iii) transactions take less time. Feline and Lakhani (2018) further defined blockchain as a technology that is a revolutionary invention that has attracted the attention of companies and governments around the world. Blockchain technology can be done at three distinct levels (networks) depending on its aim and need (Nomura Research Institute, 2016): public, private (primarily beneficial for internal networking inside a single organization), and consortium (an expanded private network suitable for networking with multiple collaborating organizations). A private blockchain system entails vital business data related to a company, such as contracts, licenses, trade secrets, and strategies. On the other hand, a consortium blockchain includes the data of that company and its business partners and other organizations, like transaction data, communication information, and contracts (Marvin, 2017). Governments can create public networks between all organizations on a larger scale (Feng et al., 2018).

16.2.2 BLOCKCHAIN AND RESTAURANTS

Although blockchain technology has been associated with cryptocurrencies, it can also be used in food service businesses. The foodservice businesses encompass a variety of business establishments, such as restaurants, bars, cafes, clubs, snack bars, delivery kitchens, caterers, institutional food services, kiosks, and mobile food services (Vetter, 2021). The food sector can significantly benefit from the implementation of blockchain technology. This technology enhances transparency and visibility across the whole supply chain by enabling data traceability from start to finish in the food retail network (Mona Hussain, 2023). Blockchain can monitor and record information about food shipments, such as the amount and type of food, and keep relevant information about the restaurant owner or food provider (Filimonau and Naumova, 2020).

For better collaboration and efficient management between different hospitality and tourism sectors, it is necessary to embrace blockchain as the latest technology that has the potential to provide essential information to swiftly adapt, accommodate, and connect the various stakeholders involved. This helps meet the diverse needs of consumers and the growing demand for fast service delivery, and the most crucial factor in this process is transparency for all parties involved (Melnychenko et al., 2019). As a result, by offering transparent, decentralized, and uncorrupted records, blockchain technology can improve food reliability and supply chain tracking (Schinckus, 2020). However, questions may arise in implementing blockchain as a solution to the traceability problem. If one is wondering how to coordinate individual activities over the Internet without secured centralized storage, the answer is: blockchain makes use of distributed databases that are chronologically organized. These blocks are connected in a straight path and are not erasable.

Since there are so many stakeholders that need to enter data in the food supply chain, the next question is: How can one validate entries without a central authority to verify that a transaction is not fraudulent or invalid? Blockchain operates on a probabilistic basis. It uses mathematical puzzles to compel data moving over a network of computers to become more visible and verifiable. To be sure, how can one ensure that only legitimate transactions are recorded into a blockchain? Wright and De Filippi (2015) answered that only when all the network's computers agree that the transaction is legitimate will a new block of data be added to the end of the blockchain. Various voting methods can be used within the network to reach a consensus. Once a block has been added to a blockchain, it becomes immutable and cannot be removed. Furthermore, the transactions it encompasses can be readily accessed and authenticated by all individuals connected to the network. The information is stored as a permanent record that can be accessed by all computers on the network to coordinate actions or verify events.

16.2.3 BLOCKCHAIN INTEGRATION WITH THE RESTAURANT

The emergence of a homogenized "global palate" or "global cuisine" in tourism brought challenges to the restaurant supply chain (George et al., 2019). Creating network connections between food operators and tourism developments is problematic, and their cooperation is complex. Thus, the hospitality and tourism sectors must be equipped with advanced solutions that, in the context of intelligent tourism, guarantee the uniqueness of their offerings and their ability to be tracked (Baralla et al., 2019). George et al. (2019) investigated the novel concept of integrating blockchain technology with restaurants. This technology enhances transparency and visibility across the whole supply chain by enabling data traceability from start to finish in the food retail network (Mona Hussain, 2023). Blockchain can monitor and record information about food shipments, such as the amount and type of food, and keep relevant information about the restaurant owner or food provider (Filimonau and Naumova, 2020). Some key features of the blockchain technology integration are discussed in the subsections.

16.2.3.1 SUPPLY CHAIN TRACKING

Till now, the lack of adequate food quality traceability systems and the unavailability of reliable food quality indices for restaurants have hindered their capacity to achieve their value chain objectives. On the other hand, customers are now more concerned about fresh and additive-free food and their expectations for cross-seasonal products. Blockchain-based traceability is most important in the food industry because it can pinpoint the source of food items, letting buyers, sellers, and customers follow their journey from the fields to production and delivery (Musamih et al., 2021).

16.2.3.2 FOOD QUALITY

Providing fresh and tasty food is an integral part (e.g., driving out the competitors) of the restaurant business. Returning customers at the restaurants and customer satisfaction are based on food quality. Restaurant managers must understand the importance of high food quality and its direct impact on customer revisit intentions (Namkung and Jang, 2007). Implementation of blockchain technology in restaurants can bring new hope to the food quality control systems which ensure traceability starting from post-harvest to the dining table with its involvement of stakeholders in the value chain (George et al., 2019). Therefore, blockchain technology implementation in restaurants will change the entire business process by providing end-to-end visibility of the food supply chain, enhancing traceability and agility, and meeting consumer demand efficiently (Galvin, 2017).

16.2.3.3 FOOD SAFETY

Implementation of blockchain technology in restaurants would be a game changer, as it saves transaction times, removes overhead and intermediary costs, fraud, and cybercrime, and increases trust in shared processes and record keeping (Galvin, 2017). For example, Walmart has assisted IBM with the implementation of blockchain technology for ensuring food safety for products like pork and mango. This ensures visibility by capturing the data from the farm to the final consumer and creates more loyalty among consumers. It also helped Walmart to reduce the overhead, intermediary, and overhead costs and increased the trust of stakeholders through the shared process (Kshetri, 2018). Therefore, food safety can be enhanced via traceability with blockchain technology implementation in restaurants. Furthermore, the implementation of blockchain technology in restaurants can automate and track food expiration dates and expiration alerts. This aspect is especially important for the restaurant industry to comply with food safety regulations (Terziyska et al., 2024).

16.2.3.4 FOOD AUTHENTICATION

The role of blockchain in the context of food authentication would offer a clear and verifiable account of the complete supply chain. Barilla et al. (2019) claimed that tourists can confirm food authenticity by looking at menus and meals in detail. It enables travelers to view food history from farm to fork beyond the typical narrative and visual representations of restaurateurs. This can be done by scanning product barcodes or QR codes on a registered blockchain device (2019). Rejeb et al. (2020) highlighted the example of customers' Kosher and Halal food demands in this context. They noted that blockchain-enabled traceability ensures quality, enabling consumers to check the claims made by producers. The other example is when a restaurant can monitor the origin of its seafood to guarantee that it is obtained sustainably.

16.2.3.5 WASTE MANAGEMENT

Blockchain can enhance inventory management by offering real-time data on the flow of items in the supply chain. Having real-time visibility enables the prevention of excessive or insufficient stock of products, minimizes waste, and optimizes resource allocation. Implementing blockchain technology can aid restaurants in minimizing food wastage (Yele and Litoriya, 2024). It also enables restaurants to monitor their inventory levels and the expiry dates of their supplies. Additionally, this leads to decreased demand due to issues arising from indirect storage or handling concerns produced by third parties (George and Al-Ansari, 2023).

16.2.3.6 PAYMENT

Daily online food orders must submit delivery information to different restaurant systems, sometimes exchanging sensitive consumer data with partners without consent, putting client data at risk. Online food purchases are possible through blockchain, which maintains sensitive customer data that only the customer and restaurant can view, protecting customer privacy (Alves, 2021). Blockchain technology can enhance the efficiency and security of food delivery and payment operations. Blockchain-based payment systems allow clients to make payments for their food orders using cryptocurrencies or other digital tokens (Wood, 2023).

16.2.3.7 LOYALTY PROGRAM

Restaurants aim to optimize loyalty programs by utilizing blockchain technology, enhancing their efficiency and user-friendliness. The decentralized nature of blockchain enables connecting different companies, enhancing the overall user experience. Customers and brands could both benefit from this new idea, which could change the way loyalty programs work and make the entire process easier and more fun (Rana et al., 2022). A restaurant could generate its digital tokens and provide guests with rewards for each transaction they complete. This can be utilized to foster consumer loyalty and establish a pattern of repeating business. Furthermore, this could aid restaurants in overseeing consumer preferences and tailoring their menu accordingly. Implementing this strategy may increase a restaurant's profitability by cultivating customer loyalty and providing enhanced service (Wood, 2023).

16.2.3.8 SUSTAINABILITY

The tangible advantages of implementing blockchain technology in the hospitality and tourism industry, in terms of economic, environmental, and societal sustainability, have not been adequately proven. There is insufficient empirical and pragmatic evidence to support the notion that blockchain technology has a significant impact on the sustainability of the hospitality and tourism business. Most studies primarily examine examples that are focused on the pilot and planning stages (Kouhizadeh et al., 2021).

The food service sector has been specifically recognized as having a shared responsibility for climate change due to the substantial amounts of energy, water, and waste needed for food production and consumption, all of which significantly increase the sector's carbon footprint (Teng and Wu, 2019). Hospitality and tourism are an industry where tourists (consumers) and the host community (producers) interact directly with each other. This influences each person's desire to behave and demand tourism in a more sustainable way (UNWTO, 2019).

The food service business (such as restaurants) is responsible for creating approximately 34% of all anthropogenic carbon emissions (Vetter, 2021). This means that businesses that serve food should admit and take responsibility for their part in climate change by taking action to stop all practices that are bad for the environment. Hence, scholars emphasize the need for food service businesses (hereby restaurants) to adopt technologies that lower carbon emissions and the earth's temperature to combat climate change (Sharma et al., 2020). In addition, blockchain technology has drawn attention recently because of its enormous potential to mitigate climate change (Schinckus, 2020; Truby et al., 2022).

Because of the lack of visibility and transparency in the supply chain, the food and beverage (F&B) (e.g., restaurants) business has had to deal with costly products (Yang et al., 2022). Blockchain technology helps food and drink companies get traceability and immutability by letting them track data back to its original, unchanged source along the entire global supply chain. This keeps data safe while it is being managed throughout the process (Galvez et al., 2018). Saberi et al. (2019) stress that the implementation of blockchain technology can stop people, governments, or organizations from unfairly hurting the environment. It can also accurately track products that do not meet standards, cut down on rework and recalls by finding more transactions between companies, and lower resource use and greenhouse gas emissions.

16.3 CHALLENGES OF BLOCKCHAIN IN RESTAURANTS

Blockchain is thought to be a modern technology and an inevitable trend in business (e.g., restaurants). Adopters are worried about the youth of the technology, problems with the law, and a lack of clear business models, all of which are stopping it from becoming widely recognized. Technology that has not been fully developed may not be strong enough in terms of data throughput, scalability, and latency. A lack of this kind of information will hurt the potential of blockchain to help the hospitality and tourism businesses stay alive (Ranta et al., 2021; Toufaily et al., 2021). Some of the most popular features of challenges are discussed in the subsections.

16.3.1 TECHNICAL CHALLENGES

1. **Lack of Interoperability:** One of the main infrastructures of implementing blockchain is interoperability, which refers to the capacity of various apps and programs to process and establish connections across different ledgers and facilitating cross-chain communication, engagement, and value transfer. To enhance the sustainability of supply chain performance, different blockchain platforms in the food sector must be able to work together seamlessly. Nevertheless, the absence of interoperability remains a significant obstacle that impedes the successful execution of blockchain technology (Perrons and Cosby, 2020; Sahebi et al., 2020; Sanka et al., 2021).

2. **Lack of Security and Privacy:** Practitioners are currently engaged in discussions regarding the significant issue of security in public blockchains. While private blockchains are seen as safer, there are still concerns regarding data security in blockchain applications. Blockchain, in contrast to conventional centralized databases, enables the disclosure of more information to external parties due to the requirement of distributing data among several peers (Makhdoom et al., 2019; Sanka et al., 2021; Spathoulas et al., 2021).
3. **Cost of Implementation:** Blockchain is a feature-dependent technology; the final cost of implementation will change based on the needs of the project. A significant amount of money may need to be spent to set up the blockchain system at first, which could affect the long-term viability of the business (Lohmer and Lasch, 2020; Thakur et al., 2019). Hence, many restaurant businesses need help to afford the costly blockchain technology.

16.3.2 ORGANIZATIONAL CHALLENGES

1. **Lack of Expertise and Human Capital:** The scarcity of blockchain expertise continues to be a significant worry within the hospitality and tourism industry, raising doubts about its long-term viability. The recent progress and growing complexity of technology have widened the disparity between the need for skilled human resources and expertise and the availability of such workers (Choi et al., 2020; Helliar et al., 2020; Kamble et al., 2019).
2. **Conflicts With The Existing Business Culture:** The hospitality and tourism industry's current business procedures may limit the success of blockchain applications. Current bureaucratic processes and mindsets can pose significant challenges to creating sustainable blockchain projects (Toufaily et al., 2021). This dictates businesses currently rely on their experience to make decisions about using blockchain technology rather than considering suggestions from modern technology.
3. **Lack Of Top Management Commitment:** Top management support is crucial for the successful implementation of blockchain technology in restaurants. Organizational leadership is crucial for promoting innovation and facilitating technology adoption. However, some managers may not fully support blockchain

technology implementation initiatives. Insufficient support from senior management can hinder the success of blockchain applications in the hospitality and tourism industry (Li et al., 2019).

4. **Lack Of Collaboration And Coordination In The Supply Chain:** Effective adoption of blockchain in the hospitality and tourism industry requires collaboration and coordination among several stakeholders including government, developers, financial players, start-ups, regulators, accountants, audit businesses, and consultants. These stakeholders are regarded as essential components of blockchain initiatives. Similarly, increased collaboration and coordination have the potential to strengthen three key areas of sustainability in the sector (Toufaily et al., 2021).

16.3.3 REGULATION

Blockchain technology takes the Internet's cross-border features beyond information exchange and communication to the world of contracts and transactions. Blockchain's decentralization reduces the need for the system to be linked to a specific legal framework.

A blockchain's nodes can be located anywhere in the world. A nation may declare that any node within its borders is subject to its laws. The government's action is almost useless if the other nodes in the network deny the regulations' application. To put it another way, governments have minimal control over what happens on a blockchain because there is no central database or party. In terms of identity, it can raise problems. However, all transactions on open blockchains are public; it is not always evident who executed transactions on a public blockchain. In the Bitcoin Blockchain, each user has a public key that cannot be traced back to a specific individual. There is no central agency that regulates the Bitcoin Blockchain, making it difficult to identify the owner of a given account. In practice, without knowing the identity of a blockchain transaction partner, it is hard to sue them for any issues that arise which includes (Naves et al., 2018):

1. **Lack of Legislation and Standardization:** Laws that protect consumers and data must be considered by every blockchain project. The roles and duties of people in the market may change if blockchain is used to create a decentralized and distributed supply chain model. Because of this, a new governing system should take all these changes into account (Sharma et al., 2021; Wang et al., 2021).

2. **Lack of Governmental Support and Effective Incentive Programs:** Governments are not proactively supporting blockchain education and training. Investors may be cautious to invest in startups that offer innovative solutions for new business models, either through direct cash investment or tax incentives (Sharma et al., 2021; Wang et al., 2021).

Even though blockchain technology is modern technology, the above-mentioned concerns are not groundless. It is important to emphasize that to maximize revenue by introducing blockchain technology to the restaurant's ecosystem, restaurants should scrutinize suitable strategies that are suitable to their business environment and consumers' needs (Ghadge and Bourlakis, 2023).

16.4 CONCLUSION

The integration of blockchain technology presents a transformative opportunity for the restaurant industry, offering innovative solutions to longstanding challenges and paving the way for enhanced efficiency, transparency, and sustainability. As emphasized by Treiblmaier (2019), blockchain's inherent features, such as immutability, transparency, and decentralization, make it well-suited for revolutionizing various aspects of restaurant operations. By leveraging blockchain technology, restaurants can address critical issues in supply chain management, food authentication, waste management, payment systems, loyalty programs, and sustainability efforts.

The benefits of blockchain technology in the restaurant sector are evident across multiple dimensions. Firstly, blockchain enables enhanced transparency and visibility throughout the food supply chain, facilitating data traceability from farm to fork (Mona Hussain, 2023). This transparency not only fosters consumer trust but also helps restaurants ensure food safety and quality standards (Filimonau and Naumova, 2020). Moreover, blockchain-based traceability systems empower consumers to make informed choices about the food they consume, whether it be verifying the authenticity of ingredients or ensuring compliance with dietary preferences such as Kosher or Halal (Rejeb et al., 2020). Additionally, blockchain technology offers significant potential for improving waste management and inventory optimization in restaurants. Real-time data tracking facilitated by blockchain enables eateries to minimize food wastage, optimize inventory levels, and streamline resource allocation (George and Al-Ansari, 2023). By reducing inefficiencies in the supply chain, restaurants can not only lower costs but also contribute to environmental sustainability by mitigating their carbon footprint (Teng and Wu, 2019).

However, the adoption of blockchain in the restaurant sector is not without challenges. Technical hurdles such as interoperability, security concerns, and high implementation costs pose significant obstacles (Sanka et al., 2021). Moreover, organizational challenges including a lack of expertise, resistance to change, and regulatory uncertainties hinder the widespread adoption of blockchain technology in restaurants (Choi et al., 2020). Overcoming these challenges will require collaborative efforts from industry stakeholders, government support, and the development of clear regulatory frameworks. Despite these obstacles, the potential benefits of blockchain technology in enhancing efficiency, transparency, and sustainability in the restaurant industry make it a promising avenue for further exploration and implementation (Melnychenko et al., 2019; Treiblmaier, 2019).

KEYWORDS

- **blockchain technology**
- **decentralization**
- **food supply chain**
- **hospitality**
- **tourism industry**
- **transparency**

REFERENCES

Alves, D. (2021). Proof-of-concept (POC) of restaurant's food requests in the Lisk blockchain/sidechain. *Journal of Physics: Conference Series, 1828*(1), 012110. https://doi.org/10.1088/1742-6596/1828/1/012110 (accessed on 8th July 2025).

Baralla, G., Ibba, S., Marchesi, M., Tonelli, R., & Missineo, S. (2019). A blockchain-based system to ensure transparency and reliability in food supply chain. In Mencagli, G. et al. (Eds.), *Euro-Par 2018: Parallel Processing Workshops. Lecture Notes in Computer Science, 11339* (pp. 393–403). Springer. https://doi.org/10.1007/978-3-030-10549-5_30 (accessed on 8th July 2025).

Choi, D., Chung, C. Y., Seyha, T., & Young, J. (2020). Factors affecting organizations' resistance to the adoption of blockchain technology in supply networks. *Sustainability, 12*(21), 8882. https://doi.org/10.3390/su12218882 (accessed on 8th July 2025).

Felin, T., & Lakhani, K. (2018). What problems will you solve with blockchain? *MIT Sloan Management*, 32–38.

Feng, Q., He, D., Zeadally, S., Khan, M. K., & Kumar, N. (2019). A survey on privacy protection in the blockchain system. *Journal of Network and Computer Applications, 126*, 45–58. https://doi.org/10.1016/j.jnca.2018.10.020 (accessed on 8th July 2025).

Filimonau, V., & Naumova, E. (2020). The blockchain technology and the scope of its application in hospitality operations. *International Journal of Hospitality Management, 87*, 102383. https://doi.org/10.1016/j.ijhm.2019.102383 (accessed on 8th July 2025).

Galvez, J. F., Mejuto, J. C., & Simal-Gandara, J. (2018). Future challenges on the use of blockchain for food traceability analysis. *TrAC Trends in Analytical Chemistry, 107*, 222–232.

Galvin, R. (2017). IBM and Walmart: Blockchain for food safety. https://blockchain578369954.wordpress.com/wp-content/uploads/2018/06/6-using-blockchain-for-food-safe-2.pdf (accessed on 8th July 2025).

George, R. V., Harsh, H. O., Ray, P., & Babu, A. K. (2019). Food quality traceability prototype for restaurants using blockchain and food quality data index. *Journal of Cleaner Production, 240*, 118021. https://doi.org/10.1016/j.jclepro.2019.118021 (accessed on 8th July 2025).

George, W., & Al-Ansari, T. (2023). Review of blockchain applications in food supply chains. *Blockchains, 1*(1), 34–57.

Helliar, C. V., Crawford, L., Rocca, L., Teodori, C., & Veneziani, M. (2020). Permissionless and permissioned blockchain diffusion. *International Journal of Information Management, 54*, 102136. https://doi.org/10.1016/j.ijinfomgt.2020.102136 (accessed on 8th July 2025).

Jang, H. W., Jung, H. S., & Cho, M. (2024). Blockchain adoption in the food and beverage industry from a behavioral reasoning perspective: Moderating roles of supply chain partnerships. *Journal of Hospitality and Tourism Technology, 15*(1), 138–155.

Jang, H. W., Yoo, J. J. E., & Cho, M. (2023). Resistance to blockchain adoption in the foodservice industry: Moderating roles of public pressures and climate change awareness. *International Journal of Contemporary Hospitality Management*.

Kamble, S., Gunasekaran, A., & Arha, H. (2018). Understanding the blockchain technology adoption in supply chains—Indian context. *International Journal of Production Research, 57*(7), 2009–2033. https://doi.org/10.1080/00207543.2018.1518610 (accessed on 8th July 2025).

Kouhizadeh, M., Saberi, S., & Sarkis, J. (2021). Blockchain technology and the sustainable supply chain: Theoretically exploring adoption barriers. *International Journal of Production Economics, 231*, 107831. https://doi.org/10.1016/j.ijpe.2020.107831 (accessed on 8th July 2025).

Kshetri, N. (2018). Blockchain's roles in meeting key supply chain management objectives. *International Journal of Information Management, 39*, 80–89.

Leng, J., Ruan, G., Jiang, P., Xu, K., Liu, Q., Zhou, X., & Liu, C. (2020). Blockchain-empowered sustainable manufacturing and product lifecycle management in Industry

4.0: A survey. *Renewable and Sustainable Energy Reviews, 132*, 110112. https://doi.org/10.1016/j.rser.2020.110112 (accessed on 8th July 2025).

Li, H., Pei, L., Liao, D., Sun, G., & Xu, D. (2019). Blockchain meets VANET: An architecture for identity and location privacy protection in VANET. *Peer-to-Peer Networking and Applications, 12*(5), 1178–1193. https://doi.org/10.1007/s12083-019-00786-4 (accessed on 8th July 2025).

Lohmer, J., & Lasch, R. (2020). Blockchain in operations management and manufacturing: Potential and barriers. *Computers & Industrial Engineering, 149*, 106789. https://doi.org/10.1016/j.cie.2020.106789 (accessed on 8th July 2025).

Makhdoom, I., Abolhasan, M., Abbas, H., & Ni, W. (2019). Blockchain's adoption in IoT: The challenges, and a way forward. *Journal of Network and Computer Applications, 125*, 251–279. https://doi.org/10.1016/j.jnca.2018.10.019 (accessed on 8th July 2025).

Marvin, R. (2017). Blockchain: The invisible technology that's changing the world. *PCMAG*. https://www.pcmag.com/news/blockchain-the-invisible-technology-thats-changing-the-world (accessed on 8th July 2025).

Melkić, S., & Čavlek, N. (2020). The impact of blockchain technology on tourism intermediation. *Tourism, 68*(2), 130–143. https://doi.org/10.37741/t.68.2.2 (accessed on 8th July 2025).

Melnychenko, S., Mazaraki, N., & Tkachuk, T. (2019). Leading trends in tourism: Blockchain in franchising. In *3rd International Conference on Social, Economic, and Academic Leadership (ICSEAL 2019)* (pp. 388–395). Atlantis Press.

Mona Hussain, A. A. H. (2023). Evaluating the efficacy and drawbacks of traceability technology in the quality assurance of the food supply chain. *Advances in Food Production, Processing, and Nutrition, 1*(1), 001–010. https://doi.org/10.17352/afppn.000001 (accessed on 8th July 2025).

Musamih, A., Salah, K., Jayaraman, R., Arshad, J., Debe, M., Al-Hammadi, Y., & Ellahham, S. (2021). A blockchain-based approach for drug traceability in healthcare supply chain. *IEEE Access, 9*, 9728–9743. https://doi.org/10.1109/access.2021.3049920 (accessed on 8th July 2025).

Namkung, Y., & Jang, S. (2007). Does food quality really matter in restaurants? Its impact on customer satisfaction and behavioral intentions. *Journal of Hospitality & Tourism Research, 31*(3), 387–409.

Nomura Research Institute. (2016). *Survey on Blockchain Technologies and Related Services. FY2015 Report*. Japan's Ministry of Economy, Trade, and Industry (METI), 1–78.

Perrons, R. K., & Cosby, T. (2020). Applying blockchain in the geoenergy domain: The road to interoperability and standards. *Applied Energy, 262*, 114545. https://doi.org/10.1016/j.apenergy.2020.114545 (accessed on 8th July 2025).

Rana, R. L., Adamashvili, N., & Tricase, C. (2022). The impact of blockchain technology adoption on tourism industry: A systematic literature review. *Sustainability, 14*(12), 7383. https://doi.org/10.3390/su14127383 (accessed on 8th July 2025).

Rejeb, A., & Rejeb, K. (2019). Blockchain technology in tourism: Applications and possibilities. *World Scientific News*, *119*, 119–144.

Rejeb, A., Keogh, J. G., Zailani, S., Treiblmaier, H., & Rejeb, K. (2020). Blockchain technology in the food industry: A review of potentials, challenges and future research directions. *Logistics, 4*(4), 27. https://doi.org/10.3390/logistics4040027 (accessed on 8th July 2025).

Saberi, S., Kouhizadeh, M., Sarkis, J., & Shen, L. (2019). Blockchain technology and its relationships to sustainable supply chain management. *International Journal of Production Research, 57*(7), 2117–2135.

Sahebi, I. G., Masoomi, B., & Ghorbani, S. (2020). Expert oriented approach for analyzing the blockchain adoption barriers in humanitarian supply chain. *Technology in Society, 63*, 101427. https://doi.org/10.1016/j.techsoc.2020.101427 (accessed on 8th July 2025).

Sanka, A. I., Irfan, M., Huang, I., & Cheung, R. C. (2021). A survey of breakthrough in blockchain technology: Adoptions, applications, challenges, and future research. *Computer Communications, 169*, 179–201.

Schinckus, C. (2020). The good, the bad and the ugly: An overview of the sustainability of blockchain technology. *Energy Research & Social Science, 69*, 101614.

Sharma, S., & Bumb, A. (2021). Marketing at the bottom of the pyramid: Overcoming the challenges through MICMAC analysis. *Journal of Global Marketing, 34*(4), 292–307. https://doi.org/10.1080/08911762.2021.1884924 (accessed on 8th July 2025).

Shirley, E. (2022). Blockchain challenges in the food and beverage industry. *Clarkston Consulting News*. https://clarkstonconsulting.com/insights/blockchain-challenges/ (accessed on 8th July 2025).

Spathoulas, G., Negka, L., Pandey, P., & Katsikas, S. (2021). Can blockchain technology enhance security and privacy in the Internet of Things? In G. A. Tsihrintzis & M. Virvou (Eds.), *Advances in Core Computer Science-Based Technologies: Papers in Honor of Professor Nikolaos Alexandris* (pp. 199–228). Springer International Publishing. https://doi.org/10.1007/978-3-030-41196-1_10 (accessed on 8th July 2025).

Teng, Y. M., & Wu, K. S. (2019). Sustainability development in hospitality: The effect of perceived value on customers' green restaurant behavioral intention. *Sustainability, 11*(7), 1987. https://doi.org/10.3390/su11071987 (accessed on 8th July 2025).

Terziyska, M., Nikolova-Alexieva, V., & Valeva, K. (2024). Application of blockchain technologies in the supply chain in food enterprises. In *BIO Web of Conferences, 102*, 04010. EDP Sciences.

Thakur, V., Doja, M. N., Ahmad, T., Khadanga, G., & Pal, M. (2019). Interoperability issues and data models in integrating land administration systems in India. In *International Conference on e-Governance and Education—Transforming Lives for e-Living*.

Tham, A., & Sigala, M. (2020). Roadblock(chain): Bit(coin)s for tourism sustainable development goals? *Journal of Hospitality and Tourism Technology, 11*(2), 203–222. https://doi.org/10.1108/jhtt-05-2019-0069 (accessed on 8th July 2025).

Thompson, B. S., & Rust, S. (2023). Blocking blockchain: Examining the social, cultural, and institutional factors causing innovation resistance to digital technology in seafood supply chains. *Technology in Society, 73*, 102235.

Toufaily, E., Zalan, T., & Dhaou, S. B. (2021). A framework of blockchain technology adoption: An investigation of challenges and expected value. *Information & Management, 58*(3), 103444. https://doi.org/10.1016/j.im.2021.103444 (accessed on 8th July 2025).

Treiblmaier, H. (2019). Toward more rigorous blockchain research: Recommendations for writing blockchain case studies. *Frontiers in Blockchain, 2*. https://doi.org/10.3389/fbloc.2019.00003 (accessed on 8th July 2025).

Truby, J., Brown, R. D., Dahdal, A., & Ibrahim, I. (2022). Blockchain, climate damage, and death: Policy interventions to reduce the carbon emissions, mortality, and net-zero implications of non-fungible tokens and bitcoin. *Energy Research & Social Science, 88*, 102499.

Vetter, D. (2021). How much does our food contribute to global warming? New research reveals all. *Forbes*. https://www.forbes.com/sites/davidrvetter/2021/03/10/how-much-does-our-food-contribute-to-global-warming-new-research-reveals-all/?sh=5b563cdd27d7 (accessed on 8th July 2025).

Vu, N., Ghadge, A., & Bourlakis, M. (2023). Blockchain adoption in food supply chains: A review and implementation framework. *Production Planning & Control, 34*(6), 506–523.

Wang, Y., Han, J. H., & Beynon-Davies, P. (2019). Understanding blockchain technology for future supply chains: A systematic literature review and research agenda. *Supply Chain Management: An International Journal, 24*(1), 62–84. https://doi.org/10.1108/scm-03-2018-0148 (accessed on 8th July 2025).

Wood, J. (2023). Blockchain–8 ways blockchain is revolutionizing the restaurant industry. *Bloc for Business*. https://business.getonbloc.com/blockchain-in-the-restaurant-industry (accessed on 8th July 2025).

World Tourism Organization. (2019). *Baseline report On the Integration of Sustainable Consumption and Production Pattern into Tourism Policies*. https://www.e-unwto.org/doi/pdf/10.18111/9789284420605?download=true (accessed on 8th July 2025).

Wright, A., & De Filippi, P. (2015). Decentralized blockchain technology and the rise of lex cryptographic. *SSRN Electronic Journal*. https://doi.org/10.2139/ssrn.2580664 (accessed on 8th July 2025).

Yang, Y., Liu, Y., Lv, X., Ai, J., & Li, Y. (2022). Anthropomorphism and customers' willingness to use artificial intelligence service agents. *Journal of Hospitality Marketing & Management, 31*(1), 1–23.

Yele, S., & Litoriya, R. (2024). Blockchain-based secure dining: Enhancing safety, transparency, and traceability in food consumption environment. *Blockchain: Research and Applications*, 100187.

CHAPTER 17

SMART TOURISM AND CRISIS RECOVERY: A STRATEGIC RESPONSE TO CHALLENGES

SANTUS KUMAR DEB[1] and BAPON CHANDRA KURI[2*]

[1]*Department of Tourism and Hospitality Management, University of Dhaka, Dhaka, Bangladesh*

[2]*Department of Tourism and Hospitality Management, Gopalganj Science & Technology University, Gopalganj, Bangladesh*

*Corresponding author

ABSTRACT

The purpose of the chapter is the adaptation of the tourism industry through digitalization and service automation in crises. This approach spikes the necessity of "Smart Tourism" as most countries are still struggling with the manmade (war, political issues) and nature-based (Covid-19, Monkeypox) crisis. Besides this, Augmented Reality (AR), Artificial Intelligence (AI), and Virtual Reality (VR) are the main paradigms of "Smart Tourism" that provide added information about the tourism products from the tourist's perspective, attitude, language, image, tone, and expectation. This study used a qualitative approach which will be followed by a focus group discussion. Data was collected from both primary and secondary sources. Additionally, the researchers accumulate information from experts in the tourism industry and IT sector to extract the outcome through thematic analysis. We find the intensive linkage between service automation and tourist involvement during the crisis. Furthermore, we illustrate a crisis recovery model to

Smart Tourism: Managing Technological Revolution in the Tourism Industry. Santus Kumar Deb (Ed.)

DOI: 10.1201/9781779641090-17

make a strategic response to challenges for the tourism industry. They skipped the contribution of smart tourism to manage challenges during a crisis. However, this pioneering study explores how the tourism industry can adapt to the crisis through "Smart Tourism."

17.1 INTRODUCTION

Faulkner (2013) has found an alarming number of crises influencing the travel and tourism industry including natural disasters, economic crises, safety and security issues, and climate migration which are the rudimentary challenges for the global tourism and hospitality sector. A crisis is typically characterized as a negative incident that harms places' or organizations' activities (Berbekova et al., 2020). According to Racherla and Hu (2009), crises are described as unexpected, unpredictable, and ambiguous events that deteriorate social, political, and environmental ambiance. However, there are many traditional approaches have been applied to manage crises. Recently, the utilization of SMART technology has been widely utilized by tourism and hospitality stakeholders to tackle the crisis effectively and efficiently. Furthermore, the importance of digital transformation of tourism becomes more evident when the world has experienced the inexorable miseries of the COVID-19 crisis (Zhong et al., 2021). The tourism industry is facing challenges due to the coronavirus crisis, despite its short-term destructive effects including ineffective crisis management strategies in the tourism industry, technological hazards, natural disasters, terrorism, economic breakdown, and the spreading of disease (Kukanja et al., 2020). This is an additional chance to reconsider the tourism sector from a critical standpoint.

Therefore, people have realized the paramount significance of digital or SMART tourism at the moment of complete lockdown. However, digital tourism is the buzzword in the field of travel and hospitality which denotes the technological involvement among all stakeholders in their relative sectors. It allows the tourists to explore all the facilities with the help of modern technology and sophistication. Most of the countries have declared themselves as digitalized nations and now they are thriving towards the smart countries. Scholars have depicted that digitalization is the basic foundation of smart rules and protocols in the tourism and travel sector (Gretzel et al., 2015). For example, Bangladesh has fulfilled the all criteria to claim itself as a digitalized country. Now, it accelerates its movement for the attainment of a smart vision and mission. This country breaks down its mission into two or more specific objectives, like smart citizen, smart community, smart business, e-governance, smart city, and smart economy (Pal et al., 2023). Therefore, countries like Bangladesh have taken pragmatic measures to become smart countries in this cutting-edge society. Likewise, UNWTO believes that if a country turns into a smart country, then "SMART TOURISM" becomes a more lucrative sector for tourists (Almeida et al., 2020). However, recent crises like natural diseases including Covid 19, monkeypox, Ebola, and SARS virus, natural disasters including Tsunami, earthquakes, sea level rises, climate migration, and manmade crises like economic, war, crime, safety security, pollution, resources crisis make the tourism industry clogged for the last few years (Paliwal et al., 2022). Even, tourists become more concerned about these issues, which dwindled their regular travel towards a particular destination. They feel more fascinated to visit a secure place with all the facilities at a time.

Therefore, the concept of "SMART tourism" opens a new door for the tourist to embrace all the facilities of a tourist destination from any location. It is truly an excellent alternative form of tourism that can facilitate tourists how to deal and cope with crisis. Smart tourism also caters to the tourists to get attached more to technological intervention before, during, and after the crisis. According to Boes et al. (2015), Cornejo Ortega and Malcolm (2020), and Gretzel (2018), a smart city raises the standard of living for their residents while also increasing the city's competitiveness. According to Cohen (2022), a smart town has evolved in all six patterns: smart citizen, smart governance, smart ecology, smart movement, smart investment, and smart residing. This chapter has proposed a smart framework that is coupled with digital transformation and SMART involvement with all stakeholders to lessen the effect of the crisis. However, it is more challenging to introduce technological adaptation in every sphere of human communication. The challenges to getting involved people in SMART tourism:

- Lack of information about Smart technologies in the peripheral area;
- Purchasing power to buy new technology;
- Lack of skill and idea in using technologies;
- Misuse of technologies, like hacking, pornography, and drug dealing;
- Social chaos like age limitation in using smartphones;
- Accessibility of Wi-Fi, net, or satellite connection;

- Slowness of the net connection;
- Availability of power or electricity in destination;
- People's attitude towards using technologies;
- Lack of skilled technological personnel;
- Technological blasting like phone blast or fire;
- Causes Chronic diseases like cancer, and ear and eyesight problems;
- Fear of social safety includes cyberbullying.

17.2 LITERATURE REVIEW

As mentioned earlier, the fragility of the tourism industry is evident in its sensitivity to several sorts of crises, including contagious disease, natural catastrophes, terrorist attacks, and social turbulence (Femenia-Serra and Ivars-Baidal, 2021; Hosie et al., 2016; Lau et al., 2022; Ritchie 2004; Rossello et al., 2020). In other words, Kim et al. (2020) depicted that the social and political environment spreads social chaos, political riot, and diseases. For example, China experienced an exorable earthquake in 2008 that caused 12.7 billion US dollars in economic damages in the Sichuan region (Yang et al., 2021). The tourism industry was highly impeded due to terrorist assaults in the US in September 2001 (Pforr, 2006). According to Pestronk (2008), among all forms of emergencies, the contagious virus has an enormous effect on the travel and tourism business. For example, Britain witnessed a mad cow disease outbreak in the 1990s, and the foot and mouth epidemic debacle the tourism industry in England and Ireland between 2000 and 2001, resulting in a 5% drop in tourism arrival (Ferguson, 2007). Furthermore, China faced an unrecoverable loss of 16.8 billion USD, and foreign tourist arrivals decreased by 60% due to the SARS epidemic in 2003, (Huang, 2004) and Asian tourism industry saw a significant negative impact on employment and the country's GDP.

In 2004, Kenya saw an outbreak of the Chikungunya virus that resulted in a huge loss for the aviation industry, and tourist arrival was completely stopped for two years. Even, one-third of the French population got infected by Chikungunya and it cost 5 million USD in tourism business. This virus had been scattered in Gujrat, costing 6 million USD in tourism rates, Malaysia 65 million USD, and Thailand 363 million USD (Vora et al., 2009). Moreover, West Africa saw the inception of the Ebola pandemic in 2014, spreading progressively to Sierra Leone, Nigeria, Liberia, and Guinea. The Ebola disease significantly impacted the economics of many

African countries, resulting in a significant downsizing in the travel industry (Dixon et al., 2014). Therefore, to investigate the severity of the crisis Sönmez and Graefe (1998) found a significant association between terrorism risks and the process of choosing a place to travel. Moreover, Dev et al. (2023) investigated the adaptability of supply chain management in the hospitality industry during the crisis. On the other hand, Neumayer (2004) concatenated crisis management into three stages—pre-crisis, crisis, and post-crisis—which was generally acknowledged in the reviewed literature and concentrated on the effects of political unrest. Furthermore, Wohlmuther (2014) investigated the roles and responsibilities of tourism in mitigating the crisis. On the other hand, Martens (2005) examined the Impact of a crisis on tourist's decision-making and risk perception.

Dobrowolski (2020) explored the reorientation of crisis management in crisis after Covid 19. However, many scholars have reported that natural disasters (Brammer, 1990), war (Cannon et al., 1994), food crisesBerke, 1998, economic crises, and climate migration Gretzel and Collier-de-Mendonça (2019 are responsible for increasing geographical and social separation around the world. Richardson (1994) debated that the growing population, urbanization, expansion of people's habitation, and excessive use of technology lead to cybercrime and digital security issues in the online podium. Therefore, this chapter seeks to solve these weaknesses and identify several crucial factors of crisis management strategy by introducing modern technological sophistication in tourism destinations. This book chapter has examined how potential crises can be minimized or eradicated with the adoption of modern technology. This study also illustrated a crisis hypothesis frame along with crisis management by utilizing smart technology which triggers the formulation of crisis resilience theory.

The introduction of technological evolvement in the tourism sector has brought a new horizon in tourist destinations (Baggio et al., 2020). However, the writers touched on topics like human capital (Boes et al., 2015; Del et al., 2015; Li, et al., 2021), governance, novelty, demographic analysis, and financial investment associated with technology (Gretzel et al., 2015). Smart destinations represent a new paradigm for destination branding, promotion, and management Ivars-Baidal et al. (2019). Currently, the majority of smart experiences in tourism are delivered through personal devices, such as smartphones and related mobile apps. As a result, travelers who are tech-savvy and open to sharing and receiving information are needed (De Sausmarez, 2013; Hystad and Keller, 2008).

To counteract the dependence on smartphones, new advancements in smart infrastructure offer substitute interfaces including smart software, smart exhibitions, and smart keys (Lee et al., 2017). Therefore, this book uses a crisis management analysis by using smart tools and tourism, which form the foundation of many smart resilience concepts, to outline the areas that a thorough crisis management framework should take into account. The chapter is prepared as follows: first, we explore the qualitative technique used and explain thematic analysis as the primary method of investigation. We then proceed with the types of challenges in the tourism industry, providing a full examination of crisis tools and the digital transformation of destinations to manage crises. In the fourth section, this chapter highlighted the crisis hypothesis model with thematic judgment. The final section discusses the thematic analysis of crisis management and SMART resilience theory and offers a brief overview of the sense-making viewpoint and its implication in the travel and tourism sector.

17.3 METHODOLOGY

This book chapter has been developed from data collected via primary and secondary sources. Thematic analysis has been applied to figure out how the digital transformation of tourism contributes to tackling the crisis. Therefore, IT experts and academicians from public universities have given vital information on how to form adaptive measures in crisis using digital tools and technology. Moreover, researchers arranged a seminar on "Smart Technology and Crisis Management." Therefore, numerous collective information has been coded to extract the crucial coping strategy during a pandemic situation of any state or region. The researcher has designed a semi-structured questionnaire for this research work. After that, the entire chapter has been evaluated by the editor of the book to verify its validity and authenticity. Several secondary sources of information were gathered from Google Scholar, newspapers, journals, books, and magazines.

17.4 KEY RISKS AND CHALLENGES

Crisis management enables a business to deal with an unexpected danger or risk situation and then swiftly return to regular business operations. However, SMART tourism can play a fundamental role in case of mitigating crises and inspire tourists to use technological tools to enjoy the essence of tourism in an alternative form. For instance, an extraordinary situation has been brought on by COVID-19 which had a significant influence on every area of the economy particularly on the travel sector and the number of foreign tourist influxes decreased by 74% in 2020 (Deb et al., 2023). In those particular circumstances, travel sectors were solely blamed for spreading the COVID-19 virus around the world. Therefore, the UNWTO has classified tourism risk into five broad categories, including environmental (deforestation, climate change, natural catastrophes, climate refugee), societal and political (crime, forced migration, war), health-related (disease, pandemic), Technological (Hacking, online bullying, transportation fault, and accident), economic breakdown (Ural, 2016). However, this chapter has illustrated numerous crises and risks that are caused by natural and man-made interventions (Table 17.1).

In conclusion, several pragmatic steps have been taken to curtail the negative impact of the crisis by arranging dialogue among the countries. The world does not see any fruitful outcomes to abate the impact of the crisis. Therefore, to measure the influence of the current crisis technological involvement is now inevitable to create awareness among the people of this cutting-edge society.

17.5 SMART TECHNOLOGICAL TOOLS FOR MANAGING CRISIS

However, the previous chapters have already discussed major SMART technological tools and their importance for the tourism and travel industry. For instance, AR (Augmented Reality) refers to additional information automatically delivered to customers while they are roaming with their SMART technology thus helping to gather more augmented data about the destination, travel, food and beverage, and most significantly about the crisis. On the other hand, VR (Virtual Reality) delineates enjoying real-life experiences with the help of technological involvement with the human body from any location across the world. Additionally, AI, Google Alerts, Fractal Analysis, Factal, TweetDeck, SumAll, Veoci, and Konnect Insights are major logistic supports to promote the digital transformation of tourism and hospitality (Sharma et al., 2023). This chapter illustrates a security system to protect SMART technologies in the crisis domain such as (Figure 17.1):

TABLE 17.1 Crisis in the Tourism and Hospitality Industry

	Manmade Crisis
Variable	**Examples**
Socio-Cultural	Culture commoditization
Political crisis	Riot, Hartal, Detention
Economic Crisis	Overpricing, fraud, embezzlement
Human resources Deficiencies	Unskilled human resources,
Service and Facilities crisis	Lack of medical, hospitality, sports, and other hospitality-related service.
Technological / Transportation	Airline, cruise, road, and rail accidents. Technological blast and breakdown.
Health Related	Pandemics
Product Deficiencies	Scarcity of resources, ineffective supply chain.
Property Damage	Construction faults, loss, and injury
Safety and Security	Vandalism, theft, hacking.
Psychological	Propaganda, rumor, negative word of mouth.
Terrorism, War, and Crime	Child trafficking, gun war, drugs.
Forced Migration	Isolated from the origin.
Pollution	Air, water, soil, deforestation, and misuse of plastic products
Tourism and Hospitality sector	Lack of knowledge, information, hotel, unfit transportation, adulterated food and beverage, and entertainment facilities.
	Natural Crisis
Variable	**Examples**
Tourist seasonality	Push and pull factors, extrinsic and intrinsic rewards
Disease	SARS, COVID-19, Monkeypox, Ebola.
Natural Disaster	Earth quick, volcano eruption, heavy rainfall, forest fire, excessively hot and cold weather, sea level rises, Strom, tsunami.
Climate Refugee	Climate migration.

Source: Authors compilation and COMCEC (2017).

1. **Strong Firewall System:** A computer network security system called a firewall that restricts unauthorized entering and leaving in the private and public network. Therefore, hackers and crackers do not get access to break the network or technological system.

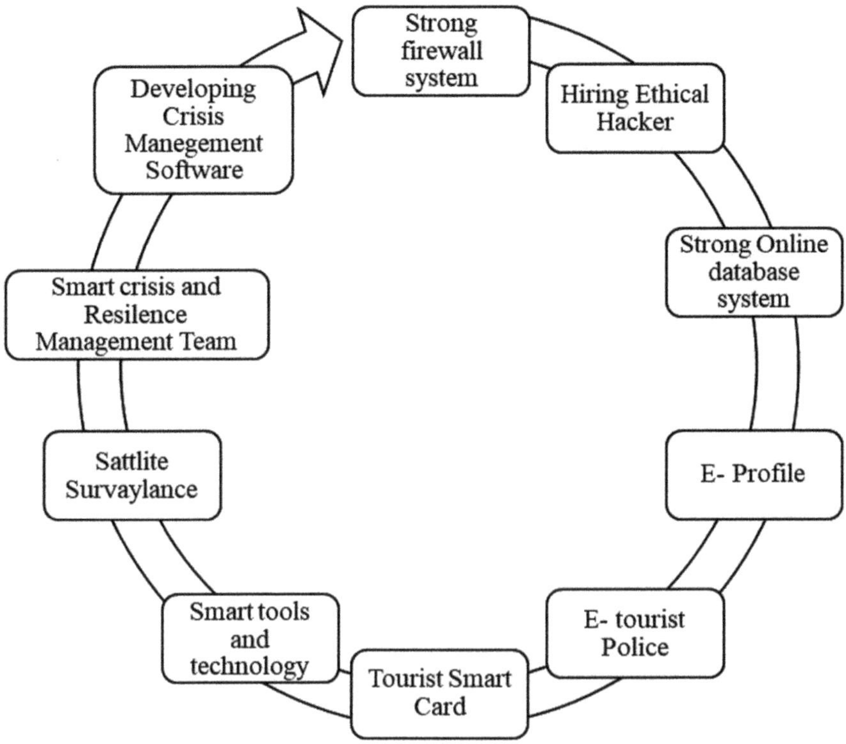

FIGURE 17.1 Smart technological tools to manage a crisis.

Source: Authors compilation.

2. **Hiring Ethical Hackers:** These types of hackers assist in building a strong security system and breach a computer network to assess or test its security, as opposed to doing so maliciously or illegally.
3. **Online Database System:** An online database allows an organization to provide employees—or even tourists and suppliers—with access to vital information. For example, this database system will store previous, current, and future crisis data for stakeholders.
4. **E-profile:** The e-profile system includes information about tourists, employees, suppliers, and other tourism stakeholders in an online database system with maximum security. For example, digital passports and e-voter IDs are the fundamental tools to identify a person.
5. **E-tourist Police:** E-tourist police combat cybercrime, provide safety to tourists, and ensure online secured information management.

In Bangladesh, one can easily communicate with police service by dialing 999 for quick response.

6. **Tourist Smart Card:** This type of card will facilitate the tourist to get access to any destination and service without any hassle. This card will store information about tourist demographic status and the purpose of traveling.
7. **SMART Tools and Technologies:** AI, VR, and AR are the more common technological tools to define Smart tourism. It also encompasses smart citizens, smart cities, and smart destinations to fulfill the embodiment of transforming tourism and hospitality in digital sophistication (Table 17.2).

TABLE 17.2 Smart Tools and Technology

Theme	Explanation
SMART City	A digitalized place where the application of smart software and instruments is used in every sector of the state, and citizens, government, business stakeholders, and society share a digital platform for e-service, decision-making, and problem-solving.
SMART infrastructure	A smart infrastructure is a system that makes better decisions about a problem by utilizing a data feedback loop.
SMART Citizen	A person who is well-iterated in using digital technological tools to get e-service, generate legal online income, and get associated with the technological environment in a smart city complex.
SMART Government	Application of creativity and technology to enable improved planning and decision-making within governing bodies.
E-medical service	The provision of medical services remotely over a communications network is known as e-medical service.
Smart Accommodation	A smart hotel has been outfitted with up-to-date technological and internet systems, to provide customers with better service and improve management, efficiency, and control from the hotel management.
Virtual Tourism	Providing real-time experience of destination by installing digital tools and technology in a simulated world.
Augmented Reality	Software-based added information enriches the tourist's knowledge about the destination and gets precautionary data before the crisis appears.
Artificial intelligence	The simulated robotic machine copies the human neural system to manipulate the environment in its own style. However, future crises can be easily spotted by installing AI in the crisis management strategy.
Other Smart tools	Smartphones and software, Google Alerts, Google Map, Fractal Analysis, Factal, TweetDeck, SumAll, Veoci, Konnect, and GPS tracking.

Source: Authors compilation.

8. **Satellite Surveillance:** A satellite surveillance system is intended to detect nuclear explosions, acquire electronic intelligence, weather forecasting, pandemic zones, take pictures, monitor the launch of strategic missiles, and provide early warning.
9. **Smart Crisis and Resilience Management Team:** This team is classified into the following categories. Figure 17.2 is designed based on a specific country context like Bangladesh.

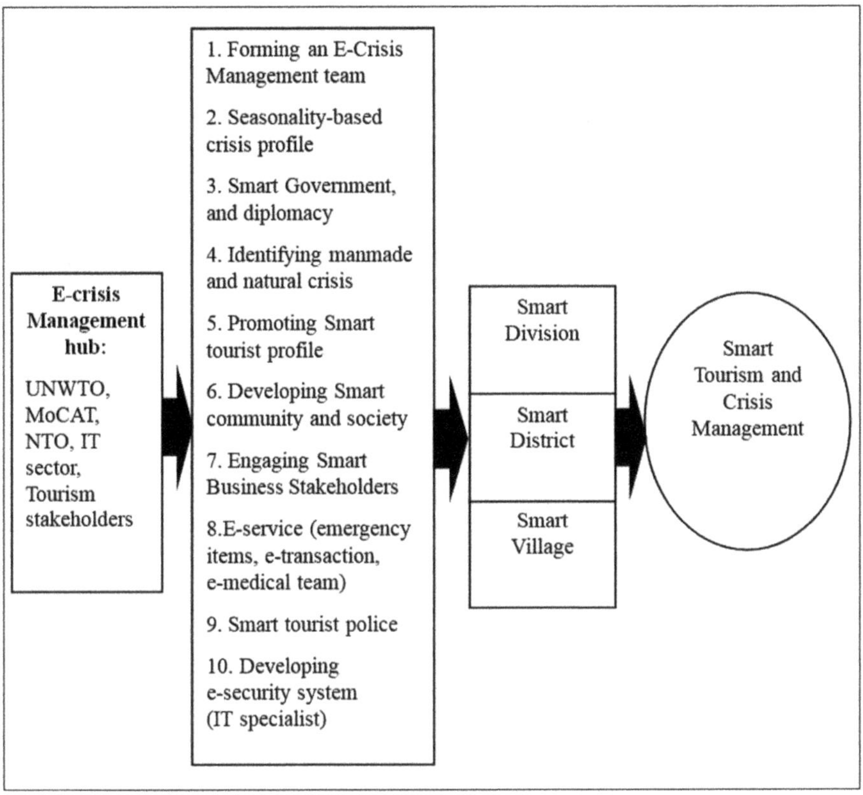

FIGURE 17.2 Crisis resilience team and smart country.

Source: Authors compilation.

10. **Developing Crisis Management Software:** This software preserves a record of event data with accurate information on what was done and what decisions were taken, allowing teams to learn and improve for the next time. Often an afterthought amid a crisis,

the Post Incident Review (PIR) is developed in the background as the crisis progresses.

However, cyber security is a vital factor in managing crises and providing safety for the technological tools used in smart crisis management systems (Opara et al., 2022). The main cause of this high susceptibility is the existence of multiple diverse and fragmented business sectors that are vulnerable to cyber-attacks and rely on a wide range of technology for production and operation (Paraskevas, 2022).

17.6 SMART TOURISM AND CRISIS HYPOTHESIS

We might soon find ourselves living in a world where travel and technology continue to converge toward the smart form. Therefore, with VR, AR, and other digital platforms, the way visitors engage with a location may alter in the future. Virtual tourism and other technological advancements restore cultural authenticity, even as they can alleviate issues like over-tourism (Van Nuenen and Scarles, 2021). Smart tourism and crisis hypothesis start with three fundamental stages:

1. **Avoidance Phase:** In the case of the avoidance phase, people try to skip the crisis and adapt to the circumstance, but it may cause a lot of struggles to accustom to the crisis. What if the people cannot adjust to the crisis? Then, they move toward the wait-and-watch or confrontation stage.
2. **Wait and Watch Phase:** On the other hand, the wait-and-watch stage deals with observing and analyzing the situation with care for a specific period, and then they decide to move toward the confrontation or avoidance phase.
3. **Confrontation Phase:** Lastly, the confrontation stage is the more crucial stage that bridges the gap between "Wait and Watch" and "Avoidance" situations. In this stage, understanding the nature of the crisis brings several solutions by utilizing SMART technologies and crisis management strategies.

Finally, training is an essential component of good natural disaster preparedness while dealing with prospective crisis hypotheses. It is necessary to enable dynamic access to this instruction, which can be accomplished by developing a mobile app. Moreover, one can use technology that allows for a quick response and make sure that community members have access to this mode of communication. Social media has repeatedly demonstrated its effectiveness in creating awareness. Therefore, it is mandatory to incorporate social media sites into the crisis management plan (Figure 17.3).

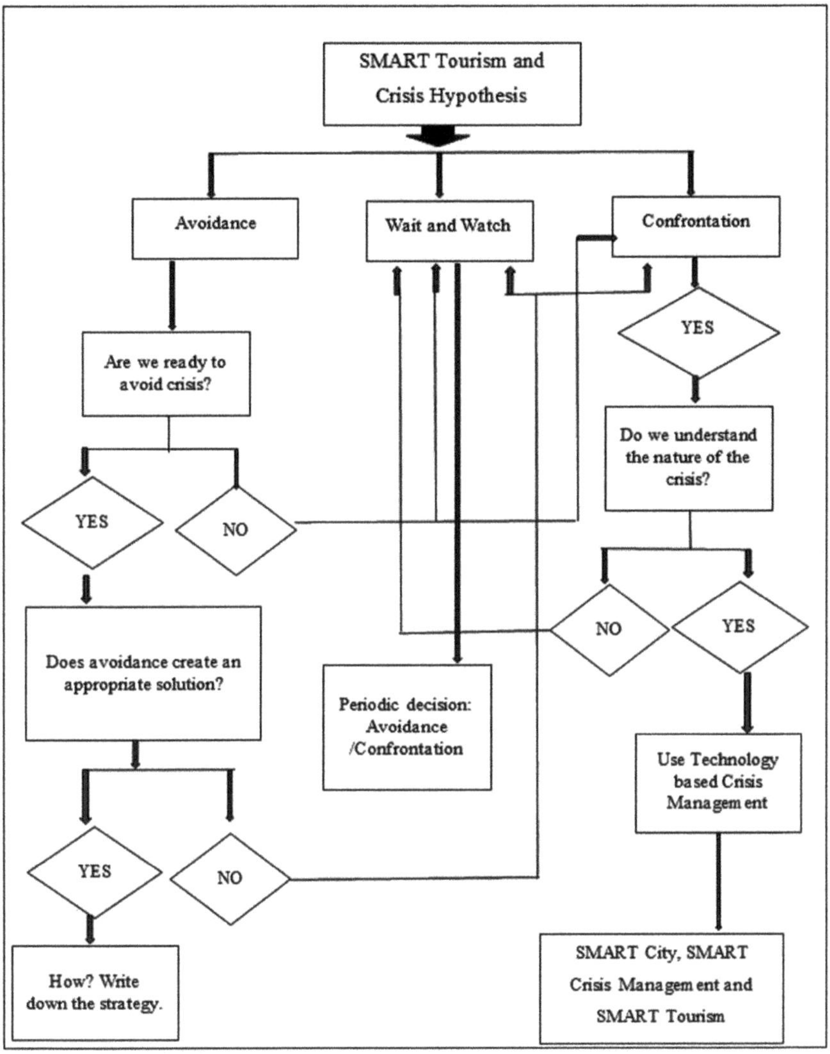

FIGURE 17.3 Crisis hypothesis.

Source: Authors compilation.

The above-mentioned crisis hypothesis triggers some valuable solutions during the pandemic situation. This smart solution can be a logistic support for stakeholders to take proper action in emergency circumstances.

17.7 EFFECTIVE TOURISM CRISIS MANAGEMENT: A FRAMEWORK OF SMART TOOLS AND TOURISM

Based on the crisis management framework by Faulkner (2013) and PATA (Wilks et al., 2004), this chapter has amplified the figure by adding more variables that are essential to implement the strategy from the technological perspective.

Table 17.3 has made a comparison among all the crisis management systems. However, the crisis management framework given by Faulkner (2013) and PATA has followed the traditional approach to deal with the challenges (Wilks et al., 2004). They avoided the application of technology in case of diminishing the crisis. This research gap has been bridged by highlighting an updated smart crisis management system that states the digital transformation of destination. Moreover, it will prioritize the risks after assessing their probability, timelines, and possible effects by installing technological sophistication in the entire system. Table 17.4 has illustrated the crisis management strategy through the digitalization process.

TABLE 17.3 Effective Crisis Management

Crisis Management by Faulkner	PATA	SMART Crisis Management (by Author)	
Pre-event	Identify	Historical Background	Climate and Demography
Prodromal	Analysis	Pre-crisis	Digitalization and Technological Transformation
Emergency	Plan	During Crisis	SMART Tourism
Intermediate	Track	Post Crisis	Upgrading SMART facilities and securities
Recovery	Control	SMART tourism resilience theory	Initiating conceptual concept
Resolution	Communicate	Knowledge management	Digital storing and sharing

Source: Faulkner (2013); Wilks et al. (2004) and Authors compilation.

TABLE 17.4 Crisis Management Process and Smart Tools

Crisis Management Process and Smart Tools			
Historical Background of the region	**Precrisis** (Digitalization and Technological involvement)	**During Crisis** (SMART Tourism: staying at Home)	**Post Crisis** (SMART Resilience system)
Land	Risk symptom analysis	Analysis of the risk	E-Reviewing System
Climate	E-insurance	Risk Hypothesis Evaluate the risk:	Storing in the Smart database system
Disease	E-Tourist response team	Monitoring	Analyzing TSA management
Geo-politics	Media and Social	Taking proactive and reactive marking strategies	Smart Monitoring System
Demography	E-medical team	Control and solve the risk	Smart resilience plan
Society	E-service	Augmented Tourism	Digital lock and safety recognition
Economic	E- Crisis Response Team	Artificial intelligence	Crisis Management Software
Disease	E-Infrastructure	E-Recreation /Entertainment	
	E-Government	Online food service	
		Emergency number and E-crisis team	
		E-service and transaction	
		E hotel booking/ transportation	
		Knowledge Management and sharing	Developing SMART tourism Resilience theory

Source: Authors compilation.

1. **Historical Background of the Country:** To understand the nature of the crisis of any region, it is essential to explore the historical background of a country like the nature of the weather, climate, disease pattern, technological advancement, demographic status, and so forth. For example, Pizer, AstraZeneca, and Moderna corporation conducted several experiments on the human body to formulate an authentic vaccination by using by analyzing historical data of the SARS virus of a particular region.
2. **Pre-crisis Strategy:** As mentioned earlier, each problem has come up with several symptoms before it appears. It is mandatory to transform society into a SMART country, which brings several opportunities like how to deal with crises by using modern technology. For instance, the e-crisis response team, e-medical team, social and electric media, e-infrastructure, and e-government are the rudimentary factors in this phase. However, smart citizens proactively take numerous measures to handle the crisis. The automated system will interconnect people with the e-crisis response team which is operated by the e-government system.
3. **Risk Symptom Analysis:** A crisis is always interconnected with several symptoms that tell us the pattern of the problem. In this phase, symptoms reflect how to deal with a crisis just before it appears and give us frequent guidelines by using a technological approach (satellite forecasting, biological analysis of flora and fauna) at the destination.
4. **During the Crisis:** This is the most crucial phase which provides real-time experiences about the crisis. However, this chapter has already discussed the crisis hypothesis in the previous section which illustrates three basic hypothetical questions: (i) Avoidance (ii) Wait and Watch; and (iii) Confrontation. Based on these three perspectives, it is nearly impossible for the tourist to visit any location. For example, in the covid 19 crisis, people were bound to confine themselves in their houses. However, WHO provided health protocols to be maintained during the situation which implies complete impediments for the tourism and hospitality industry. After that, the world successfully invented COVID-19 vaccination to mitigate the crisis. Now the question arises, how the tourism and travel industry will rejuvenate during the crisis? The answer is very simple and easy, technological involvement

will be a peremptory strategy for stakeholders in the tourism and hospitality sector. The Smart citizen can use virtual tourism, AI applications, and augmented reality opportunities to get real-time experiences of any destination or service. On the other hand, the smart hotel industry can convert their business into a major quarantine center for affluent people to enjoy their services for the time being. The food and beverage sector can get online food orders and provide them at the doorsteps of the customers. This is how tourism can revive their alternative business niche during the crisis. The e-crisis management team will actively run their services by utilizing the satellite surveillance system, this method will easily detect the customer needs during the crisis.

5. **Analysis of the Risk:** Risk analysis is the most complex phenomenon, as, it requires several crucial steps to go through. Most significantly, how? when? and where did the crisis originate? However, the historical background of the region gives some insinuation about the crisis phenomenon. For example, the COVID-19 crisis originated from China, as, this country has a notorious reputation for smuggling and eating wild animals. Smart technological tools like AI are massively used to measure the severity of the crisis.
6. **Evaluate the Risk:** This segment measures the extremity of the crisis. This stage asks the question of whether the crisis is minor, moderate, or major. Based on these three rudimentary criteria, steps be taken to mitigate or eliminate the challenge. However, any crisis might start with a minor stage and end up with major hazards, if the challenges are not handled effectively. Modern technological matrix tools and software have been applied to evaluate the risk.
7. **Monitoring:** However, most of the studies suggested that monitoring can be done during and after the crisis. The study has argued that monitoring can be done in all phases of crisis management, on the ground that digitalization in every sector makes this possible, and how all the stakeholders inclusively work to combat the crisis.
8. **Taking Proactive and Reactive Strategies:** The proactive strategies refer to taking measures before the crisis. There is a link between proactive approaches and symptom analysis. When the symptoms of a crisis have been traced in any location then the proactive action will be accelerated in that region and news should

be transferred rest of the world to make them aware of the upcoming crisis. On the other hand, reactive strategies will be implemented in a moment of crisis by adopting modern sophistication.

9. **Control the Risk:** The WHO had prescribed some essential health protocols to be maintained in the COVID-19 pandemic situation before inventing the vaccination. However, these guidelines and mandates have been disseminated around the world because of SMART technologies. This strategy works as a miracle to control the risk for the particular time being.
10. **Solve the Risk:** The situation will be exacerbated if the crisis will not be handled with care. For example, juveniles in Bangladesh initiated an e-vaccination program named "Surokka. Com" which brought the citizens into the mass jabs program.
11. **Post-crisis Management:** In this situation, tourists become more meticulous in choosing the destination. However, smart tourists want to avail all types of technological opportunities in the destination. Additionally, they will explore the destination that managed the crisis successfully. For example, during the Covid crisis, Maldives summoned tourist to visit their country to provide maximum health safety. This situation demands a smart destination that is embellished with smart hotels, transportation systems, smart stakeholders, e-visa, tourist smart cards, and so forth. Furthermore, this stage accumulates the entire analysis of pre- and post-crisis situations. This phase also accentuates on e-reviewing system in a crisis moment, storing the primary and secondary data in the data management software. Moreover, this situation also analyzes the Tourism Satellite Account (TSA) to compare the tourist arrival and the contribution to the country's GDP during and post crises. However, smart monitoring, initiating digital lock and safety recognition, and smart resilience theory are the major indicators in the post-crisis management system.
12. **Establishing Tourism Resilience Theory:** Tourism and Resilience theory's application to the tourism industry is described by Cochrane (2010). After a crisis event, a system that renews itself enters the "exploitation" phase, where an orderly system appears by utilizing the various forms of potential—social, human, natural, and financial—that have been developed over previous cycles and channeling them into new structures. However, the above-mentioned theory

gave several hypothetical intrusions for the stakeholders of the travel industry. After the covid 19 crisis, the entire theory has fallen into a question of argument which is no longer applicable for the industry until or unless this theory is incorporated with digitalization. SMART tourism resilience theory expedites the previous theory and delineates the emphasis on smart destinations, smart citizens, smart countries, and smart crisis teams linked with modern digitalization systems before, during, and post crises to dwindle or eradicate the challenge.

13. **Knowledge Management and Sharing:** After establishing the tourism resilience theory, it is mandatory to transfer, share, and store it for future purposes. The next generation and stakeholders will get a lot of information about the nature of the crisis that the world encountered and how it was handled and solved. The technological tools like a pen drive, presentation, e-documentation, and hard disk will be a logistic support to store and demonstrate this knowledge on both national and international podiums (Deb et al., 2023).

Finally, knowledge communication is critical in the early stages of a crisis management plan. It is critical not only during but also before a crisis happens. We live in the age of social media, therefore, information from this source is available and easy to obtain. It could be information about preventing or detecting a potentially disastrous incident. In most circumstances, during the second and third phases of a crisis management plan, a huge number of agencies, such as e-police forces, firefighters, civil protection, e-health services, and so on, become involved. Mobile technology could be valuable for both communication and information collecting. Modern mobile devices have high-definition cameras and can accomplish activities that were previously only possible on PCs. However, if communications networks are broken, then, communication through this source would be hampered. Therefore, conventional modes of e-communication systems should be preserved.

17.8 MANAGERIAL AND PRACTICAL IMPLICATION

Several smart tools are offered to managers and policymakers in this book chapter. Initially, it offers a way for determining and cataloging a great deal of actions suggested related to the application of smart technology and tourism to manage crises. It also highlighted a crisis management hypothesis for organizing these acts logically and cohesively. A balanced smart and crisis management framework with regions of comparable relevance was presented with several strategies that will help the tourism policymaker how tackle or adapt to crises.

Because of growing costs, the current state of the economy, staffing shortages, and ongoing challenges with the supply chain, service providers frequently have to deliver fewer services. For instance, a lot of hotels are reducing back on services like room service and regular housekeeping. During the pandemic, several travel-related businesses made layoffs or lost employees; according to reports, 62 million jobs associated with tourism were lost globally in 2020. This chapter illustrated a smart technological model to rejuvenate the industry after the COVID-19 crisis. Therefore, the hospitality sector can apply this model how to deal with crises in any emergency. Despite economic prosperity, traveler trust in a place can be negatively impacted by unforeseen occurrences such as natural or man-made catastrophes, regardless of whether the risk is genuine or imagined. Therefore, this chapter introduces tourist e-profiling and e-visa systems that will guide the government to reformulate its tax pattern. Most importantly, tourism stakeholders will get information when they go through the chapter and get a proper interpretation of financial strategy and crisis minimizing system.

17.9 CONCLUSION

However, smart tourism development can also assist tourism stakeholders in developing crisis resilience in the travel and tourism industry. Recently, it has been frequently talked about "Zombie Virus" that will be exposed, only when the Antarctic Ice area melts. Therefore, this "Smart Resilience" theory connects a six-pillar framework and discusses notions of smart destinations at the conceptual level that guide the travel industry on how to deal with future crises like the "Zombie virus" or new variant of Covid 19. However, the excessive use of digital technology could lead to negative outcomes like social marginalization, the digital divide, invasions of privacy and confidentiality, the spread of false information and political prejudice, as well as ineffective distance work and educational programs. It is recommended that authorities free up the digitalization process, improve digital literacy, and make digital services more accessible to lessen these negative effects of the crisis.

KEYWORDS

- **artificial intelligence**
- **augmented reality**
- **e-crisis management**
- **service digitalization**
- **smart destination**
- **smart tourism**
- **tourism industry**
- **virtual reality**

REFERENCES

Almeida-Santana, A., David-Negre, T., & Moreno-Gil, S. (2020). New digital tourism ecosystem: Understanding the relationship between information sources and sharing economy platforms. *International Journal of Tourism Cities, 6*(2), 335–345.

Baggio, R. (2020). Tourism destinations: A universality conjecture based on network science. *Annals of Tourism Research, 82*, 102929.

Beirmen, D. (2021). The PATA rapid response taskforce. *Tourism Cases, tourism20210001.*

Berbekova, A., & Baker, M. A. (2020). Crisis management and tourism experiences. In *The Routledge Handbook of Tourism Experience Management and Marketing* (pp. 580–589). Routledge.

Berke, J. D., Paletzki, R. F., Aronson, G. J., Hyman, S. E., & Gerfen, C. R. (1998). A complex program of striatal gene expression induced by dopaminergic stimulation. *Journal of Neuroscience, 18*(14), 5301–5310.

Blaikie, P., Cannon, T., Davis, I., & Wisner, B. (1994). *At Risk: Natural Hazards, People's Vulnerability and Disasters.* New York: Routledge.

Boes, K., Buhalis, D., & Inversini, A. (2015). Conceptualizing smart tourism destination dimensions. In *Information and Communication Technologies in Tourism: Proceedings of the International Conference in Lugano* (pp. 391–403). Springer International Publishing.

Brammer, H. (1990). Floods in Bangladesh: II. Flood mitigation and environmental aspects. *Geographical Journal*, 158–165.

Breitsohl, J., & Garrod, B. (2016). Assessing tourists' cognitive, emotional, and behavioral reactions to an unethical destination incident. *Tourism Management, 54*, 209–220.

Cherry, J. D. (2004). The chronology of the 2002–2003 SARS mini pandemic. *Pediatric Respiratory Reviews, 5*(4), 262–269.

Ciotti, M., Ciccozzi, M., Terrinoni, A., Jiang, W. C., Wang, C. B., & Bernardini, S. (2020). The COVID-19 pandemic. *Critical Reviews in Clinical Laboratory Sciences, 57*(6), 365–388.

Cochrane, J. (2010). The sphere of tourism resilience. *Tourism Recreation Research, 35*(2), 173–185.

Cohen, S., & Karatzimas, S. (2022). Analyzing smart cities' reporting: Do they report "smart"? *Journal of Public Budgeting, Accounting & Financial Management, 34*(5), 602–621.

COMCEC. (2017). *Crisis Management in the Tourism Sector: Recovery from Crisis in the OIC Member Countries*. Ankara: COMCEC Coordination Office.

Cornejo Ortega, J. L., & Malcolm, C. D. (2020). Touristic stakeholders' perceptions about the smart tourism destination concept in Puerto Vallarta, Jalisco, Mexico. *Sustainability, 12*(5), 1741.

De Sausmarez, N. (2013). Challenges to Kenyan tourism since 2008: Crisis management from the Kenyan tour operator perspective. *Current Issues in Tourism, 16*(7–8), 792–809.

Deb, S. K., Kuri, B. C., Roy, H., Roy, B., & Nafi, S. M. (2023). The adaptability of tourism and hospitality supply chain management to manage COVID-19 crisis: Bangladesh perspective. *International Journal of Agile Systems and Management, 16*(4), 484–511.

Del Chiappa, G., & Baggio, R. (2015). Knowledge transfer in smart tourism destinations: Analyzing the effects of a network structure. *Journal of Destination Marketing & Management, 4*(3), 145–150.

Dixon, M. G., & Schafer, I. J. (2014). Ebola viral disease outbreak—West Africa, 2014. *Morbidity and Mortality Weekly Report, 63*(25), 548.

Dobrowolski, Z. (2020). After COVID-19: Reorientation of crisis management in crisis. *Entrepreneurship and Sustainability Issues, 8*(2).

Faulkner, B. (2013). Towards a framework for tourism disaster management. In *Managing Tourist Health and Safety in The New Millennium* (pp. 155–176). Routledge.

Femenia-Serra, F., Neuhofer, B., & Ivars-Baidal, J. A. (2019). Towards a conceptualization of smart tourists and their role within the smart destination scenario. *The Service Industries Journal, 39*(2), 109–133.

Ferguson, L. (2007). The United Nations World Tourism Organization. *New Political Economy, 12*(4), 557–568.

Gretzel, U., & Jamal, T. (2020). Guiding principles for good governance of the smart destination. *Travel and Tourism Research Association: Advancing Tourism Research Globally, 42*, 1–10.

Gretzel, U., & Collier de Mendonça, M. (2019). Smart destination brands: Semiotic analysis of visual and verbal signs. *International Journal of Tourism Cities, 5*(4), 560–580.

Gretzel, U., & Scarpino-Johns, M. (2018). Destination resilience and smart tourism destinations. *Tourism Review International, 22*(3–4), 263–276.

Gretzel, U., Koo, C., Sigala, M., & Xiang, Z. (2015). Special issue on smart tourism: Convergence of information technologies, experiences, and theories. *Electronic Markets, 25*, 175–177.

Gretzel, U., Sigala, M., Xiang, Z., & Koo, C. (2015). Smart tourism: Foundations and developments. *Electronic Markets, 25*, 179–188.

Hosie, P., & Pforr, C. (2016). *Crisis Management in The Tourism Industry: Beating the Odds?* Routledge.

Huang, Y. (2004). The SARS epidemic and its aftermath in China: A political perspective. In *Learning from SARS: Preparing for the next disease outbreak* (pp. 116–136).
Hystad, P. W., & Keller, P. C. (2008). Towards a destination tourism disaster management framework: Long-term lessons from a forest fire disaster. *Tourism Management, 29*(1), 151–162.
Ivars-Baidal, J. A., Celdrán-Bernabeu, M. A., Femenia-Serra, F., Perles-Ribes, J. F., & Giner-Sánchez, D. (2021). Measuring the progress of smart destinations: The use of indicators as a management tool. *Journal of Destination Marketing & Management, 19*, 100531.
Ivars-Baidal, J. A., Celdrán-Bernabeu, M. A., Mazón, J. N., & Perles-Ivars, Á. F. (2019). Smart destinations and the evolution of ICTs: A new scenario for destination management? *Current Issues in Tourism, 22*(13), 1581–1600.
Kim, J., & Lee, J. C. (2020). Effects of COVID-19 on preferences for private dining facilities in restaurants. *Journal of Hospitality and Tourism Management, 45*, 67–70.
Kukanja, M., Planinc, T., & Sikošek, M. (2020). Crisis management practices in tourism SMEs during the COVID-19 pandemic. *Organizacija, 53*(4), 346–361.
Lau, V. W., Tse, D. C., Bligh, M. C., Hong, Y. Y., Kakarika, M., Chan, H. W., & Chiu, C. P. (2022). Not "my" crisis: Social identity and followers' crisis responses to COVID-19. *Analyses of Social Issues and Public Policy, 22*(2), 506–535.
Lee, J., Jun, S., Chang, T. W., & Park, J. (2017). A smartness assessment framework for smart factories using analytic network process. *Sustainability, 9*(5), 794.
Li, J., Nguyen, T. H. H., & Coca-Stefaniak, J. A. (2021). Coronavirus impacts on post-pandemic planned travel behaviors. *Annals of Tourism Research, 86*, 102964.
Martens, H. M., Feldesz, K., & Merten, P. (2016). Crisis management in tourism–A literature based approach on the proactive prediction of a crisis and the implementation of prevention measures. *Athens Journal of Tourism, 3*(2), 89–101.
Neumayer, E. (2004). The impact of political violence on tourism: Dynamic cross-national estimation. *Journal of Conflict Resolution, 48*(2), 259–281.
Ntalakos, A., Belias, D., Koustelios, A., & Tsigilis, N. (2022). Effect of COVID-19 on the tourism industry: Opportunities and threats in COVID-19 era. In *FEMIB* (pp. 107–114).
Opara, E., Wimmer, H., & Rebman, C. M. (2022). Auto-ML cyber security data analysis using Google, Azure and IBM cloud platforms. In *International Conference on Electrical, Computer and Energy Technologies (ICECET)* (pp. 1–10). IEEE.
Pal, S. K., & Sarker, P. C. (2023). SMART Bangladesh Vision 2041: Concept of a sustainable developed country. *Environmental Management and Sustainable Development, 12*(1), 67–81.
Paliwal, M., Chatradhi, N., Singh, A., & Dikkatwar, R. (2022). Smart tourism: Antecedents to Indian traveller's decision. *European Journal of Innovation Management*.
Paraskevas, A. (2022). Cybersecurity in travel and tourism: A risk-based approach. In *Handbook of e-Tourism* (pp. 1605–1628). Cham: Springer International Publishing.
Pestronk, R. M., Kamoie, B., Fidler, D., Matthews, G., Benjamin, G. C., Bryan, R. T., & Redd, S. (2008). Improving laws and legal authorities for public health emergency legal preparedness. *Journal of Law, Medicine & Ethics, 36*(S1), 47–51.
Pforr, C. (2006). *Tourism in Post Crisis is Tourism in Pre-Crisis: A Review of the Literature on Crisis Management in Tourism.* PhD Dissertation, Curtin University.

Racherla, P., & Hu, C. (2009). A framework for knowledge-based crisis management in the hospitality and tourism industry. *Cornell Hospitality Quarterly, 50*(4), 561–577.

Rajapaksha, K. (2021). COVID-19 and tourism. *Available at SSRN 3909608*.

Richardson, V. (1994). Conducting research on practice. *Educational Researcher, 23*(5), 5–10.

Ritchie, B. W. (2004). Chaos, crises and disasters: A strategic approach to crisis management in the tourism industry. *Tourism Management, 25*(6), 669–683.

Rosselló, J., Becken, S., & Santana-Gallego, M. (2020). The effects of natural disasters on international tourism: A global analysis. *Tourism Management, 79*, 104080.

Sharma, R., & Thapa, S. (2023). Cybersecurity awareness, education, and behavioral change: Strategies for promoting secure online practices among end users. *Eigenpub Review of Science and Technology, 7*(1), 224–238.

Van Nuenen, T., & Scarles, C. (2021). Advancements in technology and digital media in tourism. *Tourist Studies, 21*(1), 119–132.

Vora, K. S., Mavalankar, D. V., Ramani, K. V., Upadhyaya, M., Sharma, B., Iyengar, S., & Iyengar, K. (2009). Maternal health situation in India: A case study. *Journal of Health, Population, and Nutrition, 27*(2), 184.

Wilks, J., & Moore, S. (2004). *Tourism Risk Management for The Asia-Pacific Region: An Authoritative Guide for Managing Crises and Disasters*. Gold Coast.

Wohlmuther, C., & Wintersteiner, W. (2014). *International Handbook on Tourism and Peace.* Klagenfurt/Celovec, Austria.

Yang, J., Xie, H., Yu, G., & Liu, M. (2021). Antecedents and consequences of supply chain risk management capabilities: An investigation in the post-coronavirus crisis. *International Journal of Production Research, 59*(5), 1573–1585.

Zhong, L., Sun, S., Law, R., & Li, X. (2021). Tourism crisis management: Evidence from COVID-19. *Current Issues in Tourism, 24*(19), 2671–2682.

GLOSSARY

A

AI Literacy: Involves the skills and competencies required for using AI technologies and applications as tools, viewing them critically, understanding their context and embedded principles, and questioning their design and implementation.

Artificial Intelligence (AI): Refers to computer systems capable of performing complex tasks that historically only a human could perform, such as reasoning, making decisions, or solving problems.

Augmented Reality (AR): An interactive experience in which a real-world environment is enhanced with computer-generated visual elements, sounds, and other stimuli.

B

Big Data: Refers primarily to data sets that are too large or complex to be processed using traditional data-processing application software.

Blockchain: A system in which a record of transactions—especially those made using cryptocurrency—is maintained across computers that are linked in a peer-to-peer network.

C

Chatbots: A chatbot is a computer program that simulates human conversation with an end user. Not all chatbots are equipped with artificial intelligence (AI), but modern chatbots increasingly use conversational AI techniques, such as natural language processing (NLP), to understand user questions and automate responses.

Closed-Circuit Television (CCTV): A system that transmits television signals to a limited number of screens and is often used in stores and public places to prevent crime.

Cloud Computing: The use of hosted services—such as data storage, servers, databases, networking, and software—over the Internet.

Constructed Environment (CE): CE refers to the human-made environment designed to cater to various needs; however, its primary purpose lies in satisfying tourists.

Crisis Resilience: The ability not only to recover quickly from a crisis but to bounce back better—and even thrive. This concept is used not only for critical infrastructures (such as the supply of basic services like water, food, energy, transport, housing/shelter, communications, finance, and health), but also for the wider public, to integrate and address human and social dynamics in crises and disaster situations—including the role of the population, the media, and rescuers (staff, volunteers, and ad-hoc volunteers).

Cyber Security: The state of being protected against the criminal or unauthorized use of electronic data, or the measures taken to achieve this.

D

Data Encryption: It is essential to employ industry-standard practices for encrypting data transmitted among visitors, service providers, and destinations to effectively protect it from interception and theft.

Digital Channels: Online platforms embraced by travel companies, hotels, airlines, and tour operators that enable travelers to plan itineraries, make bookings, and reserve lodging through websites and mobile applications with ease.

Digital Infrastructure: Digital infrastructure refers to the physical and software-based systems necessary to deliver digital goods, products, and services. This includes data centers, fiber infrastructure, server hardware, personnel, IT virtualization and infrastructure software, operating systems, and more.

Digital Transactions: Transactions in which the customer authorizes the transfer of money through electronic means, with the funds flowing directly from one account to another. These accounts may be held in banks or with other entities or providers.

E

Eco-Smart Destination: A green and environmentally friendly destination equipped with all types of logistical technological support, where the community, businesspeople, and tourists are connected through technological webs, software, and tools.

E-Guides: A digital guide application that accompanies tourists during their travels by providing information about the area to visit. It is installed on technological devices and includes do's and don'ts for visitors while creating awareness about natural areas.

E-Medical Service: The provision of medical services remotely over a communications network. It involves the use of the Internet or other electronic media by patients, health workers, and the public to disseminate or provide access to health information and prescriptions.

Emotional Intelligence (EI): The ability to manage one's own emotions and understand the emotions of others. There are five key elements to EI: self-awareness, self-regulation, motivation, empathy, and social skills, all of which help enhance employee efficiency and overall firm performance.

E-Profile: A system that includes information about tourists, employees, suppliers, and other tourism stakeholders in an online database with maximum security. For example, digital passports and e-voter IDs are fundamental tools used to identify a person.

E-Safety: Also called "internet safety," "online safety," or "web safety," e-safety is defined as the safe and responsible use of technology. This includes the use of the Internet as well as other means of communication via electronic media (such as text messages, gaming devices, email, etc.).

F

Fourth Industrial Revolution (4IR): The current and evolving environment in which disruptive technologies and trends—such as the Internet of Things (IoT), robotics, virtual reality (VR), and artificial intelligence (AI)—are transforming the way people live and work.

G

Geocaching: The process of hiding and locating small containers using GPS coordinates. The main goal of geocaching is to encourage people to spend more time outdoors while introducing them to the world of technology.

Geographic Information System (GIS): A computer system for capturing, storing, checking, and displaying data related to positions on Earth's surface.

H

HDM Method: A human-driven design approach that focuses on the early phases of design, is strongly future-oriented, and aims to contribute to innovation for a sustainable society and improved quality of life.

HEAT-D Framework: A sophisticated, scientifically based framework developed through research on tourism and climate change to track destination maintenance costs and manage the long-term effects of climate change.

HVAC System: Heating, ventilation, and air conditioning; the use of various technologies to control the temperature, humidity, and purity of the air in an enclosed space.

I

Information and Communication Technologies (ICTs): Refers to all communication technologies, including the Internet, wireless networks, cell phones, computers, software, middleware, video conferencing, social networking, and other media applications and services.

Interactive Guides: Augmented reality (AR) makes standard guidebooks interactive through the creation of interactive guides. Applications such as *AR City* overlay real-time information on restaurants, sights, and landmarks to provide users with interactive, context-aware guides.

Internet of Things (IoT): Devices with sensors, processing ability, software, and other technologies that connect and exchange data with other devices and systems over the Internet or other communication networks. IoT helps provide customized services to tourists and guests in airlines, hotels, cruises, and tourism destinations. In addition, the use of IoT in the tourism and hospitality industry helps reduce costs, enhance productivity, improve efficiency, and offer customized and differentiated services.

L

Language Translation: By using language translation features, augmented reality (AR) makes cross-cultural communication easier. Apps such as *Google Translate* can use AR to translate text instantly, allowing travelers to view menus, signage, and other written content in their native language.

Local Area Network (LAN): A collection of devices connected in one physical location, such as a building, office, or home. A LAN can range in size from a small home network with one user to a large enterprise network with thousands of users and devices in an office or school.

M

Mobile Applications: A mobile app (or mobile application) is a software application developed specifically for use on small, wireless computing devices, such as smartphones and tablets, rather than desktop or laptop computers. These applications were originally popularized by Apple Inc. and its App Store, which offers thousands of applications for the iPhone, iPad, and iPod Touch.

N

Near Field Communication (NFC): A set of short-range wireless technologies, typically requiring a distance of 4 cm or less to initiate a connection. NFC allows for the exchange of small payloads of data between an NFC tag and an Android-powered device, or between two Android-powered devices.

Neuroscience: The study of the nervous system—from its structure to function, development to degeneration, in both health and disease. It encompasses the entire nervous system, with a primary focus on the brain. Incredibly complex, the brain defines who we are and what we do.

P

Participatory Sensing: An approach to data collection and interpretation in which individuals—acting alone or in groups—use their personal mobile devices and web services to systematically explore aspects of their environment, ranging from health to culture. This paradigm involves using mobile devices to create a collaborative sensor network to collect, share, and analyze data.

Personalization of Services: Services such as recommendations, trip planning, and targeted marketing based on client data to provide individualized experiences. This enhances client satisfaction, loyalty, and engagement by offering services aligned with personal preferences.

Q

Quick Response (QR) Code: A machine-readable code consisting of an array of black and white squares, typically used to store URLs or other information readable by a smartphone camera. QR codes are utilized for contactless ticketing, reservations, and check-ins at various establishments, including museums, events, and accommodations.

R

Radio Frequency Identification (RFID): Involves small devices that use radio frequencies to transfer data, mainly for tracking and identifying objects, animals, and people. This is a method of automatic identification and data capture.

Robotics: A branch of engineering and computer science that involves the conception, design, manufacture, and operation of robots. In the travel industry, robotic technology is used to automate and improve services such as airport cleaning, restaurant customer service, hotel luggage handling, information provision, and more.

S

Satellite Surveillance: A satellite surveillance system is designed to detect nuclear explosions, acquire electronic intelligence, forecast weather, identify pandemic zones, capture images, monitor strategic missile launches, and provide early warning.

Satellite System: Systems through which content is delivered via satellite transmission for display on television receivers or video monitors located in private dwellings.

Smart Buildings: A smart building uses technology to enable the efficient and economical use of resources while creating a safe and comfortable environment for occupants.

Smart Citizen: A person well-versed in using digital technological tools to access e-services, generate legal online income, and engage with the technological environment within a smart city complex.

Smart City: A digitalized environment where smart software and instruments are applied across all sectors of the state, involving citizens, government, business stakeholders, and society.

Smart Destinations: A smart destination considers economic, social, and environmental sustainability in order to enhance tourism benefits while minimizing negative impacts. In a smart destination, tourist experiences and services are supported and developed through the use of smart technologies such as mobile applications, IoT, AI, and big data. It is a destination that leverages innovation and solutions to address challenges such as seasonality, sustainable mobility, and tourist flow management—all while improving both the visitor and resident experience.

Smart Economy: A concept relevant to both the present and the future, as it refers to policies that stimulate innovation, investment, and creativity, combined with scientific research, advanced technology, and environmental responsibility through the lens of sustainability. A smart economy is built on technological innovation, resource efficiency, sustainability, and high social welfare as key success drivers.

Smart Environments: It is an ecosystem of communicating items that can coordinate, provide services, and control complex information (e.g., sensors, actuators, information devices, and other network-connected devices). This environment links computers and other smart devices to everyday settings and tasks. Smart environments include smart homes, smart cities, and smart manufacturing.

Smart Government: The application of creativity and technology to improve planning and decision-making within governing bodies. Smart government uses frontier technologies—led by artificial intelligence (AI)—to deliver services tailored to specific citizen demands. This automated system provides assistance at every stage. Moreover, smart government implies simple, moral, accountable, responsive, and transparent governance.

Smart Grids: It is an electricity network based on digital technology that supplies electricity to consumers through two-way digital communication. This system enables monitoring, analysis, control, and communication within the supply chain to improve efficiency, reduce energy consumption and costs, and maximize transparency and reliability. The smart grid was introduced to overcome the limitations of conventional electrical grids by using smart net meters.

Smart Hotels and Accommodations: Internet of things (IoT) devices in hotels provide seamless, customized experiences for guests—ranging from connected amenities and smart room controls to enhanced guest security, energy-efficient operations, personalized room settings, and automated check-ins.

Smart Infrastructure: A system that makes better decisions about problems by utilizing a data feedback loop. Smart infrastructure involves the integration of advanced technology, data analytics, and connectivity into various components of physical infrastructure to improve efficiency, resilience, and sustainability.

Smart Living: Technological advancements that give people the opportunity to benefit from new ways of living. It involves original and innovative solutions aimed at making life more efficient, controllable, economical, productive, integrated, and sustainable.

Smart Mobility: The use of alternative modes of transportation in place of—or in addition to—owning a gasoline-powered automobile, such as ride-sharing, car-sharing, public transportation, walking, cycling, and more. Smart mobility connects various elements of technology and transportation and involves rethinking the infrastructure used in daily life and business.

Smart Resource Management: Digital technologies—such as the Internet of Things (IoT) and data analytics—used by Tourism 4.0 to manage resources intelligently. This includes maximizing water and energy use, as well as reducing waste in lodging and tourist attractions.

Smart Services and Experiences: Smart services—such as guided tours, interactive exhibits, and location-based information services—can be developed through the Internet of Things (IoT). Visitors can participate in engaging activities, access relevant information about attractions, and receive personalized suggestions.

Smart Tourism: Tourism that uses interconnected physical and digital technologies to enhance the overall guest experience. Travel experiences are digitally enabled and connected to technology, allowing guests to efficiently navigate foreign cities, book activities online, and manage their entire itinerary through tools such as artificial intelligence (AI), the Internet of Things (IoT), big data, and 5G.

Smart Transportation: IoT-enabled vehicles and buses allow for route planning, maintenance monitoring, and real-time tracking. This results in greater safety, reduced traffic, more efficient transportation operations, and improved navigation for tourists.

Social Media Network: Social networking refers to the use of internet-based social media platforms to stay connected with friends, family, colleagues, or customers. It can serve a social purpose, a business purpose, or both—through platforms such as Facebook, X (formerly Twitter), Instagram, and Pinterest.

T

Tailored Experiences: Experiences customized to meet individual customer needs. Using user data and preferences, mobile apps can offer tailored suggestions for activities, restaurants, and attractions. By considering unique interests, this personalization enhances the overall user experience.

TDM Method: The term "Technology-Driven" refers to a management philosophy that emphasizes the development of new products or services based on a firm's technical capabilities rather than proven demand—essentially, making keys first and then looking for locks to open.

Total Quality Management (TQM): A management system for a customer-focused firm that involves all employees in ongoing improvement. TQM is a continuous process of identifying, reducing, or eliminating errors in manufacturing.

Tourism 4.0: A concept representing the transformative digitalization that is revolutionizing the tourism sector. It emphasizes how digitization, connectivity, and data-driven insights are replacing traditional methods and aligns closely with the broader concept of Industry 4.0.

Tourist Smart Card: A smart card that enables tourists to access destinations and services without hassle. These cards integrate multiple services and activities offered at discounted rates—through a chip, QR code, or mobile application. Additionally, the card stores information about the tourist's demographic profile and travel purpose.

V

Virtual Private Network (VPN): A VPN establishes a digital connection between a user's computer and a remote server owned by a VPN provider, creating a point-to-point tunnel that encrypts personal data, masks IP addresses, and allows users to bypass website blocks and firewalls on the internet.

Virtual Reality (VR): The use of computer modeling and simulation to enable a person to interact with an artificial three-dimensional (3D) visual or other sensory environment. VR applications immerse users in computer-generated surroundings that simulate reality through the use of interactive devices—such as goggles, headsets, gloves, or bodysuits—that send and receive information.

Virtual Tourism: A tourism niche that leverages technology to enable travelers to engage in activities, visit locations, and experience destinations without leaving their homes. Virtual tourism can be an excellent tool for converting prospects into clients through a unique and more engaging approach than that of competitors.

W

Wearables and Smart Accessories: Travelers frequently use wearables with Internet of Things (IoT) capabilities for augmented reality, activity tracking, health monitoring, interactive experiences, and navigation—all of which contribute to a more informed and connected travel experience.

Wi-Fi: A wireless networking technology that uses radio waves to provide high-speed internet access. Wi-Fi is a set of wireless network protocols based on the IEEE 802.11 family of standards, commonly used for local area networks (LANs) and internet access, allowing adjacent digital devices to exchange data via radio signals.

Wireless Environmental Monitoring: Wireless environmental monitoring solutions offer a range of cost-effective data-gathering and administrative benefits. These systems can assist ecotourism management by providing accurate, actionable data for the maintenance and improvement of ecosystems.

INDEX

T

www.ingramcontent.com/pod-product-compliance
Ingram Content Group UK Ltd.
Pitfield, Milton Keynes, MK11 3LW, UK
UKHW022200190926
475702UK00003B/124

* 9 7 8 1 7 7 9 6 4 1 0 8 3 *